OPTICAL FIBER COMMUNICATIONS

Principles and Practice

Kruger Brentt
Publishers

OPTICAL FIBER COMMUNICATIONS

Principles and Practice

Harper Berry
Betty Price
Marcus Ramirezh

Kruger Brentt
Publishers

2023

Kruger Brentt Publishers UK. LTD.
Company Number 9728962

Regd. Office: 68 St Margarets Road, Edgware, Middlesex HA8 9UU

ISBN: 978-1-78715-169-7

For information on all our publications visit our website at http://krugerbrentt.com/

PREFACE

An optical Fiber is a thin, flexible, transparent Fiber that acts as a waveguide, or "light pipe", to transmit light between the two ends of the Fiber. Optical fibers are widely used in Fiber-optic communications, which permits transmission over longer distances and at higher bandwidths (data rates) than other forms of communication. Fibers are used instead of metal wires because signals travel along them with less loss and are also immune to electromagnetic interference. Fiber optic data transmission systems send information over fiber by turning electronic signals into light. Light refers to more than the portion of the electromagnetic spectrum that is near to what is visible to the human eye.

Fiber optics plays an important role in the field of communication to transmit voice, television and digital data signals fro one place to another. The transmission of light along the thin cylindrical glass fiber by total internal reflection was first demonstrated by John Tyndall in 1870 and the application of this phenomenon in the field of communication is tried only from 1927. Today the applications of fiber optics are also extended to medical field in the form of endoscopes and to instrumentation engineering in the form of optical sensors. The basic principle of optical fiber in the transmission of optical signal is total internal reflection. The basic principle of optical fiber in the transmission of optical signal is total internal reflection.

Fiber optics (optical fibers) are long, thin strands of very pure glass about the size of a human hair. They are arranged in bundles called optical cables and used to transmit signals over long distances. The present book contains 13-chapters namely Introduction, optical fiber waveguides, characteristics of optical transmission, optical fibers and cables, optical fiber connections: joints, couplers and isolators, optical sources 1: the laser, optical sources 2: the light-emitting diode, optical detectors, optical amplification, wavelength conversion and regeneration, integrated optics and photonics, optical fiber systems 1: intensity modulation/direct detection, optical fiber systems 2: coherent and phase modulated, optical networks. This book provides problems and techniques of design and utilisation of optical fiber systems. It explores the principles for understanding and applying optical fiber technology to sophisticated modern telecommunication systems. The subject matter is presented in a lucid and succinct style so that even those who do not have a prerequisite knowledge of the subject can easily understand it. This book is an excellent reference work for anyone involved with optical fiber communications, electronic devices and other circuitry applications, such as electrical and technical engineers.

We are grateful to all those persons as well as various books, manuals, periodicals, magazines, journals etc. that helped in the preparation of this book. In spite of the best efforts, it is possible that some errors may have occurred into the compilation and editing of the book. Further queries, constructive suggestions and criticisms for the improvement of the book are always welcomed and shall be thankfully acknowledged.

Harper Berry

Betty Price

Marcus Ramirezh

CONTENTS

CHAPTER 1
INTRODUCTION

For decades, electronic signals have been sent effectively via normal 'hard-wired' connections or by the use of different kinds of radio links which had their own downfalls. On the contrary, optic fiber links, whether utilized for video or audio links over long or short ranges, offer some unique advantages as compared to the standard wired cables.

Presently, the growth in information technology has had increased use of the current telecommunication systems. Often, optical fiber communication plays a significant role in the development of telecommunication systems with high quality and speed. Nowadays, optical fiber applications majorly involve telecommunication systems with an inclusion of internet and local area networks (LAN) to achieve high signalling rates.

In optical fiber technology, an optical fiber link is utilized to transfer analog or digital data in light frequency form via a cable with a highly reflective central core. The role of the highly reflective central core is to act as a light guide for the transfer of light through it through continuous reflections across its characteristic reflective walls. As is illustrated in the block diagram below, the optical fiber communication module mainly comprises a transmitter (Tx) circuit and a receiver (Rx) module.

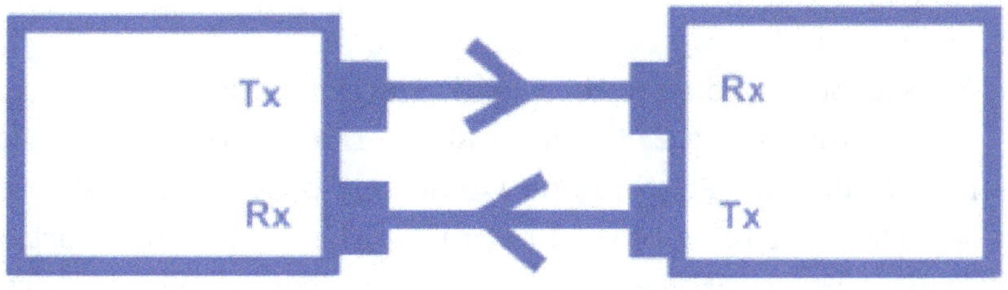

Fig. 1. A simple receiver-transmitter block diagram

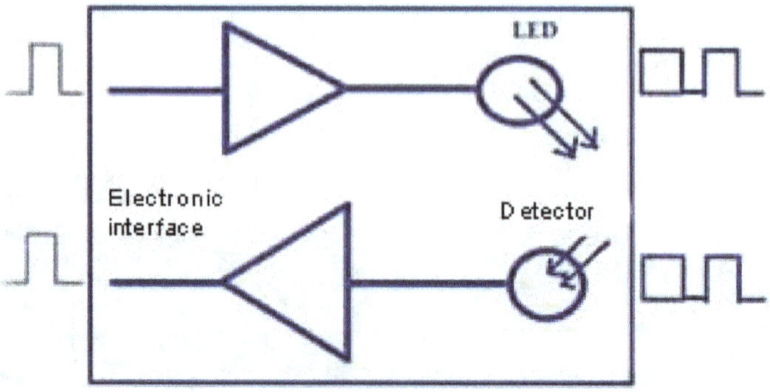

Fig. 2. Block diagram for a typical optical transmitter and receivers

The transmitter is located on one end of the fiber cable while the receiver is located on the other sides. As is common, a transceiver- a module that includes a transmitter and receiver- is employed in most systems. The input to the transmitter comprises an electrical signal which converts into an optical signal from either a light-emitting diode (LED) or laser diode. As required, the light signal from the transmitter is linked to the fiber cable with the use of a connector and broadcasted via the cable. Subsequently, the light signal from the fiber terminal can be linked to a receiver, anywhere a detector converts the light to an electrical signal, after which it is suitably conditioned for use by the receiving equipment. The figure below depicts a block diagram for a typical optical transmitter and receivers.

HISTORY OF OPTICAL FIBER COMMUNICATION

First developed in the 1970s, fiber-optics have revolutionized the telecommunications industry and have played a major role in the advent of the Information Age. Because of its advantages over electrical transmission, optical fibers have largely replaced copper wire communications in core networks in the developed world.

In 1880 Alexander Graham Bell and his assistant Charles Sumner Tainter created a very early precursor to fiber-optic communications, the Photophone, at Bell's newly established Volta Laboratory in Washington, D.C. Bell considered it his most important invention. The device allowed for the transmission of sound on a beam of light.

On June 3, 1880, Bell conducted the world's first wireless telephone trans-mission between two buildings, some 213 meters apart. Due to its use of an atmospheric transmission medium, the Photophone would not prove practical until advances in laser and optical fiber technologies permitted the secure transport of light. The Photophone's first practical use came in military communication systems many decades later.

In 1954 Harold Hopkins and Narinder Singh Kapany showed that rolled fiber glass allowed light to be transmitted. Initially it was considered that the light can traverse in only straight medium. Junichi Nishizawa, a Japanese scientist at Tohoku University, proposed the use of optical fibers for communications in 1963. Nishizawa invented the PIN diode and the static induction transistor, both of which contributed to the development of optical fiber communications.

In 1966 Charles K. Kao and George Hockham at STC Laboratories (STL) showed that the losses of 1,000 dB/km in existing glass (compared to 5–10 dB/km in coaxial cable) were due to contaminants which could potentially be removed.

Optical fiber was successfully developed in 1970 by Corning Glass Works, with attenuation low enough for communication purposes (about 20 dB/km) and at the same time GaAs semiconductor lasers were developed that were compact and therefore suitable for transmitting light through fiber optic cables for long distances.

In 1973, Optelecom, Inc., co-founded by the inventor of the laser, Gordon Gould, received a contract from APA for the first optical communication systems. Developed for Army Missile Command in Huntsville, Alabama, it was a laser on the ground and a spout of optical fiber played out by missile to transmit a modulated signal over five kilometers.

After a period of research starting from 1975, the first commercial fiber-optic communications system was developed which operated at a wavelength around 0.8 μm and used GaAs semiconductor lasers. This first-generation system operated at a bit rate of 45 Mbit/s with repeater spacing of up to 10 km. Soon on 22 April 1977, General Telephone and Electronics sent the first live telephone traffic through fiber optics at a 6 Mbit/s throughput in Long Beach, California.

In October 1973, Corning Glass signed a development contract with CSELT and Pirelli aimed to test fiber optics in an urban environment: in September 1977, the second cable in this test series, named COS-2, was experimentally deployed in two lines (9 km) in Turin, for the first time in a big city, at a speed of 140 Mbit/s.

The second generation of fiber-optic communication was developed for commercial use in the early 1980s, operated at 1.3 μm and used InGaAsP semiconductor lasers. These early systems were initially limited by multi mode fiber dispersion, and in 1981 the single-mode fiber was revealed to greatly improve system performance, however practical connectors capable of working with single mode fiber proved difficult to develop.

Canadian service provider SaskTel had completed construction of what was then the world's longest commercial fiber optic network, which covered 3,268 km (2,031 mi) and linked 52 communities.[11] By 1987, these systems were operating at bit rates of up to 1.7 Gb/s with repeater spacing up to 50 km (31 mi).

The first transatlantic telephone cable to use optical fiber was TAT-8, based on Desurvire optimised laser amplification technology. It went into operation in 1988.

Third-generation fiber-optic systems operated at 1.55 μm and had losses of about 0.2 dB/km. This development was spurred by the discovery of Indium gallium arsenide and the development of the Indium Gallium Arsenide photodiode by Pearsall. Engineers overcame earlier difficulties with pulsespreading at that wavelength using conventional InGaAsP semiconductor lasers. Scientists overcame this difficulty by using dispersion-shifted fibers designed to have minimal dispersion at 1.55 μm or by limiting the laser spectrum to a single longitudinal mode.

These developments eventually allowed third-generation systems to operate commercially at 2.5 Gbit/s with repeater spacing in excess of 100 km (62 mi).

The fourth generation of fiber-optic communication systems used optical amplification to reduce the need for repeaters and wavelength-division multiplexing to increase data capacity. These two improvements caused a revolution that resulted in the doubling of system capacity every six months starting in 1992 until a bit rate of 10 Tb/s was reached by 2001. In 2006 a bit-rate of 14 Tbit/s was reached over a single 160 km (99 mi) line using optical amplifiers.

The focus of development for the fifth generation of fiber-optic communications is on extending the wavelength range over which a WDM system can operate. The conventional wavelength window, known as the C band, covers the wavelength range 1.53–1.57 μm, and dry fiber has a low-loss window promising an extension of that range to 1.30–1.65 μm. Other developments include the concept of "optical solutions", pulses that preserve their shape by counteracting the effects of dispersion with the nonlinear effects of the fiber by using pulses of a specific shape.

In the late 1990s through 2000, industry promoters, and research companies such as KMI, and RHK predicted massive increases in demand for communications bandwidth due to increased use of the Internet, and commercialization of various bandwidth-intensive consumer services, such as video on demand. Internet protocol data traffic was increasing exponentially, at a faster rate than integrated circuit complexity had increased under Moore's Law. From the bust of the dot-com bubble through 2006, however, the main trend in the industry has been consolidation of firms and offshoring of manufacturing to reduce costs. Companies such as Verizon and AT&T have taken advantage of fiber-optic communications to deliver a variety of high-throughput data and broadband services to consumers' homes.

The general system

An optical fiber can be understood as a dielectric waveguide, which operates at optical frequencies. The device or a tube, if bent or if terminated to radiate energy, is called a **waveguide**, in general. Figure 3 depicts a bunch of fiber optic cables.

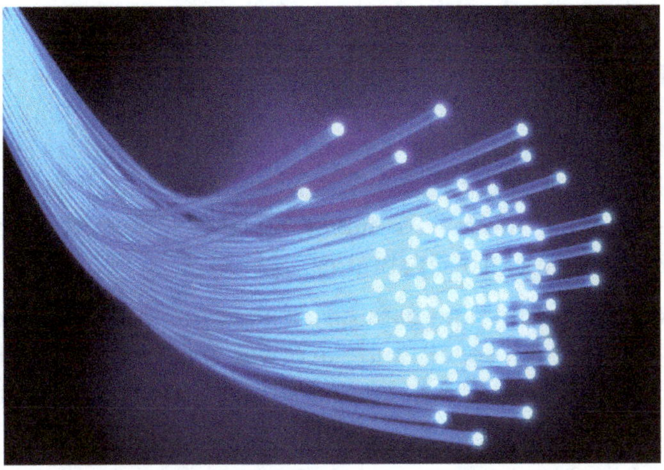

Fig. 3. Bunch of fiber optic cables

ELEMENTS OF A FIBER OPTIC COMMUNICATION SYSTEM

There are three main elements of a fiber optic communication system. They are

- Compact Light Source
- Low loss Optical Fiber
- Photo Detector

Accessories like connectors, switches, couplers, multiplexing devices, amplifiers, and splices are also essential elements in this communication system.

Compact Light Source

Depending on the applications like local area networks and long-haul communication systems, the light source requirements vary. The requirements of the sources include power, speed, spectral linewidth, noise, ruggedness, cost, temperature, and so on. Two components are used as light sources: light-emitting diodes (LEDs) and laser diodes.

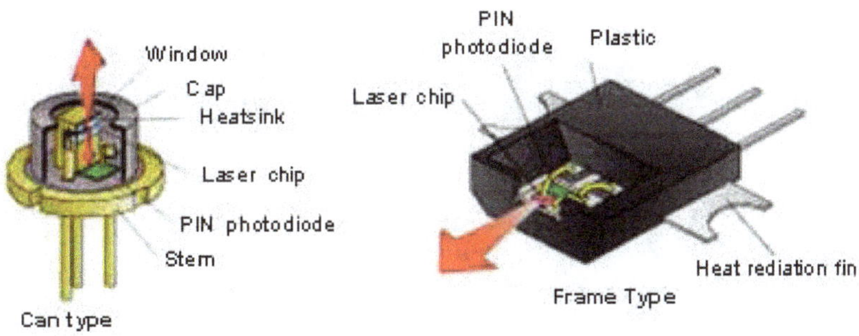

Fig. 4. Laser Diode Structure

Laser Diodes

The light-emitting diodes are used for short distances and low data rate applications due to their low bandwidth and power capabilities. Two such LED structures include Surface and Edge Emitting Systems. The surface-emitting diodes are simple in design and are reliable, but due to their broader line width and modulation frequency limitation edge-emitting diodes are mostly used. Edge emitting diodes have high power and narrower linewidth capabilities.

For longer distances and high data rate transmission, Laser Diodes are preferred due to its high power, high speed, and narrower spectral linewidth characteristics. But these are inherently non-linear and more sensitive to temperature variations.

Nowadays many improvements and advancements have made these sources more reliable. A few of such comparisons of these two sources are given below. Both these sources are modulated using either direct or external modulation techniques.

LOW LOSS OPTICAL FIBER

Optical fiber is a cable, which is also known as a cylindrical dielectric waveguide made of low-loss material. Optical fiber also considers the parameters like the environment in which it is operating, the tensile strength, durability, and rigidity. The Fiber optic cable is made of high-quality extruded glass (si) or plastic, and it is flexible. The diameter of the fiber optic cable is in between 0.25 to 0.5mm (slightly thicker than a human hair).

FIBER OPTIC CABLE

A Fiber Optic Cable consists of four parts.

- Core
- Cladding
- Buffer
- Jacket

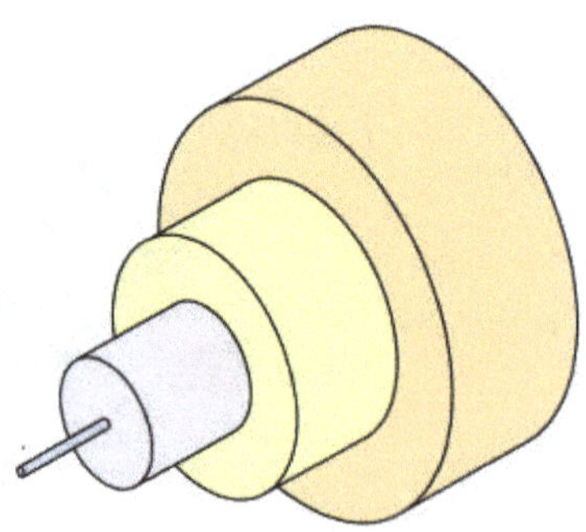

Fig. 5. Fiber Optic Cable

Core

The core of a fiber cable is a cylinder of plastic that runs all along the fiber cable's length and offers protection by cladding. The diameter of the core depends on the application used. Due to internal reflection, the light traveling within the core reflects from the core, the cladding boundary. The core cross-section needs to be a circular one for most of the applications.

Cladding

Cladding is an outer optical material that protects the core. The main function of the cladding is that it reflects the light back into the core. When light enters through the core (dense material) into the cladding(less dense material), it changes its angle, and then reflects back to the core.

Buffer

The main function of the buffer is to protect the fiber from damage and thousands of optical fibers arranged in hundreds of optical cables. These bundles are protected by the cable's outer covering which is called a jacket.

Jacket

Fiber optic cable jackets are available in different colors that can easily make us recognize the exact color of the cable we are dealing with. The color yellow clearly signifies a single-mode cable, and the orange color indicates multimode.

Types of Optical Fibers

Single-Mode Fibers: Single-mode fibers are used to transmit one signal per fiber; these fibers are used in telephone and television sets. Single-mode fibers have small cores.

Multi-Mode Fibers: Multimode fibers are used to transmit many signals per fiber; these signals are used in computer and local area networks that have larger cores.

Photo Detectors

The purpose of photodetectors is to convert the light signal back to an electrical signal. Two types of photodetectors are mainly used for optical receivers in optical communication systems: PN photodiode and avalanche photodiode. Depending on the application's wavelengths, the material composition of these devices vary. These materials include silicon, germanium, InGaAs, etc.

FIBRE OPTIC TRANSMISSION SYSTEM

Any fibre optic data transmission system will comprise a number of different elements. There are three major elements (marked in bold), and a further one that is vital for practical systems:

- Transmitter (light source)
- Fibre optic cable
- Optical repeater
- Receiver (Detector)

The different elements of the system will vary according to the application. Systems used for lower capacity links, possibly for local area networks will employ somewhat different techniques and components to those used by network providers that provide extremely high data rates over long distances. Nevertheless the basic principles are the same whatever the system.

In the system the transmitter of light source generates a light stream modulated to enable it to carry the data. Conventionally a pulse of light indicates a "1" and the absence of light indicates "0". This light is transmitted down a very thin fibre of glass or other suitable material to be presented at the receiver or detector. The detector converts the pulses of light into equivalent electrical pulses. In this way the data can be transmitted as light over great distances.

Fibre optic transmitter

Although the original telecommunications fibre optic systems would have used large lasers, today a variety of semiconductor devices can be used. The most commonly used devices are light emitting diodes, LEDs, and semiconductor laser diodes.

The simplest transmitter device is the LED. Its main advantage is that it is cheap, and this makes it ideal for low cost applications where only short runs are needed. However they have a number of drawbacks. The first is that they offer a very low level of efficiency.

Only about 1% of the input power enters the optical fibre, and this means that high power drivers would be needed to provide sufficient light to enable long distance transmissions to be made. The other disadvantage of LEDs is that they produce what is termed incoherent light that covers a relatively wide spectrum. Typically the spectral width is between 30 and 60 nm. This means that any chromatic dispersion in the fibre will limit the bandwidth of the system.

In view of their performance, LEDs are used mainly in local-area-network applications where the data rates are typically in the range 10-100 Mb/s and transmission distances are a few kilometres.

Where higher levels of performance are required, i.e. it is necessary that the fibre optic link can operate over greater distances and with higher data rates, then lasers are used. Although more costly, they offer some significant advantages. In the first instance they are able to provide a higher output level, and in addition to this the light output is directional and this enables a much higher level of efficiency in the transfer of the light into the fibre optic cable. Typically the coupling efficiency into a single mode fibre may be as high as 50%.

A further advantage is that lasers have a very narrow spectral bandwidth as a result of the fact that they produce coherent light. This narrow spectral width enables the lasers to transmit data at much higher rates because modal dispersion is less apparent. Another advantage is that semiconductor lasers can be modulated directly at high frequencies because of short recombination time for the carriers within the semiconductor material.

Laser diodes are often directly modulated. This provides a very simple and effective method of transferring the data onto the optical signal. This is achieved by controlling current applied directly to the device. This in turn varies the light output from the laser. However for very high data rates or very long distance links, it is more effective to run the laser at a constant output level (continuous wave). The light is then modulated using an external device. The advantage of using an external means of modulation is that it increases the maximum link distance because an effect known as laser chirp is eliminated. This chirp broadens the spectrum of the light signal and this increases the chromatic dispersion in the fibre optic cable.

Fibre optic cable

The full details and description of fibre optic cabling are found in a separate article / tutorial on this area of the website. In essence a fibre optic cable consists of core, around which is another layer referred to as the cladding. Outside of this there is a protective outer coating.

The fibre optic cables operate because their cladding has a refractive index that is slightly lower than that of the core. This means that light passing down the core undergoes total internal reflection when it reaches the core / cladding boundary, and it is thereby contained within the core of the optical fibre.

Repeaters and amplifiers

There is a maximum distance over which signals may be transmitted over fibre optic cabling. This is limited not only by the attenuation of the cable, but also the distortion of the light signal along the cable. In order to overcome these effects and transmit the signals over longer distances, repeaters and amplifiers are used.

Opto-electric repeaters may be used. These devices convert the optical signal into an electrical format where it can be processed to ensure that the signal is not distorted and then converted back into the optical format. It may then be transmitted along the next state of the fibre optic cable.

An alternative approach is to use an optical amplifier. These amplifiers directly amplify the optical signal without the need to convert the signal back into an electrical format. The amplifiers consist of a length of fibre optic cable that is doped with a rare earth mineral named Erbium. The treated fibre cable is then illuminated or pumped with light of a shorter wavelength from another laser and this serves to amplify the signal that is being carried.

In view of the much reduced cost of fibre optic amplifiers over repeaters, amplifiers are far more widely used. Most repeaters have been replaced, and amplifiers are used in virtually all new installations these days.

Receivers

Light travelling along a fibre optic cable needs to be converted into an electrical signal so that it can be processed and the data that is carried can be extracted. The component that is at the heart of the receiver is a photo-detector. This is normally a semiconductor device and may be a p-n junction, a p-i-n photo-diode or an avalanche photo-diode. Photo-transistors are not used because they do not have sufficient speed.

Once the optical signal from the fibre optic cable has been applied to the photo-detector and converted into an electrical format it can be processed to recover the data which can then be passed to its final destination.

Fibre optic transmission of data is generally used for long distance telecommunications network links and for high speed local area networks. Currently fibre optics is not used for the delivery of services to homes, although this is a long term aim for many telcos. By using optical fibre cabling here, the available bandwidth for new services would be considerably higher and the possibility of greater revenues would increase. Currently the cost oft his is not viable, although it is likely to happen in the medium term.

Advantages and Disadvantages of Optical Fiber

Optical fiber is rising in both telecommunication and data communication due to its unsurpassed advantages: faster speed with less attenuation, less impervious to electromagnetic interference (EMI), smaller size and greater information carrying capacity. The unceasing bandwidth needs, on the other hand, are also yielding significant growth in optical fiber demands. Let's take a review of common fiber optic cable types, explore the advantages and disadvantage of optical fiber, and learn tips on selecting fiber optic cable.

Optical fiber uses light pulses instead of electrical pulses to transmit information, thus delivers hundreds of times higher bandwidth than traditional electrical systems. Fiber optic cable can be protected by sheathing and armor to make it resistant to harsh environmental conditions. Hence it is widely adopted in commercial business, governments, military and many other industries for voice, video and data transmission (Fig. 6).

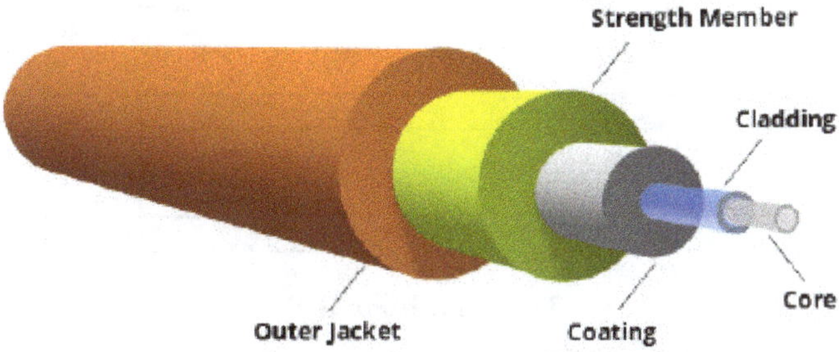

Fig. 6: Optical fiber structure

COMMON FIBER OPTIC CABLE TYPES

Generally, there are three types of fiber optic cables: the two glass optical fiber—single mode fiber optic cable and multimode optical fiber, as well as plastic optical fiber (POF).

Single Mode Fiber Optic Cable

The "mode" in fiber optic cable refers to the path in which light travels. Single mode fiber has a smaller core diameter of 9 microns (8.3 microns to be exact) and only allows a single wavelength and pathway for light to travel, which greatly decreases light reflections and lowers attenuation. Slightly more expensive than its multimode counterparts, single mode fiber optic cable is often used in network connections over long lengths (Fig. 7).

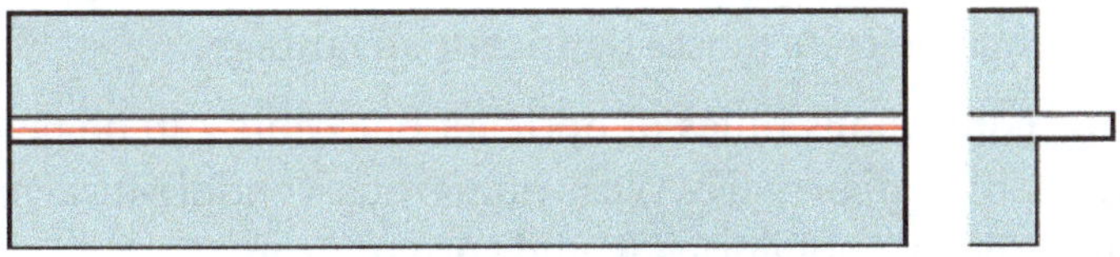

Fig. 7: Single mode fiber

Multimode Fiber Optic Cable

Multimode optical fiber has a larger core diameter than that of single mode fiber optic cable, which allows multiple pathways and several wavelengths of light to be transmitted. Multimode optical fiber is available in two sizes, 50 microns and 62.5 microns. It is commonly used for short distances, including patch cable applications such as fiber to the desktop or patch panel to equipment, data and audio/video applications in LANs. According to the fiber refractive index distribution, multimode fiber can be divided into two types: Step-Index Multimode Fiber vs Graded-Index Multimode Fiber (Fig. 8).

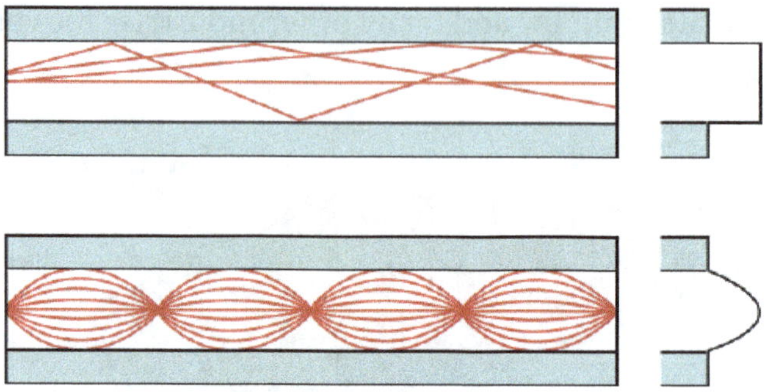

Fig. 8: Multimode fiber

PLASTIC OPTICAL FIBER (POF)

POF is a large core step-index optical fiber with a typical diameter of 1 mm. The large size enables it to easily couple lots of light from sources and connectors that do not need to be high precision. So typical connector costs are 10-20% as much as for glass fibers and termination is simple. Being plastic, it is more durable and can be installed in minutes with minimal tools and training. For applications do not require high bandwidth over great distances, POF is more competitive, making it a viable option for desktop LAN connections and low speed short links.

Though optical fiber has speed and bandwidth advantages over copper cable, it also contains some drawbacks. Here are the advantages and disadvantages of optical fiber cable.

Advantages of Optical Fiber

Greater bandwidth & faster speed—Optical fiber cable supports extremely high bandwidth and speed. The large amount of information that can be transmitted per unit of optical fiber cable is its most significant advantage.

Cheap—Long, continuous miles of optical fiber cable can be made cheaper than equivalent lengths of copper wire. With numerous vendors swarm to compete for the market share, optical cable price would sure to drop.

Thinner and light-weighted—Optical fiber is thinner, and can be drawn to smaller diameters than copper wire. They are of smaller size and light weight than a comparable copper wire cable, offering a better fit for places where space is a concern.

Higher carrying capacity—Because optical fibers are much thinner than copper wires, more fibers can be bundled into a given-diameter cable. This allows more phone lines to go over the same cable or more channels to come through the cable into your cable TV box.

Less signal degradation—The loss of signal in optical fiber is less than that in copper wire.

Light signals—Unlike electrical signals transmitted in copper wires, light signals from one fiber do not interfere with those of other fibers in the same fiber cable. This means clearer phone conversations or TV reception.

Long lifespan—Optical fibers usually have a longer life cycle for over 100 years.

Disadvantages of Optical Fiber

Low power—Light emitting sources are limited to low power. Although high power emitters are available to improve power supply, it would add extra cost.

Fragility—Optical fiber is rather fragile and more vulnerable to damage compared to copper wires. You'd better not to twist or bend fiber optic cables too tightly.

Distance—The distance between the transmitter and receiver should keep short or repeaters are needed to boost the signal.

How to Select the Right Optical Fiber Cable?

Optical fiber cable has gained much momentum in communication networks, and there emerges a dazzling array of vendors competing to manufacture and supply fiber optic cables. When selecting optical fiber, you'd better start with a reliable vendor and then consider the selection criteria. Here's a guide to clarify some of the confusions about choosing fiber optic cable.

Check Manufacturer Qualification

The major optical cable manufacturers should be granted ISO9001 quality system certification, ISO4001 international environment system certification, the ROHS, the relevant national and international institutions certification such as the Ministry of Information Industry, UL certification and etc.

Fiber Mode: Single Mode or Multimode

Single mode fiber is often used for long distances while multimode optical fiber is commonly used for short range. Moreover, the system cost and installation cost change with different fiber modes. You can refer to Single Mode vs Multimode Fiber: What's the Difference? and then decide which fiber mode you need.

Optical Cable Jackets: OFNR, OFNP, or LSZH

The standard jacket type of optical cable is OFNR, which stands for "Optical Fiber Non-conductive Riser". Besides, optical fibers are also available with OFNP, or plenum jackets, which are suitable for use in plenum environments such as drop-ceilings or raised floors. Another jacket option is LSZH. Short for "Low Smoke Zero Halogen", it is made from special compounds which give off very little smoke and no toxic when put on fire. So always refer to the local fire code authority to clarify the installation requirement before choosing the jacket type.

Optical Fiber Internal Construction: Tight Pack or Breakout or Assembly or Loose Tube

Tight pack cables are also known as distribution style cables, features that all buffered fibers under a single jacket with strength members for Enclosure to Enclosure and Conduit under Grade installations. Breakout fiber cable or fan out cable is applicable for Device to Device applications with tough and durable advantages. Assembly or zip cord construction is often used for making optic patch cables and short breakout runs. While loose tube construction is a Telco standard used in the telecommunications industry.

INDOOR VS. OUTDOOR

The choice greatly depends on your application. The major difference between indoor and outdoor fiber cable is water blocking feature. Outdoor cables are designed to protect the fibers from years of exposure to moisture. However, nowadays there have been cables with both dry water-blocked outdoor feature and indoor designs. For example, in a campus environment, you can get cables with two jackets: an outer PE jacket that withstands moisture and an inner PVC jacket that is UL-rated for fire retardancy.

Fiber Count

Both indoor and outdoor fiber cable have a vast option of fiber count ranging from 4-144 fibers. If your fiber demand exceeds this range, you can custom the fiber count for indoor or outdoor optical cable. Unless you are making fiber patch cords or hooking up a simple link with two fibers, it is highly recommended to get some spare fibers.

CONCLUSION

Optical fiber provides a fast, constant and stable Internet connection that allows a lot of data to be transmitted over incredible distances. As data demands become enormous, fiber optic cabling is the sure way to go for network flexibility and stability.

CHAPTER

OPTICAL FIBER WAVEGUIDES 2

INTRODUCTION

An optical waveguide is a spatially inhomogeneous structure for guiding light, i.e. for restricting the spatial region in which light can propagate. Usually, a waveguide contains a region of increased refractive index, compared with the surrounding medium (called *cladding*). However, guidance is also possible with other physical mechanisms, e.g. by the use of reflections at metallic interfaces, or with photonic crystal structures. Some waveguides also involve plasmonic effects at metals.

Most waveguides exhibit two-dimensional guidance, thus restricting the extension of guided light in two dimensions and permitting propagation essentially only in one dimension. An example is the *channel waveguide* shown in Figure 1. The most important type of two-dimensional waveguide is the optical fiber. There are also one-dimensional waveguides, called planar waveguides.

WAVEGUIDE FABRICATION

There are many different techniques for fabricating **dielectric** waveguides. Some examples:

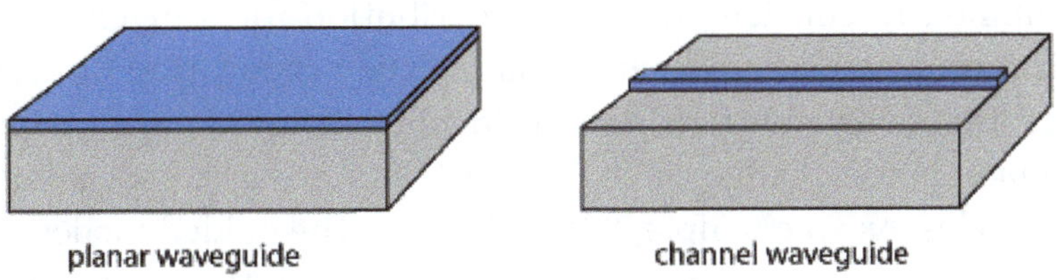

Fig 1: Two different kinds of waveguides. Planar waveguides guide light only in the vertical direction, whereas channel waveguides guide in two dimensions.

- Planar waveguides can be fabricated on various crystal and glass materials with epitaxy or with polishing methods. The waveguide may be made on the top of the device (as shown on the left side of Figure 1), but it can also be placed between other solid layers.

- Channel waveguides on semiconductor, crystal and glass materials can be made with lithographic methods in combination with, e.g., epitaxy, ion exchange, or thermal indiffusion. It is possible to make a *buried waveguide* by growing an additional layer on top of the waveguide. That may lead to lower propagation losses and a more symmetric mode profile.
- Optical fibers can be fabricated by drawing from a ***preform***, which is a large glass rod with a built-in refractive index profile. Fibers can again be drawn into waveguides of further reduced dimensions, in the extreme case resulting in ***nanofibers***.
- Waveguides can be written into transparent media (e.g. glasses or crystals, and even in polymers) with focused and pulsed laser beams, exploiting laser-induced breakdown and related phenomena. In glasses, the affected volume often exhibits a somewhat increased refractive index, which can be directly used for guiding light. In crystals, the refractive index may be decreased; one then has to treat some region around the desired waveguide region.

The trade-offs between different fabrication techniques can be complicated. They can involve aspects such as cost, flexibility and reproducibility of manufacturing, propagation losses, possible side effects on the material (e.g. via heating or indiffused materials), optimum **mode size** and symmetry for coupling to other waveguides, etc.

WAVEGUIDE MODES

For waveguides with large extensions, geometrical optics are often used for describing the propagation of injected light. Such a description, however, becomes invalid when interference effects occur, and this is particularly the case for very small waveguide dimensions. In that case, a wave description of the light is required – normally on the basis of Maxwell's equations, often simplified with approximating assumptions.

It is common to consider the field distribution for a given optical frequency and polarization in a plane perpendicular to the propagation direction. Of special interest are those distributions which do not change during propagation, apart from a common phase change. Such field distributions are associated with so-called waveguide modes. As an example, Figure 2 shows the guided modes of a multimode fiber. Each mode has a so-called propagation constant, the imaginary part of which quantifies the phase delay per unit propagation distance. A fiber also has a large number of unguided modes (cladding modes), which are not restricted to the vicinity of the fiber core.

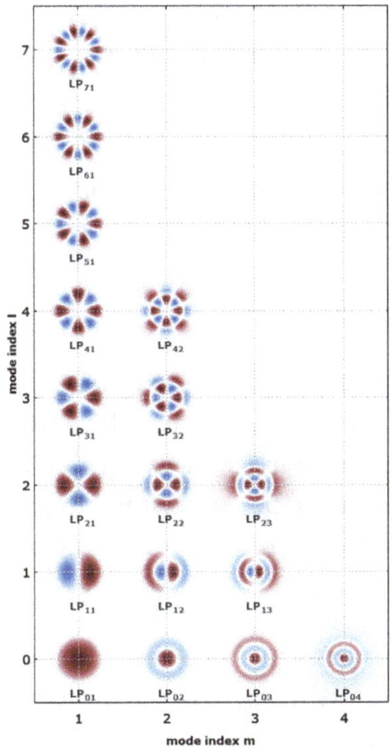

Figure 2: Electric field amplitude profiles for all the guided modes of an optical fiber. The two colors indicate different signs of electric field values. The lowest-order mode (l = 1, m = 0, called LP_{01} mode) has an intensity profile which is similar to that of a Gaussian beam. In general, light launched into a multimode fiber will excite a superposition of different modes, which can have a complicated shape.

Any initial field distribution, which may describe light injected at the beginning of the waveguide, can be decomposed into a linear combination of the field distributions of the guided waveguide modes, plus some function which can not be expressed as such a combination. The latter part corresponds to light which can not be guided. Depending on the type of waveguide, the not guided light may propagate in the cladding or may be reflected. The propagation of the guided part is easily calculated, using a linear combination of the waveguide modes with local expansion coefficients calculated from the propagation constants of the modes.

A waveguide with a small transverse spatial extension and/or a small refractive index difference (small numerical aperture) may be able to guide only a single transverse mode (for a given optical frequency and polarization) and no higher-order modes; it is then called a single-mode waveguide ('! single-mode fibers). The field distribution after a certain propagation distance then always resembles the constant mode field distribution, independent of the initial field distribution, provided that the unguided modes have been lost (e.g. in the cladding). Multimode waveguides are those supporting several or even many guided modes (sometimes many thousands).

Some types of waveguides (e.g. the channel waveguide on the right side of Figure 1) exhibit modes with strongly asymmetric intensity profiles. It also happens that guided modes exist only for one polarization direction, or that the modes for different polarization directions have very different properties.

Various properties such as the propagation losses, the bend sensitivity (for fibers), the propagation constant and the chromatic dispersion can substantially depend on the type of guided mode.

WAVEGUIDE DISPERSION

Confinement of light in a waveguide leads to **wave vectors** which are tilted against the propagation direction. This affects the phase delay per unit length and thus the **chromatic dispersion** properties (**waveguide dispersion**). For example, the dispersion of a **photonic crystal fiber** with small **mode area** can be anomalous in the visible spectral region, although the silica material would have normal dispersion.

Plasmonic Waveguides for Nano Optics

For various applications, for example in the context of photonic integrated circuits, it is of great interest to strongly localize light in waveguides to dimensions far below the optical wavelength. Here, dielectric waveguides exhibit serious limitations. For example, although nanofibers can have diameters far below the wavelength, the electric field distributions of light guided in nanometer-scale fibers extend far beyond dielectric structure. Therefore, new waveguide technologies based on other physical guiding mechanisms are investigated. A promising field is that of nanoplasmonics, where nanometer-scale metallic structures embedded in dielectric materials are used. In that way, it is possible to obtain much more localized field distributions than possible with dielectric structures alone. However, the propagation losses are typically quite high. Additional challenges are to efficiently couple light into such structures and to realize various passive and active photonic components such as strong bends, couplers, filters, amplifiers and detectors.

RAY THEORY TRANSMISSION

Total internal reflection

To consider the propagation of light within an optical fiber utilizing the ray theory model it is necessary to take account of the refractive index of the dielectric medium. The refractive index of a medium is defined as the ratio of the velocity of light in a vacuum to the velocity of light in the medium.

A ray of light travels more slowly in an optically dense medium than in one that is less dense, and the refractive index gives a measure of this effect. When a ray is incident on the interface between two dielectrics of differing refractive indices (e.g. glass–air), refraction occurs, as illustrated in Fig. 3(a). It may be observed that the ray approaching the interface is propagating in a dielectric of refractive index *n* and is at an angle ö to the normal at the surface of the interface. If the dielectric on the other side of the interface has a refractive index *n*which is less than n_1, then the refraction is such that the ray path in this lower index medium is at an angle to the normal, where is greater than φ. The angles of incidence and refraction are related to each other and to the refractive indices of the dielectrics by Snell's law of refraction, which states that:

$$n_1 \sin \varphi_1 = n_2 \sin \varphi_2$$

Or

$$\frac{\sin \varphi_1}{\sin \varphi_1} = \frac{n_2}{n_2} \quad (1)$$

It may also be observed in Fig. 3(a) that a small amount of light is reflected back into the originating dielectric medium (partial internal reflection). As *n*is greater than *n*, the angle of refraction is always greater than the angle of incidence. Thus when the angle of refraction is 90° and the refracted ray emerges parallel to the interface between the dielectrics, the angle of incidence must be less than 90°. This is the limiting case of refraction and the angle of incidence is now known as the critical angle öc, as shown in Fig. 3(b). From Eq. (1) the value of the critical angle is given by

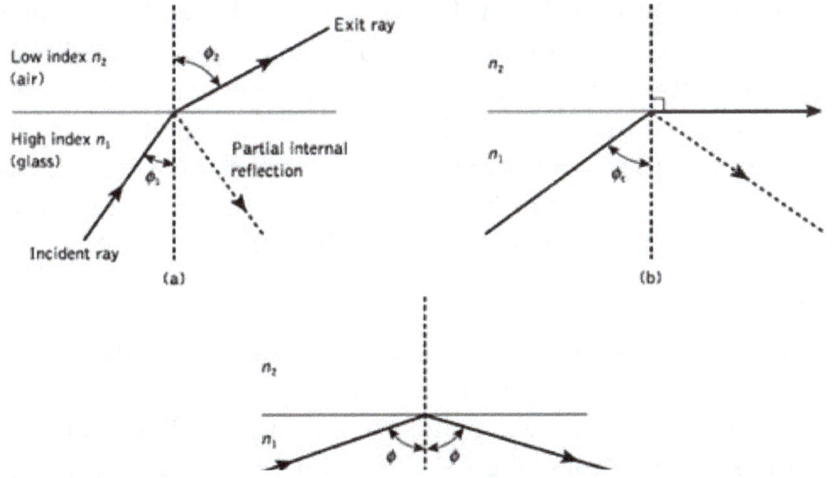

Fig. 3. Light rays incident on a high to low refractive index interface (e.g. glass air): refraction, (b) the limiting case of refraction showing the critical ray at angle j_c (c) total internal reflection where $j > j_c$

$$\sin \phi_c = \frac{n_2}{n_1} \quad (2)$$

At angles of incidence greater than the critical angle the light is reflected back into the originating dielectric medium (total internal reflection) with high efficiency (around 99.9%). Hence, it may be observed in Fig. 3(c) that total internal reflection occurs at the inter- face between two dielectrics of differing refractive indices when light is incident on the dielectric of lower index from the dielectric of higher index, and the angle of incidence of the ray exceeds the critical value. This is the mechanism by which light at a sufficiently shallow angle (less than 90° " may be considered to propagate down an optical fiber with low loss.

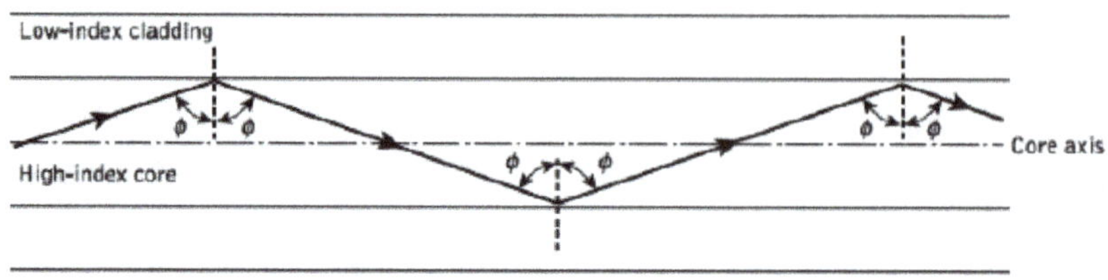

Fig. 4. Transmission of a light ray in an optical fiber

Figure 4 illustrates the transmission of a light ray in an optical fiber via a series of total internal reflections at the interface of the silica core and the slightly lower refractive index silica cladding. The ray has an angle of incidence ö at the interface which is greater than the critical angle and is reflected at the same angle to the normal.

The light ray shown in Fig. 4 is known as a meridional ray as it passes through the axis of the fiber core. This type of ray is the simplest to describe and is generally used when illustrating the fundamental transmission properties of optical fibers. It must also be noted that the light transmission illustrated in Figure 1.3 assumes a perfect fiber, and that any discontinuities or imperfections at the core–cladding interface would probably result in refraction rather than total internal reflection, with the subsequent loss of the light ray into the cladding.

Acceptance angle

Having considered the propagation of light in an optical fiber through total internal reflection at the core–cladding interface, it is useful to enlarge upon the geometric optics approach with reference to light rays entering the fiber. Since only rays with a sufficiently shallow grazing angle (i.e. with an angle to the normal greater than φ_c) at the core–cladding interface are transmitted by total internal reflection, it is clear that not all rays entering the fiber core will continue to be propagated down its length.

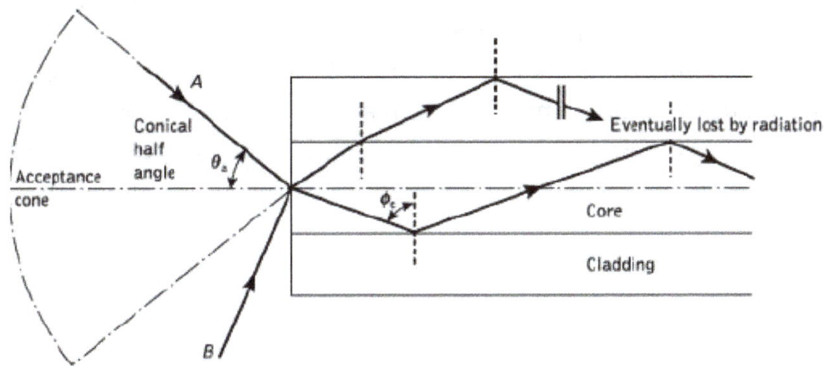

Fig. 5. Illustrates a meridional ray A at the critical angle j c within the fiber at the core-cladding interface

The geometry concerned with launching a light ray into an optical fiber is shown in Fig. 5, which illustrates a meridional ray *A* at the critical angle öc within the fiber at the core–cladding interface. It may be observed that this ray enters the fiber core at an angle φa to the fiber axis and is refracted at the air–core interface before transmission to the core–cladding interface at the critical angle. Hence, any rays which are incident into the fiber core at an angle greater than èa will be transmitted to the core–cladding interface at an angle less than φ_c, and will not be totally internally reflected. This situation is also illustrated in Fig. 5, where the incident ray *B* at an angle greater than θ_a is refracted into the cladding and eventually lost by radiation. Thus for rays to be transmitted by total internal reflection within the fiber core they must be incident on the fiber core within an acceptance cone defined by the conical half angle èa.

Hence θ_a is the maximum angle to the axis at which light may enter the fiber in order to be propagated, and is often referred to as the acceptance angle for the fiber.

If the fiber has a regular cross-section (i.e. the core–cladding interfaces are parallel and there are no discontinuities) an incident meridional ray at greater than the critical angle will continue to be reflected and will be transmitted through the fiber. From symmetry considerations it may be noted that the output angle to the axis will be equal to the input angle for the ray, assuming the ray emerges into a medium of the same refractive index from which it was input.

Numerical aperture

The acceptance angle for an optical fiber was defined in the preceding section. However, it is possible to continue the ray theory analysis to obtain a relationship between the acceptance angle and the refractive indices of the three media involved, namely the core, cladding and air. This leads to the definition of a more generally used term, the numerical aperture of the fiber. It must be noted that within this analysis, as with the preceding discussion of acceptance angle, we are concerned with meridional rays within the fiber. Figure 6 shows a light ray incident on the fiber core at an angle θ_1 to the fiber axis which is less than the acceptance angle for the fiber θa. The ray enters the fiber from a medium (air) of refractive index n_0, and the fiber core has a refractive index n_1, which is slightly greater than the cladding refractive index n_2.

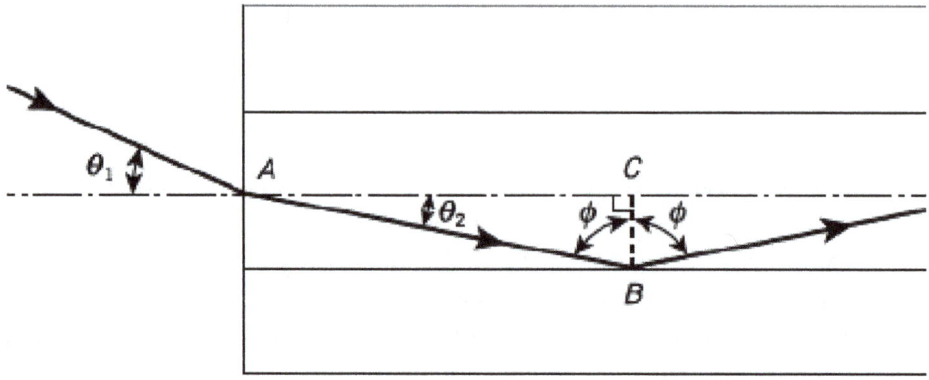

Fig. 6. Light ray incident on the fiber core at an angle q1 to the fiber axis

Assuming the entrance face at the fiber core to be normal to the axis, then considering the refraction at the air–core interface and using Snell's law given by Eq. (1.1):

$$n_0 \sin \phi_1 = n_1 \sin \phi_2 \tag{3}$$

Considering the right-angled triangle *ABC* indicated in Figure 6, then:

$$\phi = \frac{\pi}{2} - \theta_2 \tag{4}$$

where ϕ is greater than the critical angle at the core–cladding interface. Hence Eq. (3) becomes:

$$n_0 \sin \theta_1 = n_1 \cos \phi \tag{5}$$

Using the trigonometrical relationship $\sin_2 \phi + \cos_2 \phi = 1$, Eq. (5) may be written in the form:

$$n_0 \sin \theta_1 = n_1 (1 - \sin^2 \phi)^{1/2} \tag{6}$$

When the limiting case for total internal reflection is considered, ö becomes equal to the critical angle for the core–cladding interface and is given by Eq. (2). Also in this limiting case è1 becomes the acceptance angle for the fiber èa. Combining these limiting cases into Eq. (6) gives:

$$n_0 \sin \theta_a = (n_1^2 - n_2^2)^{1/2} \tag{7}$$

Equation (7), apart from relating the acceptance angle to the refractive indices, serves as the basis for the definition of the important optical fiber parameter, the numerical aperture (*NA*). Hence the *NA* is defined as:

$$NA = n_0 \sin \theta_a = (n_1^2 - n_2^2)^{1/2} \tag{8}$$

Since the *NA* is often used with the fiber in air where n_0 is unity, it is simply equal to $\sin \theta_a$. It may also be noted that incident meridional rays over the range $0 < \theta_1 < \theta_a$ will be propagated within the fiber.

SKEW RAYS

In the preceding sections we have considered the propagation of meridional rays in the optical waveguide. However, another category of ray exists which is transmitted without passing through the fiber axis. These rays, which greatly outnumber the meridional rays, follow a helical path through the fiber, as illustrated in Fig. 7, and are called skew rays.

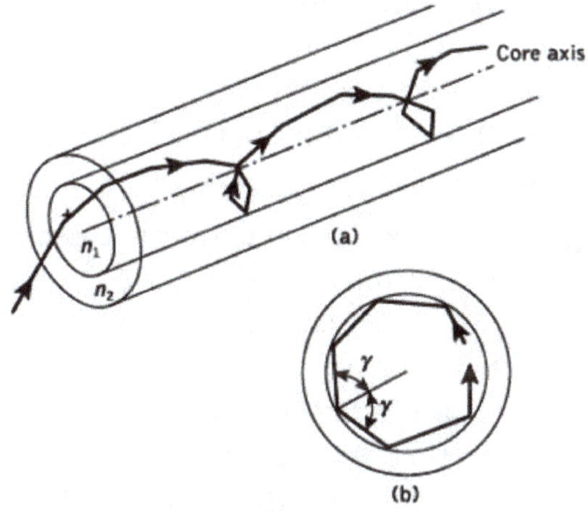

Fig. 7 A helical path through the fiber: (a) skew ray path down the fiber, (b) cross-sectional view of the fiber

It is not easy to visualize the skew ray paths in two dimensions, but it may be observed from Fig. 7(b) that the helical path traced through the fiber gives a change in direction of 2γ at each reflection, where ã is the angle between the projection of the ray in two dimensions and the radius of the fiber core at the point of reflection. Hence, unlike meridional rays, the point of emergence of skew rays from the fiber in air will depend upon the number of reflections they undergo rather than the input conditions to the fiber.

When the light input to the fiber is non uniform, skew rays will therefore tend to have a smoothing effect on the distribution of the light as it is transmitted, giving a more uniform output. The amount of smoothing is dependent on the number of reflections encountered by the skew rays. A further possible advantage of the transmission of skew rays becomes apparent when their acceptance conditions are considered

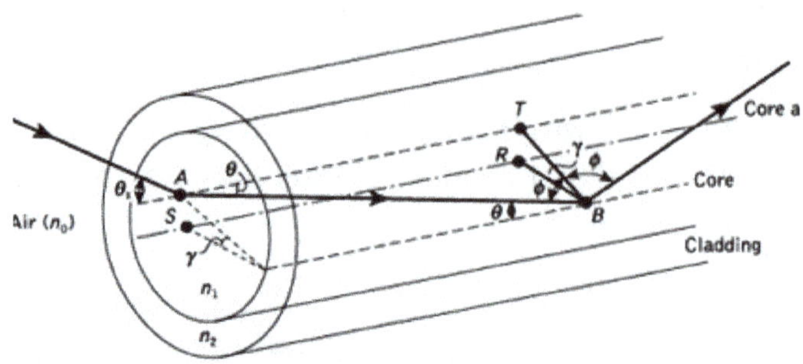

Fig. 8. Skew ray is shown incident on the fiber core at the point A, at an angle q[s] to the normal at the fiber end face

In order to calculate the acceptance angle for a skew ray it is necessary to define the direction of the ray in two perpendicular planes. The geometry of the situation is illustrated in Fig. 8 where a skew ray is shown incident on the fiber core at the point *A*, at an angle θ_s to the normal at the fiber end face. The ray is refracted at the air–core interface before traveling to the point *B* in the same plane. The angles of incidence and reflection at the point *B* are ϕ, which is greater than the critical angle for the core-cladding interface.

When considering the ray between *A* and *B* it is necessary to resolve the direction of the ray path *AB* to the core radius at the point *B*. As the incident and reflected rays at the point *B* are in the same plane, this is simply cos ϕ. However, if the two perpendicular planes through which the ray path *AB*traverses are considered, then γ is the angle between the core radius and the projection of the ray onto a plane *BRS* normal to the core axis, and ϕ is the angle between the ray and a line *AT* drawn parallel to the core axis. Thus to resolve the ray path *AB* relative to the radius *BR* in these two perpendicular planes requires multiplication by cos γ and sin θ

SINGLE-MODE WAVEGUIDE CONDITIONS IN OPTICAL FIBERS

Telecommunication applications predominantly use optical waveguides to transmit large amounts of data from one point to another. An optical waveguide's mode structure plays a significant role in determining the wave propagation in optical applications.

Depending on the mode's structure, optical waveguides transmit a single signal (single-mode) or a multitude (multimode) of signals. Since single-mode waveguide conditions are different from multimode conditions, optical waveguides need to be specifically designed for each type of mode.

Optical fibers used in telecommunication transmission systems are the derived versions of optical waveguides. In this article, we will focus our discussion on single-mode optical fibers, but first, let's briefly discuss optical waveguides in general.

OPTICAL WAVEGUIDES

Optical waveguides are the basic element in any optical circuit. They are often applied to integrated optical circuits, telecommunication systems, and biomedical imaging. Optical waveguides are classified into various types:

Depending on the Structure

- Planar
- Strip
- Fiber waveguide

Depending on the Material

- Glass
- Polymer
- Semiconductor

Depending on the Refractive Index Distribution

- Step
- Graded index

Depending on the Mode Structure

- Single-mode
- Multimode

Let's take a closer look at single-mode optical waveguides.

SINGLE-MODE OPTICAL WAVEGUIDES

In fiber optics, optical waveguides are commonly used to guide the propagation of electromagnetic signals and they only support a single transmission mode.

Single-Mode Optical Fibers

An optical fiber is an example of a non-planar optical waveguide. There are also two categories of optical fibers: single-mode optical fibers and multimode optical fibers. Single-mode optical fibers consist of a "core" placed at the center, through which electromagnetic signals are guided down to a target location. The core is enclosed inside "cladding". Cladding utilizes total internal reflection to prevent signal loss or spreading from the core. In commercial optical fibers, two protective layers cover the core and cladding unit.

Single-Mode Waveguide Conditions in Optical Fibers

Modal dispersion in optical waveguides is detrimental to their functioning. Single-mode optical waveguides are a solution to reduce modal dispersion. Single-mode optical waveguides are designed so that they prevent the existence of higher-order waveguide modes. Properly selecting waveguide parameters is critical to cutting off the higher-order waveguide modes in optical waveguides.

The V-Number

All optical waveguides are characterized by a parameter called the V-number, otherwise known as normalized frequency. The V-number is important in optical fibers to achieve single-mode waveguide conditions. The V-number is a unitless quantity and is given by equation (1), where γ is the wavelength of the electromagnetic signal, a is the core radius, and n_1 and n_2 are the refractive indices of core and cladding, respectively:

$$V = \frac{2\pi a}{\lambda}\sqrt{n_1^2 - n_2^2}$$

In optical fibers, the relationship to be satisfied for single-mode waveguide conditions is:

Optical fibers support the single propagation mode, LP01, when the V-number is less than 2.405. When the V-number is greater than 2.405, optical fibers support the next linearly polarized mode.

If the refractive index values are closer to each other, then thin optical fibers can establish single-mode wave propagation through them. The alignment of the core size is very important to achieve single-mode waveguide conditions in optical fibers. Cadence's software can help you design optical waveguides that support only single-mode propagation.

Subscribe to our newsletter for the latest updates. If you're looking to learn more about how Cadence has the solution for you, talk to us and our team of experts.

PHASE VELOCITY, GROUP VELOCITY, AND SIGNAL VELOCITY

Frequency dispersion in groups of gravity waves on the surface of deep water. The red dot moves with the phase velocity, and the green dots propagate with the group velocity. In this deep-water case, the phase velocity is twice the group velocity. The red dot overtakes two green dots when moving from the left to the right of the figure.

The velocity of a wave can be defined in many different ways, partly because there many different kinds of waves, and partly because we can focus on different aspects or components of any given wave. The ambiguity in the definition of "wave velocity" often leads to confusion, and we frequently read stories about experiments purporting to demonstrate "superluminal" propagation of electromagnetic waves (for example). Invariably, after looking into the details of these experiments, we find the claims of "superluminal communication" are simply due to a failure to recognize the differences between phase, group, and signal velocities.

In the simple case of a pure traveling sinusoidal wave we can imagine a "rigid" profile being physically moved in the positive x direction with speed v as illustrated below.

Clearly the wave function depends on both time and position. At any fixed instant of time, the function varies sinusoidally along the x axis, whereas at any fixed location on the x axis the function varies sinusoidally with time. One complete cycle of the wave can be associated with an "angular" displacement of 2p radians. The angular frequency w of a wave is the number of radians per unit time at a fixed position, whereas the wave number k is the number of radians per unit distance at a fixed time. (If we prefer to speak in terms of cycles rather than radians, we can use the wavelength l = 2p /k and the frequency n = w/2p .) In terms of these parameters we can express a pure travelling wave as the function

$$A(t,x) = A_0 \cos(kx - wt)$$

where the "amplitude" A_0 is the maximum of the function. (We use the cosine function rather than the sine merely for convenience, the difference being only a matter of phase.) The minus sign denotes the fact that if we hold t constant and increase x we are moving "to the right" along the function, whereas if we focus on a fixed spatial location and allow time to increase, we are effectively moving "to the left" along the function (or rather, it is moving to the right and we are stationary). Reversing the sign gives A_0 cos(kx + wt), which is the equation of a wave propagating in the negative x direction. Note that the function A(t,x) is the fundamental solution of the (one-dimensional) "wave equation"

Since w is the number of radians of the wave that pass a given location per unit time, and 1/k is the spatial length of the wave per radian, it follows that w/k = v is the speed at which the shape of the wave is moving, i.e., the speed at which any fixed phase of the cycle is displaced. Consequently this is called the *phase velocity* of the wave, denoted by v_p. In terms of the cyclical frequency and wavelength we have v_p = ln.

If we imagine the wave profile as a solid rigid entity sliding to the right, then obviously the phase velocity is the ordinary speed with which the actual physical parts are moving. However, we could also imagine the quantity "A" as the position along a transverse space axis, and a sequence of tiny massive particles along the x axis, each oscillating vertically in accord with A_0 cos(kx -wt). In this case the wave pattern propagates to the right with phase velocity v_p, just as before, and yet no material particle has any lateral motion at all. This illustrates that the phase of a traveling wave form may or may not correspond to a particular physical entity. It's entirely possible for a wave to "precess" through a sequence of material entities, none of which is moving in the direction of the wave. In a sense this is similar to the phenomenon of aliasing in signal processing. What we perceive as a coherent wave may in fact be simply a sequence of causally disjoint processes (like the individual spring-mass systems) that happen to be aligned spatially and temporally, either by chance or design, so that their combined behavior exhibits a wavelike pattern, even though there is no actual propagation of energy or information along the sequence.

Since a general wave (or wavelike phenomenon) need not embody the causal flow of any physical effects, there is obviously there is no upper limit on the possible phase velocity of a wave. However, even for a "genuine" physical wave, i.e., a chain of sequentially dependent events, the phase velocity does not necessarily correspond to the speed at which energy or information is propagating. This is partly a semantical issue, because in order to actually convey information, a signal cannot be a simple periodic wave, so we must consider non-periodic signals, making the notion of "phase" somewhat ambiguous. If the wave profile never exactly repeats itself, then arguably the "period" of the signal must be the entire signal. On this basis we might say that the velocity of the signal is unambiguously equal to the "phase velocity", but in this context the phase velocity could only be defined as the speed of the leading (or trailing) edge of the overall signal.

In practice and common usage, though, we tend to define the "phase" of a signal with respect to the intervals between consecutive local maxima (or minima, or zero crossings). To illustrate, consider a signal consisting of two superimposed sine waves with slightly different frequencies and wavelengths, i.e., a signal with the amplitude function

As most people know from experience, the combination of two slightly unequal tones produces a "beat", resulting from the tones cycling in and out of phase with each other. Using a well-known trigonometric identity we can express the two components of this signal in the form

Therefore, adding the two terms of A(x,t) together, the products of sines cancel out, and we can express the overall signal as

This can be somewhat loosely interpreted as a simple sinusoidal wave with the angular velocity w, the wave number k, and the modulated amplitude 2cos(Dkx – Dwt). In other words, the amplitude of the wave is itself a wave, and the phase velocity of this modulation wave is v =Dw/Dk. A typical plot of such a signal is shown below for the case w = 6 rad/sec, k = 6 rad/meter, Dw = 0.1 rad/sec, Dk = 0.3 rad/meter.

The "phase velocity" of the internal oscillations is w/k = 1 meter/sec, whereas the amplitude envelope wave (indicated by the dotted lines) has a phase velocity of Dw/Dk = 0.33 meter/sec. As a result, if we were riding along with the envelope, we would observe the internal oscillations moving forward from one group to the next.

The propagation of information or energy in a wave always occurs as a change in the wave. The most obvious example is changing the wave from being absent to being present, which propagates at the speed of the leading edge of a wave train. More generally, some modulation of the frequency and/or amplitude of a wave is required in order to convey information, and it is this modulation that represents the signal content. Hence the actual speed of content in the situation described above is Dw/Dk. This is the phase velocity of the amplitude wave, but since each amplitude wave contains a group of internal waves, this speed is usually called the group velocity.

Physical waves of a given type in a given medium generally exhibit a characteristic group velocity as well as a characteristic phase velocity. This is because within a given medium there is a fixed relationship between the wave number and the frequency of waves. For example, in a transparent optical medium the refractive index n is defined as the ratio c/v_p where c is the speed of light in vacuum and v_p is the phase velocity of light in that medium. Now, since $v_p = w/k$, we have $w = kc/n$. Bearing in mind that the refractive index is typically a function of the frequency (resulting in the "dispersion" of colors seen in prisms, rainbows, etc.), we can take the derivative of w as follows

Hence any modulation of an electromagnetic wave in this medium will propagate at the group velocity

In a medium whose refractive index is constant, independent of frequency (such as the vacuum), we have $dn/dk = 0$ and therefore the group velocity equals the phase velocity. On the other hand, in most commonly observable transparent media (such as air, water, glass, etc.) at optical frequencies have a refractive indices that increase slightly as a function of wave number and (therefore) frequency. This is why the high frequency (blue) components of a beam of white light are deflected more than the low frequency (red) components as they pass through a glass prism. It follows that the group velocity of light in such media (called dispersive) is less than the phase velocity.

It is quite possible for the phase velocity of a perfectly monochromatic wave of light, assuming such a thing exists, to exceed the value of c, because it conveys no information. In fact, the concept of a perfectly monochromatic beam of light is similar to the idea of a "free photon", in the sense that neither of them has any physical significance or content, because a photon must be emitted and absorbed, just as a beam of light cannot be infinite in extent and duration, but must always have a beginning and an end, which introduces a range of spectral components to the signal. Any modulation will propagate at the group velocity, which, in dispersive media, is always less than c.

An example of an actual physical application in which we must be careful to distinguish between the phase and the group velocity is the case of electromagnetic waves propagating through a hollow magnetic conductor, often called a *waveguide*. A waveguide imposes a "cutoff frequency" w_0 on any propagating electromagnetic waves based on the geometry of the tube, and will not sustain waves of any lower frequency. This is roughly analogous to how the pipes in a Church organ will sustain only certain resonant patterns. As a result, the dominant wave pattern of a propagating wave with a frequency of w will have a wave number k given by

$$k = \frac{1}{c}\sqrt{\omega^2 - \omega_0^2}$$

Since (as we've seen) the phase velocity is w/k, this implies that the (dominant) phase velocity in a waveguide with cutoff frequency w_0 is

$$v_p = \frac{c}{\sqrt{1 - \left(\frac{\omega_0}{\omega}\right)^2}}$$

Hence, not only is the phase velocity generally greater than c, it approaches infinity as w approaches the cutoff frequency w_0. However, the speed at which information and energy actually propagates down a waveguide is the group velocity, which (as we've seen) is given by dw/dk. Taking the derivative of the preceding expression for k with respect to w gives

$$\frac{dk}{d\omega} = \frac{\omega}{c\sqrt{\omega^{2-}\omega_0^2}}$$

so the group velocity in a waveguide with cutoff frequency w_0 is

$$v_g = \frac{d\omega}{dk} = c\sqrt{1 - \left(\frac{\omega_0}{\omega}\right)^2}$$

which of course is always less than or equal to c.

Unfortunately we frequently read in the newspapers about how someone has succeeded in transmitting a wave with a *group* velocity exceeding c, and we are asked to regard this as an astounding discovery, overturning the principles of relativity, etc. The problem with these stories is that the group velocity corresponds to the actual signal velocity only under conditions of normal dispersion, or, more generally, under conditions when the group velocity is less than the phase velocity.

In other circumstances, the group velocity does not necessarily represent the actual propagation speed of any information or energy. For example, in a regime of anomalous dispersion, which means the refractive index *decreases* with increasing wave number, the preceding formula shows that what we called the group velocity exceeds what we called the phase velocity. In such circumstances the group velocity no longer represents the speed at which information or energy propagates.

To see why the group velocity need not correspond to the speed of information in a wave, notice that in general, by superimposing simple waves with different frequencies and wavelengths, we can easily produce a waveform with a group velocity that is arbitrarily great, even though the propagation speeds of the constituent waves are all low. A snapshot of such a case is shown below. In this figure the sinusoidal wave denoted as "A" has a wave number of k_A = 2 rad/meter and an angular frequency of w_A = 2 rad/sec, so it's individual phase velocity is v_A = 1 meter/sec. The sinusoidal wave denoted as "B" has a wave number of k_B = 2.2 rad/meter and an angular frequency of w_B = 8 rad/sec, so it's individual phase velocity is v_B = 3.63 meters/sec.

The sum of these two signals is denoted as "A+B" and, according to the formulas given above, it follows that this sum can be expressed in the form 2cos(kx-wt)cos(Dkx-Dwt) where k = 5, w = 2.1, Dk = 0.1, and Dw = 3. Consequently, the "envelope wave" represented by the second factor has a phase velocity of 30 meters/sec. Nevertheless, it's clear that no information can be propagating faster than the phase speeds of the constituent waves A and B.

Indeed if we follow the midpoint of a "group" of A+B as it proceeds from left to right, we find that when it reaches the right hand side it consists of the sum of peaks of A and B that entered at the left long before the current "group" had even "appeared". This is just one illustration of how simple interfering phase effects can be miss-construed as ultra-high-speed signals. In fact, by simply setting k_A to 2.2 and k_B to 2.0, we can cause the "groups" of A+B to precess from *right to left*, which might mistakenly be construed as a signal propagating *backwards in time*!

Needless to say, we have *not* succeeded in arranging for a message to be received before it was sent, nor even in transmitting a message superluminally. Examples of this kind merely illustrate that the "group velocity" does not always represent the speed at which real information (or energy) is moving. This stands to reason, because the two cosine factors of the carrier/modulation waveform are formally identical, so we can't arbitrarily declare one of them to represent the carrier and the other to represent the modulation. Both are required, so we shouldn't expect the information speed to be any greater than the lesser of the two phase speeds, nor should it exceed the lesser of the phase speeds of the two components A and B.

Furthermore, we already know that the transmission of information via an individual wave such as A will propagate at the speed of an incremental disturbance of A, which depends on how dw and dk are related to each other. In the example above we arbitrarily selected increments of w and k, but our ability to do this in a physical context would depend on a great deal of flexibility in the wave propagation properties of the medium. This is where the ingenuity of the experimenter can be deployed to arrange various exotic substances and fields in such a way as to permit the propagation of waveforms with the desired properties. The important point to keep in mind is that none of these experiments actually achieves meaningful superluminal transfer of information.

Incidentally, since we can contrive to make the "groups" propagate in either direction, it's not surprising that we can also make them stationary. Two identical waves propagating in opposite directions at the same speed are given by

Superimposing these two waves propagating (with synchronized nodes) in opposite directions yields a standing pure wave

GOOS–HAENCHEN SHIFT

Introduction

The following section explains the Goos–Haenchen shift in detail. Lateral displacement is known as the Goos–Haenchen shift

Theory

- The phase change incurred with the total internal reflection of a light beam on a planar dielectric interface may be understood from physical observation.
- It shows that the reflected beam is shifted laterally from the trajectory predicted by simple ray theory analysis, as illustrated in Figure 9
- The geometric reflection appears to take place at a virtual reflecting plane which is parallel to the dielectric interface in the lower index medium, as indicated in Figure 9.
- Utilizing wave theory it is possible to determine this lateral shift although it is very small and difficult to observe.
- This concept provides an important insight into the guidance mechanism of dielectric optical waveguides.

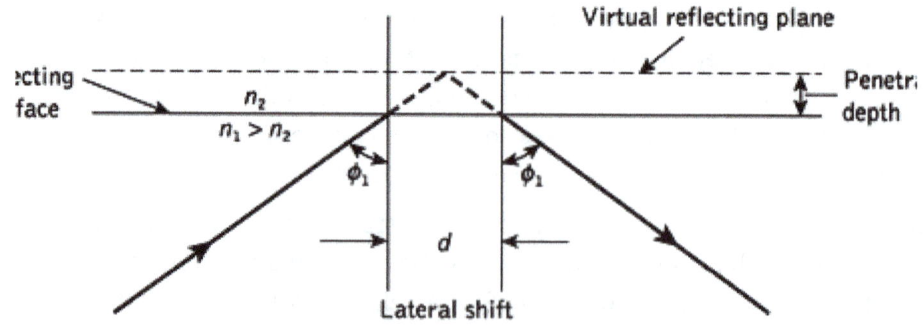

Fig. 9. Goos–Haenchen shift

CYLINDRICAL FIBER

Introduction

Optical fiber transmission takes place through guided modes. The modes are determined from the Eigen values of second-order differential equations and their boundary conditions, by an analysis similar to that used for propagation in cylindrical waveguides

Modes

- The exact solution of Maxwell's equations for a cylindrical homogeneous core dielectric waveguide involves much algebra and yields a complex result
- The cylindrical waveguide is bounded in two dimensions rather than one.
- Thus two integers, l and m, are necessary in order to specify the modes
- For the cylindrical waveguide consider TE_{lm} and TM_{lm} modes
- These modes correspond to meridional rays traveling within the fiber.
- Hybrid modes where Ez and Hz are nonzero also occur within the cylindrical waveguide.
- These modes are designated HE_{lm} and EH_{lm} depending upon whether the components of H or E make the larger contribution to the transverse (to the fiber axis) field.
- These fibers satisfy the weakly guiding approximation [Ref. 16] where the relative index difference Ä<<1. This corresponds to small grazing angles θ in Eq. (1).

  $$B_z = n_1 k \cos\theta \quad (1)$$
- Ä is usually less than 0.03 (3%) for optical communications fibers

Degenerate mode

- For weakly guiding structures with dominant forward propagation, mode theory gives dominant transverse field components.
- Hence approximate solutions for the full set of HE, EH, TE and TM modes may be given by two linearly polarized components.
- These linearly polarized (LP) modes are not exact modes of the fiber except for the fundamental (lowest order) mode.
- As Δ in weakly guiding fisbers is very small, then HE–EH mode pairs occur which have almost identical propagation constants.
- Such modes are said to be degenerate.
- The super-positions of these degenerating modes characterized by a common

propagation constant correspond to particular LP modes regardless of their HE, EH, TE or TM field configurations.

- This linear combination of degenerate modes obtained from the exact solution produces a useful simplification in the analysis of weakly guiding fibers.
- The relationship between the traditional HE, EH, TE and TM mode designations and the LP_{lm} mode designations is shown in Table 1. The mode subscripts l and m are related to the electric field intensity profile for a particular LP mode
- There are in general 2l field maxima around the circumference of the fiber core and m field maxima along a radius vector

Table 1 Correspondence between the lower order in linearly polarized modes and the traditional exact modes from which they are formed

Linearly polarized	Exact
LP_{01}	HE_{11}
LP_{11}	HE_{21}, TE_{01}, TM_{01}
LP_{21}	HE_{31}, EH_{11}
LP_{02}	HE_{12}
LP31	HE_{41}, EH_{21}
LP_{12}	HE_{22}, TE_{02}, TM_{02}
LP_{lm}	HE_{2m}, TE_{0m}, TM_{0m}
LP_{lm} ($l \neq 0$ or 1)	$HE_{l+1.m}$, $EH_{l-1.m}$

- The subscript l in the LP notation now corresponds to HE and EH modes with labels l_1 and $l + 1$ respectively
- The electric field intensity profiles for the lowest three LP modes, together with the electric field distribution of their constituent exact modes, are shown in Figure 10.
- It may be observed from the field configurations of the exact modes that the field strength in the transverse direction (Ex or Ey) is identical for the modes which belong to the same LP mode.
- Hence the origin of the term 'linearly polarized

MODE COUPLING

Introduction

In mode coupling the frequencies are close together and radial and geographical shapes are similar. It affects the transmission properties

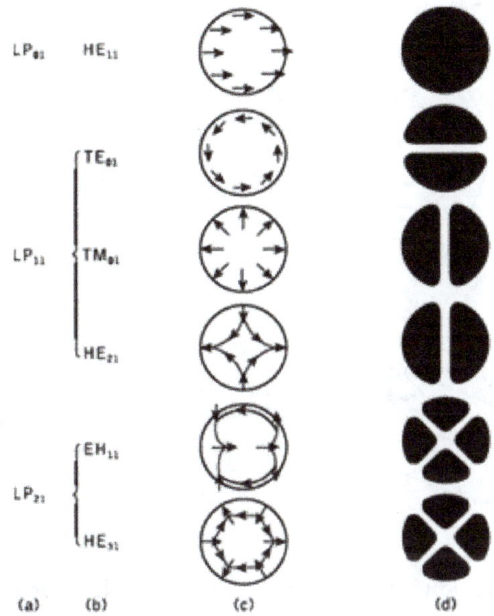

Fig. 10. (a) LP mode designations; (b) exact mode designations; (c) electric field distribution of the exact modes; (d) intensity distribution of Ex for the exact modes

Mode coupling theory

- Waveguide perturbations such as deviations of the fiber axis from straightness, variations in the core diameter, irregularities at the core–cladding interface and refractive index variations may change the propagation characteristics of the fiber (fig 11)
- These will have the effect of coupling energy traveling in one mode to another depending on the specific perturbation.
- It may be observed that in both cases the ray no longer maintains the same angle with the axis.
- In electromagnetic wave theory this corresponds to a change in the propagating mode for the light.
- Thus individual modes do not normally propagate throughout the length of the fiber without large energy transfers to adjacent modes, even when the fiber is exceptionally good quality and is not strained or bent by its surroundings.

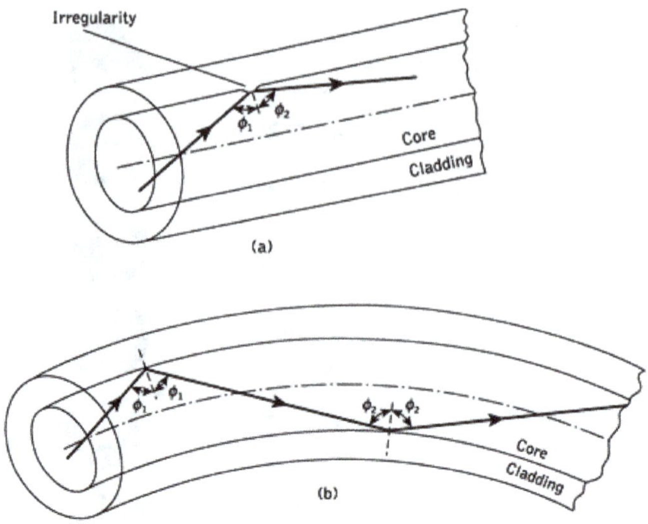

Fig. 11. Ray theory illustrations showing two of the possible fiber perturbations which give mode coupling: (a) irregularity at the core-vladding interface and (b) fiber bend

- This mode conversion is known as mode coupling or mixing. It is usually analyzed using coupled mode equations which can be obtained directly from Maxwell's equations.
- Mode coupling affects the transmission properties of fibers in several important ways, a major one being in relation to the dispersive properties of fibers over long distances

STEP INDEX FIBERS

Introduction

The refractive index of the step index fiber s uniform throughout the fiber

Fiber structure

- The optical fiber considered in the preceding sections with a core of constant refractive index n_1 and a cladding of a slightly lower refractive index n_2 is known as step index fiber.
- This is because the refractive index profile for this type of fiber makes a step change at the core-cladding interface, as indicated in Fig. 12, which illustrates the two major types of step index fiber.
- The refractive index profile may be defined as:

$$n(r) = \begin{cases} n_1 \; r < a \text{ (core)} \\ n2 \; r \geq a \text{ (cladding)} \end{cases}$$

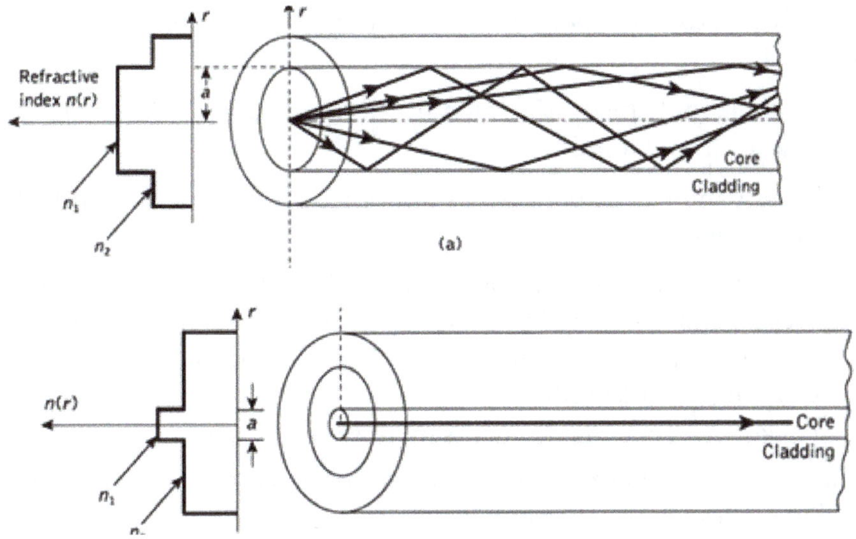

Fig. 12. The refractive index profile and ray transmission in step index fibers: (a) multimode step index; (b) single-mode step index fiber

- Figure 12(a) shows a multimode step index fiber with a core diameter of around 50 μm or greater, which is large enough to allow the propagation of many modes within the fiber core.
- This is illustrated in Figure 12(a) by the many different possible ray paths through the fiber.
- Figure 12(b) shows a single-mode or mono mode step index fiber which allows the propagation of only one transverse electromagnetic mode (typically HE11), and hence the core diameter must be of the order of 2 to 10 μm.
- The propagation of a single mode is illustrated in Figure 12(b) as corresponding to a single ray path only (usually shown as the axial ray) through the fiber.
- The single-mode step index fiber has the distinct advantage of low intermodal dispersion (broadening of transmitted light pulses), as only one mode is transmitted, whereas with multimode step index fiber considerable dispersion may occur due to the differing group velocities of the propagating modes.
- This in turn restricts the maximum bandwidth attainable with multimode step index fibers, especially when compared with single-mode fibers

The single-mode process has the following properties.

- It (almost) eliminates delay spread.
- It is more difficult to splice and to exactly align two fibers together.
- It is more difficult to couple all photonic energy from a source into it.

- It is difficult to study propagation with ray theory: Maxwell's equations are required
- It is suitable for transmitting modulated signals at 40 Gb/s (or higher) and up to 200 km without amplification.
- For lower bandwidth applications multimode fibers have several advantages over single-mode fibers. These are
    a. The use of spatially incoherent optical sources (e.g. most light-emitting diodes) which cannot be efficiently coupled to single-mode fibers;
    b. Larger numerical apertures, as well as core diameters, facilitating easier coupling to optical sources;
    c. Lower tolerance requirements on fiber connectors.
- Multimode step index fibers allow the propagation of a finite number of guided modes along the channel.
- The number of guided modes is dependent upon the physical parameters (i.e. relative refractive index difference, core radius) of the fiber and the wavelengths of the transmitted light which are included in the normalized frequency V for the fiber.
- There is a cutoff value of normalized frequency Vc for guided modes below which they cannot exist.
- Mode propagation does not entirely cease below cutoff.
- Modes may propagate as unguided or leaky modes which can travel considerable distances along the fiber.
- It is the guided modes which are of paramount importance in optical fiber communications as these are confined to the fiber over its full length.
- This allows an estimate of the number of guided modes propagating in a particular multimode step index fiber.
- In an ideal multimode step index fiber with properties (i.e. relative index difference, core diameter) which are independent of distance, there is no mode coupling, and the optical power launched into a particular mode remains in that mode and travels independently of the power launched into the other guided modes.
- The majority of these guided modes operates far from cutoff, and is well confined to the fiber core.
- Thus most of the optical power is carried in the core region and not in the cladding.
- The properties of the cladding (e.g. thickness) do not therefore significantly affect the propagation of these modes.

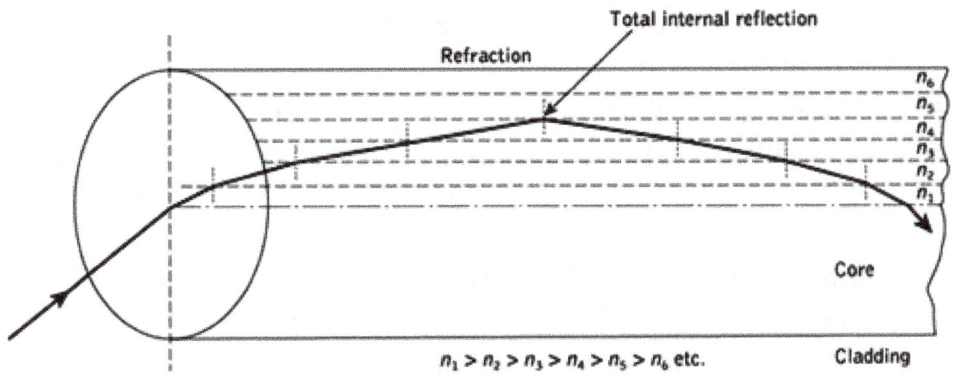

Fig. 13. Overall curved ray path

GRADED INDEX FIBERS

Introduction

Modal dispersion can be reduced by grading the refractive index a maximum value at its center to a minimum value at the core-cladding boundary. The fiber is then called a graded-index fiber

Graded index fibers

- Graded index fibers do not have a constant refractive index in the core but a decreasing core index n(r) with radial distance from a maximum value of n1 at the axis to a constant value n_2 beyond the core radius a in the cladding.
- This index variation may be represented as:

$$n(r) = \begin{cases} n_1\,(1-2\Delta(r/a)^{\alpha})^{1/2} & r < a \text{ (core)} \\ {}_1\,(1-2\Delta)^{1/2} = n_2 r \geq a \text{ (cladding)} \end{cases} \quad 1$$

- Where Δ is the relative refractive index difference and α is the profile parameter which gives the characteristic refractive index profile of the fiber core.
- Equation (1) which is a convenient method of expressing the refractive index profile of the fiber core as a variation of á, allows representation of the step index profile when $\alpha = \infty$, a parabolic profile when $\alpha = 2$ and a triangular profile when $\alpha = 1$.
- The meridional rays shown appear to follow curved paths through the fiber core.

- The gradual decrease in refractive index from the center of the core creates much refraction of the rays as they are effectively incident on a large number or high to low index interfaces
- This mechanism is illustrated in Figure 13 where a ray is shown to be gradually curved, with an ever increasing angle of incidence, until the conditions for total internal reflection are met, and the ray travels back towards the core axis, again being continuously refracted
- Multimode graded index fibers exhibit far less intermodal dispersion than multimode step index fibers due to their refractive index profile.
- Considering ray theory, the rays traveling close to the fiber axis have shorter paths when compared with rays which travel into the outer regions of the core the near axial rays are transmitted through a region of higher refractive index and therefore travel with a lower velocity than the more extreme rays.

SINGLE-MODE FIBERS

Introduction

The index of refraction of single mode fibers is a step index. The core diameter is within 3-10, µm. These fibers are used for higher data rate transmission and long haul communications

Single-mode fibers

- The advantage of the propagation of a single mode within an optical fiber is that the signal dispersion caused by the delay differences between different modes in a multimode fiber may be avoided.
- For the transmission of a single mode the fiber must be designed to allow propagation of only one mode, while all other modes are attenuated by leakage or absorption
- For single-mode operation, only the fundamental LP_{01} mode can exist.
- Hence the limit of single-mode operation depends on the lower limit of guided propagation for the LP_{11} mode.
- The cutoff normalized frequency for the LP_{11} mode in step index fibers occurs at $Vc = 2.405$.
- Thus single-mode propagation of the LP_{01} mode in step index fibers is possible over the range:

  $0 \leq V < 2.405$ (1)
- There is no cutoff for the fundamental mode.
- It must be noted that there are in fact two modes with orthogonal polarization

over this range, and the term single-mode applies to propagation of light of a particular polarization.

- Also, it is apparent that the normalized frequency for the fiber may be adjusted to within the range given in Eq. (1) by reduction of the core radius, and possibly the relative refractive index difference, for single-mode fibers, is usually less than 1%.
- It is possible to achieve single-mode operation with a slightly larger core diameter by reducing the relative refractive index difference of the fiber.
- Both these factors create difficulties with single-mode fibers.
- The small core diameters pose problems with launching light into the fiber and with field jointing, and the reduced relative refractive index difference presents difficulties in the fiber fabrication process.
- Graded index fibers may also be designed for single-mode operation and some specialist fiber designs do adopt such non step index profiles.
- However, it may be shown that the cutoff value of normalized frequency V_c to support a single mode in a graded index fiber is given by:

  $$V_c = 2.405(1+2/\alpha)^{1/2} \quad (2)$$
- It is possible to determine the fiber parameters which give single-mode operation.
- The critical value of normalized frequency for the parabolic profile graded index fiber is increased by a factor of $\sqrt{2}$ on the step index case
- The maximum V number which permits single-mode operation can be increased still further when a graded index fiber with a triangular profile is employed
- Significantly larger core diameter single-mode fibers may be produced utilizing index profile.
- With single-mode fibers with low relative refractive index differences and low V values is that the electromagnetic field associated with the LP_{10} mode extends appreciably into the cladding
- With V values less than 1.4, over half the modal power propagates in the cladding
- Thus the exponentially decaying evanescent field may extend significant distances into the cladding.
- It is therefore essential that the cladding is of a suitable thickness, and has low absorption and scattering losses in order to reduce attenuation of the mode.

- The total fiber cross-section for single-mode fibers is of a comparable size to multimode fibers for reducing the additional losses resulting from microbending
- Single-mode fiber design which allows the V value to be increased above 2.405 is the W fiber
- Use of such two-step cladding allows the loss threshold between the desirable and undesirable modes to be substantially increased.

THE GAUSSIAN APPROXIMATION

Introduction

Gaussian approximation for the transverse field distribution is very much simpler. Gaussian approximation give good result for fibers operated near the cutoff wavelength of the second-order mode.

Gaussian Approximation for optical fiber:

- The field shape of the fundamental guided mode within a single-mode step index fiber for two values of normalized frequency.
- It has the form of a Bessel function (J0(r)) in the core region matched to a modified Bessel function (K0(r)) in the cladding.
- Depending on the value of the normalized frequency, a significant proportion of the modal power is propagated in the cladding region.
- Hence, even at the cutoff value (i.e. Vc) only about 80% of the power propagates within the fiber core
- It may be observed from Figure 14 that the shape of the fundamental LP_{01} mode is similar to a Gaussian shape, which allows an approximation of the exact field distribution by a Gaussian function.
- The approximation may be investigated by writing the scalar wave equation:

$$\nabla^2\Psi + n^2k^2\Psi = 0 \tag{1}$$

Where

- k is the propagation vector
- n(x, y) is the refractive index of the fiber, which does not generally depend on z, the coordinate along the fiber axis.
- It should be noted that the time dependence exp (jωt) has been omitted from the scalar wave equation to give the reduced wave equation in Eq. (1).
- This representation is valid since the guided modes of a fiber with a small refractive index difference (i.e. Δ H=1) have one predominant transverse field component, for example Ey.

- By contrast Ex and the longitudinal component are very much smaller
- The field of the fundamental guided mode may therefore be considered as a scalar quantity and need not be described by the full set of Maxwell's equations.
- Hence Eq. (1) may be written as:

$$\nabla^2 + n^2 k^2 \phi = 0 \quad 2$$

- Where ϕ represents the dominant transverse electric field component
- The near-Gaussian shape of the predominant transverse field component of the fundamental mode has been demonstrated for fibers with a wide range of refractive index distributions.
- This proves to be the case not only for the LP_{01} mode of the step index fiber, but also for the modes with fibers displaying arbitrary graded refractive index distributions.
- Therefore, the predominant electric field component of the single guided mode may be written as the Gaussian function

$$\phi = \left(\frac{2}{\pi}\right)^{1/2} \frac{1}{\omega_0} \exp(-r^2/\omega_0^2)\exp(-j\beta z) \quad (3)$$

Where

- radius parameter $r^2 = x^2 y^2$,
- ω_0 is a width parameter which is often called the spot size or radius of the fundamental mode and
- β is the propagation constant of the guided mode field.
- If it is accepted that Eq. (3) is to a good approximation the correct shape, then the parameters β and ω_0 may be obtained either by substitution or by using a variational principle.
- Using the latter technique, solutions of the wave equation, Eq. (2), are claimed to be functions of the minimum integral:

$$J = \int_v [(\nabla\phi) \cdot [(\nabla\phi^*) - \Pi^2 k^2 \phi\phi^*] dV = \min \quad (4)$$

- Where the asterisk indicates complex conjugation.
- The integration range in Eq. (4) extends over a large cylinder with the fiber at its axis.
- The length of the cylinder L is arbitrary and its radius is assumed to tend towards infinity.

- Use of variational calculus indicates that the wave equation Eq. (2) is the Euler equation of the variational expression given in Eq. (4).
- Hence, the functions that minimize J satisfy the wave equation.

EQUIVALENT STEP INDEX

Introduction

It is the method to obtain approximate value of the cutoff wavelength and spot size for graded index single mode fiber

ESI for single mode fiber

- It is another strategy to obtain approximate values for the cutoff wavelength and spot size in graded index single-mode fibers
- Such ESI methods allow the propagation characteristics of single-mode fibers to be represented by a few parameters.
- Several different suggestions have been advanced for the choice of the core radius a ESI, and the relative index difference ÄESI, of the ESI fiber which lead to good approximations for the spot size (and hence joint and bend losses) for the actual graded index fiber.
- They are all conceptually related to the Gaussian approximation
- An early proposal for the ESI method involved transformation of the basic fiber parameters following

  $$a_s + Xa \qquad V_s = YV \qquad NA_s = (Y/X)NA \tag{1}$$

- Where the subscript s is for the ESI fiber and X, Y are constants which must be determined.
- These ESI fiber representations are only valid for a particular value of normalized frequency V and hence there is a different X, Y pair for each wavelength.
- The transformation can be carried out on the basis of either compared radii or relative refractive index differences.
- Figure 15 compares the refractive index profiles and the electric field distributions for two graded index fibers ($\alpha = 2, 4$) and their ESI fibers.
- It may be observed that their fields differ slightly only near the axis.
- An alternative ESI technique is to normalize the spot size ù0 with respect to an optimum effective fiber core radius $a_{eff.}$
- Latter quantity is obtained from the experimental measurement of the first minimum (angle θ_{min}) in the diffraction pattern using transverse illumination of the fiber immersed in an index-matching fluid. Hence:

$a_{eff} = 3.832/k \sin \theta_{min}$ (2)

Where $k = 2\Pi/\lambda$.

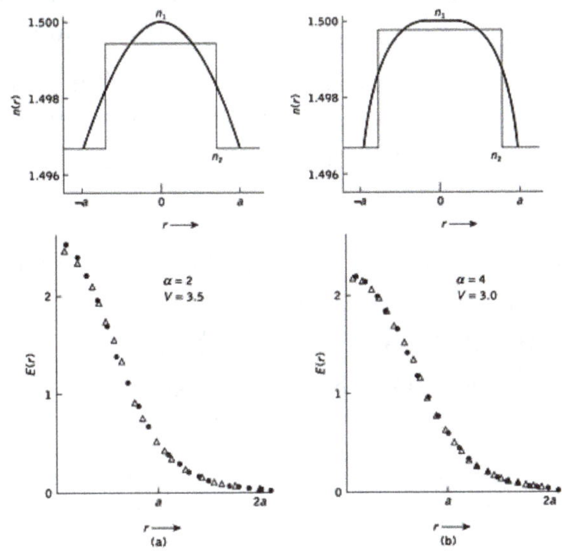

Fig. 15. Refractive index distribution n(r) and electric field distribution R(r) for graded index fibers and their ESI fiber for (a) a = 2, V = 3.5; (b) a = 4, V = 3.0

- In order to obtain the full comparison with single-mode step index fiber, the results may be expressed in terms of an effective normalized frequency V_{eff} which relates the cutoff frequencies/wavelengths for the two fibers:

$$V_{eff} = 2.405(V/V_c) = 2.405(\lambda_c/\lambda) \quad (3)$$

- The technique provides a dependence of ω_0/a_{eff} on V_{eff} which is almost identical for a reasonably wide range of profiles which are of interest for minimizing dispersion (i.e. $1.5 < V_{eff} < 2.4$).

PHOTONIC CRYSTAL FIBERS

Introduction

The following section explain the Photonic Crystal Fiber in detail

Introduction to Photonic crystal fibers:

- A new class of micro structured optical fiber containing a fine array of air holes running longitudinally downs the fiber cladding
- Since the microstructure within the fiber is often highly periodic due to the fabrication process, these fibers are usually referred to as photonic crystal fibers (PCFs), or sometimes just as holey fibers
- In PCFs two distinct guidance mechanisms arise

- PCFs have been used to realize various optical components and devices including long period gratings, multimode interference power splitters, tunable coupled cavity fiber lasers, fiber amplifiers, multichannel add/ drop filters, wavelength converters and wavelength de-multiplexers
- Increased control over the homogeneity of the fiber structures together with the use of highly purified silicon as the base material has now lowered these losses to a level of a very few decibels per kilometer for most PCF types

Index-guided microstructures:

- Although the principles of guidance and the characteristics of index-guided PCFs are similar to those of conventional fiber, there is greater index contrast since the cladding contains air holes with a refractive index of 1 in comparison with the normal silica cladding index of 1.457 which is close to the germanium-doped core index of 1.462.
- A fundamental physical difference between index-guided PCFs and conventional fibers arises from the manner in which the guided mode interacts with the cladding region.
- Whereas in a conventional fiber this interaction is largely first order and independent of wavelength, the large index contrast combined with the small structure dimensions cause the effective cladding index to be a strong function of wavelength.
- For short wavelengths the effective cladding index is only slightly lower than the core index and hence they remain tightly confined to the core.
- At longer wavelengths the mode samples more of the cladding and the effective index contrast is larger.
- This wavelength dependence results in large number of unusual optical properties which can be tailored.
- The high index contrast enables the PCF core to be reduced from around 8 μm in conventional fiber to less than 1 μm, which increases the intensity of the light in the core and enhances the nonlinear effects.
- Two common index-guided PCF designs are shown diagrammatically in Figure 16.
- In both cases a solid-core region is surrounded by a cladding region containing air holes.
- The cladding region in Figure 16(a) comprises a hexagonal array of air holes while in Figure 16(b) the cladding air holes are not uniform in size and do not extend too far from the core.

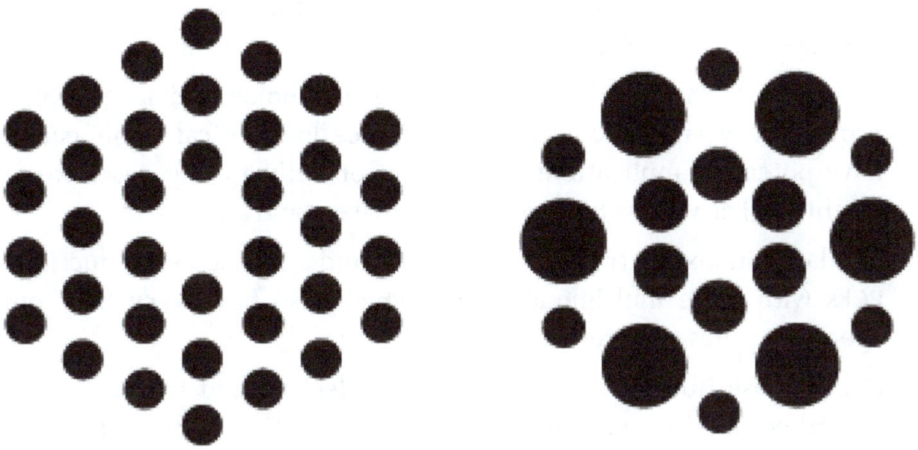

Fig. 16. Two common index-guided PCF designs

- It should be noted that the hole diameter d and hole to hole spacing or pitch ∧ are critical design parameters used to specify the structure of the PCF.
- In a silica PCF with the structure depicted in Figure 16(a) when the air fill fraction is low (i.e. $d/\wedge < 0.4$), then the fiber can be single-moded at all wavelengths.
- This property, which cannot be attained in conventional fibers, is particularly significant for broadband applications such as wavelength division multiplexed transmission
- As PCFs have a wider range of optical properties in comparison with standard optical fibers, they provide for the possibility of new and technologically important fiber devices.
- When the holey region covers more than 20% of the fiber cross-section index-guided PCFs display an interesting range of dispersive properties which could find application as dispersion-compensating or dispersion-controlling fiber components.
- In such fibers it is possible to produce very high optical nonlinearity per unit length in which modest light intensities can induce substantial nonlinear effects.
- While several kilometers of conventional fiber are normally required to achieve 2R data regeneration it was obtained with just 3.3 m of large air-filling fraction PCF.
- Filling the cladding holes with polymers or liquid crystals allows external fields to be used to dynamically vary the fiber properties.

- The temperature sensitivity of a polymer within the cladding holes may be employed to tune a Bragg grating written into the core.
- By contrast, index-guided PCFs with small holes and large hole spicing provide very large mode area (and hence low optical nonlinearities) and have potential applications in high-power delivery (e.g. laser welding and machining) as well as high-power fiber lasers and amplifiers.
- The large index contrast between silica and air enables production of such PCFs with large multimoded cores which also have very high numerical aperture values (greater than 0.7).
- Hence these fibers are useful for the collection and transmission of high optical powers in situations where signal distortion is not an issue.
- Finally, it is apparent that PCFs can be readily spliced to conventional fibers, thus enabling their integration with existing components and subsystems

CHAPTER

CHARACTERISTICS OF OPTICAL TRANSMISSION 3

INTRODUCTION

One of the important properties of optical fiber is signal attenuation. It is also known as fiber loss or signal loss. The signal attenuation of fiber determines the maximum distance between transmitter and receiver. The attenuation also determines the number of repeaters required, maintaining repeater is a costly affair.

Another important property of optical fiber is distortion mechanism. As the signal pulse travels along the fiber length it becomes broader. After sufficient length the broad pulses starts overlapping with adjacent pulses. This creates error in the receiver. Hence the distortion limits the information carrying capacity of fiber.

The factors which affect the performance of optical fibers as a transmission medium were not dealt with in detail. These transmission characteristics are of utmost importance when the suitability of optical fibers for communication purposes is investigated. The transmission characteristics of most interest are those of attenuation (or loss) and bandwidth.

The huge potential bandwidth of' optical communications helped stimulate the birth of the idea that a dielectric waveguide made of glass could be used to carry wideband telecommunication signals. However, at the time the idea may have seemed somewhat ludicrous as a typical block of glass could support optical transmission for at best a few tens of meires before it was attenuated to an unacceptable level.

Nevertheless, careful investigation of' the attenuation showed that it was largely due to absorption in the glass, caused by impurities such as iron, copper, manganese and other transition metals which occur in the third row of the periodic table. Hence, research was stimulated towards a new aerteration of 'pure' glasses for use in optical fiber communications.

A major breakthrough came in 1970 when the first fiber with an attenuation below 20 dB km-2 was reported. This level of attenuation was seen as the absolute minimum that had to be achieved before an optical fiber system could in any way compete economically with existing communication systems. Since 1970 tremendous improvements have been made, leading to silica-based glass fibers with losses of less than 0.2 dB km-2 in the laboratory.

Hence, comparatively low loss fibers have been incorporated into optical communication systems throughout the world. Moreover, as the fundamental lower limits for attenuation in silicate glass fibers have virtually been achieved, activities are increasing in relation to the investigation of other material systems which may exhibit substantially lower losses when operated at longer wavelengths. Such mid-infrared (and possibly far-infrared) transmitting fibers could eventually provide for extremely long-haul repeaterless communication assuming that, in addition to the material considerations, the optical source and detector requirements can be satisfactorily met.

The other characteristic of primary importance is the bandwidth of the fiber. This is limited by the signal dispersion within the fiber, which determines the number of bits of information transmitted in a given time period. Therefore, once the attenuation was reduced to acceptable levels attention was directed towards the dispersive properties of fibers Again, this has led to substantial improvements, giving wideband fiber bandwidths of many tens of gigahertz over a number of kilometres.

In order to appreciate these advances and possible future developments, the optical transmission characteristics of fibers must be considered in greater depth. Therefore, in this chapter we discuss the mechanisms within optical fibers which give rise to the major transmission characteristics mentioned previously (attenuation and dispersion), whilst also considering other, perhaps less obvious, effects when light is propagating down an optical fiber (modal noise, polarization and nonlinear phenomena).

We begin the discussion of attenuation with calculation of the total losses incurred in optical fibers. The various attenuation mechanisms (material absorption, linear scattering, nonlinear scattering, fiber bends). The primary focus within these sections is on silica-based glass fibers. However, consideration is given to other material systems which may be employed for mid-infrared and far-infrared optical transmission. Together with the associated limitations on fiber bandwidth.

Attenuation

The attenuation or transmission loss of optical fibers has proved to be one of the most important factors in bringing about their wide acceptance in telecommunications. As channel attenuation largely determined the maximum transmission distance prior to signal restoration, optical fiber communications became especially attractive when the transmission losses of fibers were reduced below those of the competing metallic conductors (less than 5 dB km-2).

Signal attenuation within optical fibers, as with metallic conductors, is usually expressed in the logarithmic unit of the decibel. The decibel, which is used for comparing two power levels, may be defined for a particular optical wavelength as the ratio of the input (transmitted) optical power Pi into a fiber to the output (received) optical power Po from the fiber as:

Number of decibels (dB) = 10 log Pi/Po (1)

This logarithmic unit has the advantage that the operations of multiplication and division reduce to addition and subtraction, whilst powers and roots reduce to multiplication and division. However, addition and subtraction require a conversion to numerical values which may be obtained using the relationship:

Pi/Po = 10 (dB/10) (2)

In optical fiber communications the attenuation is usually expressed in decibels per unit length- (i.e. dB/ km) following:

dBL = 10 log Pi/Po (3)

where dB is the signal attenuation per unit length in decibels and L is the fiber length

Example 1

When the mean optical power launched into an 8 km length of fiber is 120 μW, the mean optical power at the fiber output is 3 μW.

Determine:

- the overall signal attenuation or loss in decibels through the fiber assuming there are no connectors or splices;
- the signal attenuation per kilometre for the fiber.
- the overall signal attenuation for a 10 km optical link using the same fiber with splices at I km intervals, each giving an attenuation of I dB;
- the numerical input/output power ratio in (c).

Solution:

- Using Eq. (1), the overall signal attenuation in decibels through the fiber is:

  Signal attenuation = 10 log Pi/Po = 10 log 120 × 10-6/3 × 10-6
  = 10 log 40 = 16.0 dB
- The signal attenuation per kilornetre for the fiber may be simply obtained by dividing the result in (a) by the fiber length which corresponds to it using Eq. (3) where,

  adB L = 16.0 dB

  hence,

adB = 16.0 / 8

= 2.0 dBkm-1

- As adB = 2 dB km-1 the loss incurred along 10 km of the fiber is given by

  adB L = 2x10 = 20 dB

However, the link also has nine splices (at 1 km intervals) each with an attenuation of 1 dB. Therefore, the loss due to the splices is 9 dB.

Hence, the overall signal attenuation for the link is:

Signal attenuation = 20 + 9

= 29 dB

- To obtain a numerical value for the input/output power raLiO, Eq. (2) may be used where :

$$Pi/Po = 1029/10 = 794.3$$

A number of mechanisms are responsible for the signal attenuation within optical fibers. These mechanisms are influenced by the material composition, the preparation and purification technique, and the waveguide structure. They may be categorized within several major areas which include material absorption, material scattering (linear and nonlinear scattering), curve and microbending losses, mode coupling radiation losses and losses due to leaky modes. There are also losses at connectors and splices, as illustrated in Example 1. However, in this chapter we are interested solely in the characteristics of the fiber. It is instructive to consider in some detail the loss mechanisms within optical fibers in order to obtain an understanding of the problems associated with the design and fabrication of low loss waveguides.

MATERIAL ABSORPTION LOSSES IN SILICA GLASS FIBERS

Material absorption is a loss mechanism related to the material composition and the fabrication process for the fiber, which results in the dissipation of some of the transmitted optical power as heat in the waveguide. The absorption of the light may be intrinsic (caused by the interaction with one or more of the major components of the glass) or extrinsic (caused by impurities within the glass).

Intrinsic absorption

An absolutely pure silicate glass has little intrinsic absorption due to its basic material structure in the near-infrared region. However, it does have two major intrinsic absorption mechanisms at optical wavelengths which leave a low intrinsic absorption window over the 0.8 to 1.7 μm wavelength range, which shows a possible optical attenuation against wavelength characteristic for absolutely pure glass. It may be observed that there is a fundamental absorption edge, the peaks of which are centred in the ultraviolet wavelength region. This is due to the stimulation of electron transitions within the glass bv higher energy excitations. The tail of this peak may extend into the window region at the shorter wavelengths.

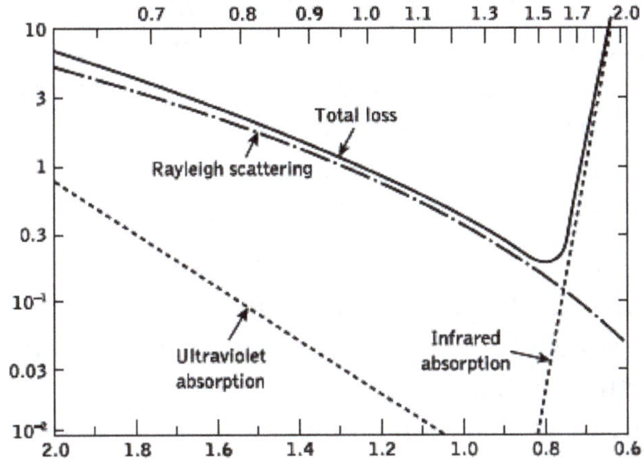

Fig. 1 The attenuation spectra for the intrinsic loss mechanisms in pure GeO2–SiO2 glass

- Intrinsic absorption occurs when material is in absolutely pure state, no density variation and in homogeneities. Thus intrinsic absorption sets the fundamental lower limit on absorption for any particular material.
- Intrinsic absorption results from electronic absorption bands in UV region and from atomic vibration bands in the near infrared region.
- The electronic absorption bands are associated with the band gaps of amorphous glass materials. Absorption occurs when a photon interacts with an electron in the valene band and excites it to a higher energy level. UV absorption decays exponentially with increasing wavelength (l).
- In the IR (infrared) region above 1.2 μm the optical waveguide loss is determined by presence of the OH ions and inherent IR absorption of the constituent materials. The inherent IR absorption is due to interaction between the vibrating band and the electromagnetic field of optical signal this results in transfer of energy from field to the band, thereby giving rise to absorption, this absorption is strong because of many bonds present in the fiber.

Extrinsic absorption

In practical optical fibers prepared by conventional melting techniques, a major source of signal attenuation is extrinsic absorption from transition metal element impurities. It may be noted that certain of these impurities, namely chromium and copper, in their worst valence state can cause attenuation in excess of 1 dB km-1 in the near-infrared region. Transition element contamination may be reduced to acceptable levels (i.e. one part in 1010) by glass refining techniques such as vapour-phase oxidation, which largely eliminates the effects of these metallic impurities.

However, another major extrinsic loss mechanism is caused by absorption due to water (as the hydroxyl or OH ion) dissolved in the glass. These hydroxyl groups are bonded into the glass structure and have fundamental stretching vibrations which occur at wavelengths between 2.7 and 4.2 m depending on group position in the glass network. The fundamental vibrations give rise to overtones appearing almost harmonically at 1.38, 0.95 and 0.72 pm. This shows the absorption spectrum for the hydroxyl group in silica. Furthermore, combinations between the overtones and the fundamental SiO2 vibration occur at 1.24, 1.13 and 0.88 µm.

It may also be observed in Figure 2 that the only significant absorption band in the region below a wavelength of 1µ m is the second overtone at 0.95 µm which causes attenuation of about I dB km-1 for one part per million (ppm) of hydroxyl. At longer wavelengths the first overtone at 1.38 µm and its sideband at 1.24 µm are strong absorbers giving attenuation of about 2 dB km-1 ppm and 4 dB km-1 pprn respectively. Since most resonances are sharply peaked, narrow windows exist in the longer wavelength region around 1.3 and 1.55 µm which are essentially unaffected by OH absorption once the impurity level has been reduced below one part in 10'. It may be observed that the lowest attenuation for this fiber occurs at a wavelength of 1.55 µm and is 0.2 dB km-1. This is approaching the minimum possible attenuation of around 0. 18 dB km-1 at this wavelength.

Extrinsic absorption results from the presence of impurities. Transition-metal impurities such as Fe, Cu, Co, Ni, Mn, and Cr absorb strongly in the wavelength range 0.6~1.6um. Their amount should be reduced to below 1 part per billion to obtain a loss level below 1dB/km. Such high-purity silica can be obtained by using modern techniques.

The main source of extrinsic absorption in state-of-the-art silica fibers is the presence of water vapors. A vibrational resonance of the OH ion occurs near 2.73um. Its harmonic and combination tones with silica produce absorption at the 1.39 µm, 1.24 µm and 0.9 µm wavelengths. The three spectral peaks seen in figure occur near these wavelengths and are due to the presence of residual water vapor in silica.+

In new kind of glass fiber, known as dry fiber, the OH ion concentration is reduced to such low levels that the 1.39 µm peak almost disappears. This is show in Fig. 2. Such fibers are used to transmit WDM (Wavelength Division Multiplexer) signals over the entire 1.30 µm to 1.65 µm wavelength range.

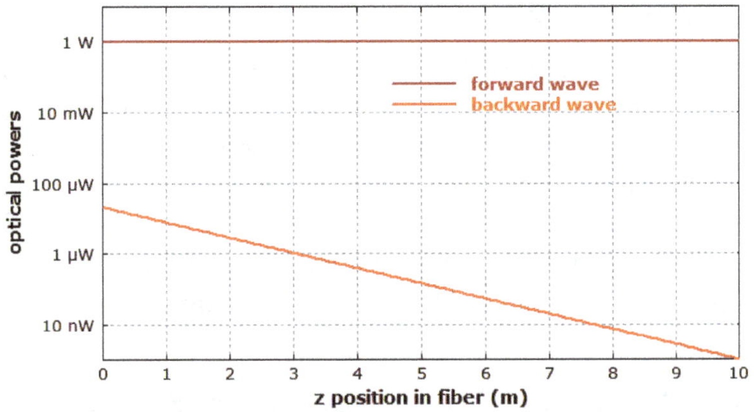

Fig. 2. The OH ion concentration and dispersion of Dry Fiber

SCATTERING LOSSES

Linear

Linear scattering mechanisms cause the transfer of some or all of the optical power contained within one propagating mode to be transferred linearly (proportionally to the mode power) into a different mode. This process tends to result in attenuation of the transmitted light as the transfer may be to a leaky or radiation mode which does not continue to propagate within the fiber core, but is radiated from the fiber. It must be noted that as with all linear processes there is no change of frequency on scattering.

Linear scattering may be categorized into two major types: Rayleigh and Mie scattering. Both result from the nonideal physical properties of the manufactured fiber which are difficult and, in certain cases, impossible to eradicate at present.

Rayleigh scattering

Rayleigh scattering is the dominant intrinsic loss mechanism in the low absorption window between the ultraviolet and infrared absorption tails. It results from inhornogeneities of a random nature occurring on a small scale compared with the wavelength of the light. These inhomogeneities manifest themselves as refractive index fluctuations and arise from density and compositional variations which are frozen into the glass lattice on cooling. The compositional variations may be reduced by improved fabrication, but the index fluctuations caused by the freezingin of density inhomogeneities are fundamental and cannot be avoided. The subsequent scattering due to the density fluctuations, which is in almost all directions, produces an attenuation proportional to 1/l4 following the Rayleigh scattering formula. For a single component glass this is given by:

gR = 8P3/3l4 n8p2bcKTF (4)

where gR is the Rayleigh scattering coefficient, l is the optical wavelength, n is the refractive index of the medium, p is the average photoelastic coefficient, c is the isothermal compressibility at a fictive temperature TF, and K is Boltzmann's constant. The fictive temperature is defined as the temperature at which the glass can reach a state of thermal equilibrium and is closely related to the anneal temperature. Furthermore, the Rayleigh scattering coefficient is related to the transmission loss factor (transmissivity) of the fiber l following the relation:

l = exp (-lRL) (5)

where L is the length of the fiber. It is apparent front Eq. (4) that the fundamental component of Rayleigh scattering is strongly reduced by operating at the longest Possible wavelength. The theoretical attenuation due to Rayleigh scattering in silica at wavelengths of 0.63, 1.00 and 1.30 Am, is 5.2, 0.8 and 0.3 dB km-1 respectively. These theoretical results are in reasonable agreement with experimental work. For instance, a low reported value for Rayleigh scattering in silica at a wavelength of 0.6328 pin is 3.9 dB km-1. However, values of 4.8 dB km-1 and 5.4 dB km-1 have also been reported.

Mie Scattering (Linear Scattering)

Mie scattering is named after German physicist Gustav Mie. This theory describes scattering of electromagnetic radiation by particles that are comparable in size to a wavelength (larger than 10% of wavelength).

For particles much larger, and much smaller than the wavelength of scattered light there are simple and excellent approximations that suffice.

For glass fibers, Mie scattering occurs in inhomogeneities such as core-cladding refractive index variations over the length of the fiber, impurities at the core-cladding interface, strains or bubbles in the fiber, or diameter fluctuations.

Mie scattering can be reduced by carefully removing imperfections from the glass material, carefully controlling the quality and cleanliness of the manufacturing process.

In commercial fibers, the effects of Mie scattering are insignificant. Optical fibers are manufactured with very few large defects. (larger than 10% of wavelength)

The scattering created by such inhomogeneities is mainly in the forward direction and is called Mie scattering. Depending upon the fiber material, design and manufacture, Mie scattering can cause significant losses. The inhomogeneities may be reduced by:

- removing imperfections due to the glass manufacturing process;
- carefully controlled extrusion and coating of the fiber;

- increasing the fiber guidance by increasing the relative refractive index difference.

By these means it is possible to reduce Mic scattering to insignificant levels.

Nonlinear

Optical waveguides do not always behave as completely lincar channels whose increase in output optical power is directlv proportional to the input optical power. Several nonlinear effects occur, which in the case of scattering cause disproportionate attenuation, usually at high optical power levels. This nonlinear scattering causes the optical power from one mode to be transferred in either the forward or backward direction to the same, or other modes, at a different frequency. It depends critically upon the optical power density within the fiber and hence only becomes significant above threshold power levels.

The most important types of nonlinear scattering within optical fibers are stimulated Brillouin and Raman scattering, both of which are usually only observed at high optical power densities in long single-mode fibers. These scattering mechanisms in fact give optical gain but with a shift in frequency, thus contributing to attenuation for light transmission at a specific wavelength. However, it may be noted that such nonlinear phenomena can also be used to give optical amplification in the context of integrated optical techniques.

Stimulated Brillouin scattering

Brillouin scattering is an effect caused by the c(3) nonlinearity of a medium, specifically by that part of the nonlinearity which is the delayed nonlinear response related to acoustic phonons. An incident photon can be converted into a scattered photon of slightly lower energy, usually propagating in the backward direction, and a phonon. The coupling of optical fields and acoustic waves occurs via electrostriction.

Spontaneous and Stimulated Brillouin Scattering

Brillouin scattering can occur spontaneously even at low optical powers, then reflecting the thermally generated phonon field. Typically, only a very small amount of the incident light is scattered in that way.

For higher optical powers, there can be a stimulated effect, where the optical fields substantially contribute to the phonon population. Above a certain threshold power of a light beam in a medium, stimulated Brillouin scattering can reflect most of the power of an incident beam. This process involves a strong nonlinear optical gain for the back-reflected wave: an originally weak counterpropagating wave at the suitable optical frequency can be strongly amplified. Here, the two counter-propagating waves generate a traveling refractive index grating; the higher the reflected power, the stronger the index grating and the higher the effective reflectance.

Brillouin Frequency Shift and Bandwidth

The optical frequency of the reflected beam is slightly lower than that of the incident beam; the frequency difference nB corresponds to the frequency of emitted phonons. This so-called Brillouin frequency shift is set by a phase-matching requirement. For pure backward Brillouin scattering, the Brillouin shift can be calculated from the refractive index n, the acoustic velocity va, and the vacuum wavelength l:

$$\nu_B = \frac{2n\upsilon_a}{\lambda}$$

(For Brillouin scattering in fibers, the effective refractive index must be used.) The Brillouin frequency shift depends on the material composition and to some extent the temperature and pressure of the medium. Such dependencies are exploited for fiber-optic sensors.

Brillouin scattering occurs only in a quite limited bandwidth, e.g. of the order of 100 MHz (which is a very tiny fraction of the optical frequency) in case of silica fibers. It depends on the damping time of the involved acoustic wave (the phonon lifetime), but can also be inhomogeneously increased, e.g. when the temperature of the active fiber in a fiber amplifier varies along the length.

Forward and Backward Brillouin Scattering

In optical fibers, Brillouin scattering occurs essentially only in backward direction. However, rather weak forward Brillouin scattering is also possible due to effects of the acoustic waveguide.

Optical Phase Conjugation

An important application of stimulated Brillouin scattering is optical phase conjugation. There are, for example, phase-conjugate mirrors for high-power Q-switched lasers which make it possible that the thermal distortions occurring in forward and backward direction in the laser crystal compensate each other.

Stimulated Brillouin Scattering in Optical Fibers

Stimulated Brillouin scattering (SBS) is frequently encountered when narrow-band optical signals (e.g. from a single-frequency laser) are amplified in a fiber amplifier, or just propagated through a passive fiber. While the material nonlinearity of e.g. silica is actually not very high, the typically small effective mode area and long propagation length strongly favor nonlinear effects.

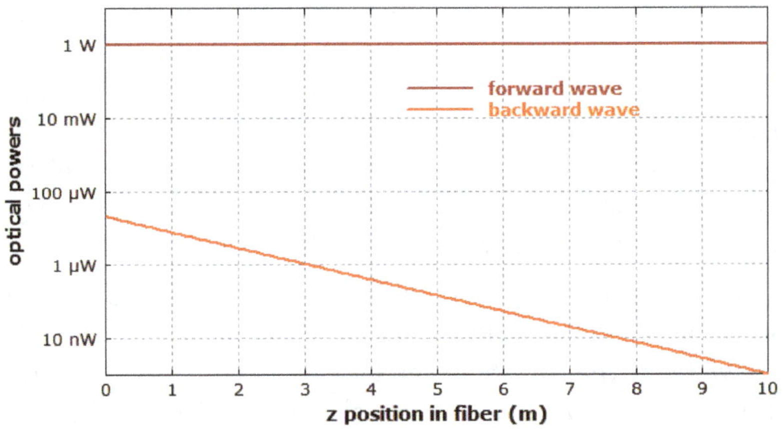

Fig. 3: Pump power (propagating from left to right, red curve) and resulting Brillouin signal power (right to left, orange curve) in a 10 m long fiber. The pump input power is 1 W.

Figure 3 shows what happens when a monochromatic light wave is injected into a 10 m long fiber. The counterpropagating Brillouin-shifted wave starts from quantum fluctuations with a very low optical power, but grows rapidly. Still, it stays far smaller than the input power of 1 W.

For a somewhat increased pump power of 1.8 W, the Brillouin gain (as measured in decibels) is nearly doubled, and the Brillouin wave becomes far stronger.

For a further increased pump power, the power of the Brillouin wave would become comparable to the pump power. In that case, substantial pump depletion occurs. For high SBS gain, that does not lead to a stable situation, but to chaotic fluctuations of the powers.

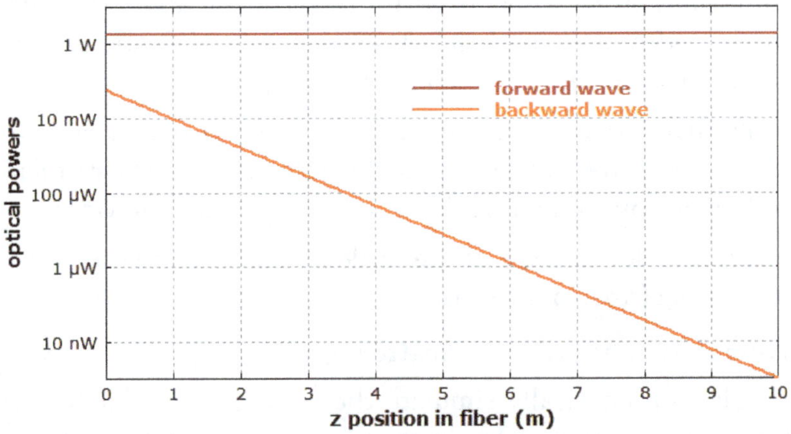

Figure 4: Same as Fig. 2, but with 1.8 W pump power.

If the fiber is many kilometers long, milliwatt powers can be sufficient to cause substantial Brillouin scattering. However, one then has to take into account propagation losses, which are substantial of such fiber lengths. The affect both the pump wave and the Brillouin wave.

For silica fibers, the Brillouin frequency shift is of the order of 10–20 GHz, and the Brillouin gain has an intrinsic bandwidth of typically 50–100 MHz, which is determined by the strong acoustic absorption (short phonon lifetime of the order of 10 ns). However, the Brillouin gain spectrum may be strongly "smeared out" by various effects, such as transverse variations of the acoustic phase velocity or longitudinal temperature variations. Accordingly, the peak gain may be strongly reduced, leading to a substantially higher SBS threshold.

The Brillouin threshold of optical fibers for narrow-band continuous-wave light typically corresponds to a Brillouin gain of the order of 90 dB. (With additional laser gain in an active fiber, the threshold can be lower.) For trains of ultrashort pulses, the SBS threshold is determined not by a peak power, but rather by a power spectral density, as explained in a Spotlight article.

Note that for short laser pulses the spatial overlap of counterpropagating pulses is limited. Therefore, much less than the total fiber length may be relevant for calculating the Brillouin gain, and further increases of peak power by pulse shortening (i.e., for constant pulse energy) do not further increase the Brillouin gain.

SBS introduces the most stringent power limit for the amplification and the passive propagation of narrow-band optical signals in fibers. In order to raise the Brillouin threshold, various methods have been developed. For example, one may increase the bandwidth of the light beyond the Brillouin gain bandwidth, reduce the fiber length, concatenate fibers with slightly different Brillouin shift, or (in high-power active fiber devices) exploit the longitudinally varying temperature. There are also attempts to reduce the overlap of guided optical and acoustic waves, or to introduce significant propagation losses for the acoustic wave. To some extent, SBS problems can be reduced via basic amplifier design modifications, concerning e.g. the doping concentration, effective mode area and pump propagation direction.

On the other hand, the Brillouin gain can be used for operating a Brillouin fiber laser. Such devices are often made as fiber ring lasers. Due to low resonator loss, they can have a relatively low pump threshold and a very small linewidth.

The temperature dependence of the Brillouin shift can be used for temperature and pressure sensing (fiber-optic sensors).

Numerical Simulation of Brillouin Scattering

It is possible to numerically simulate the propagation of light (e.g. in optical fibers) under the influence of Brillouin scattering. However, this is technically quite difficult, as it involves counterpropagating light waves and depends on their detailed frequency spectrum. Also, computation times can be fairly long.

However, in most practical cases it is not necessary to do full-blown numerical simulations for answering the relevant questions. The core question is usually whether or not Brillouin scattering will have substantial effects on life propagation, i.e., cause substantial nonlinear back-scattering – but not what exactly what would happen in that regime where it becomes substantial. That core question can usually be answered in a quite simple way. One simulates the propagation without taking into account Brillouin scattering and then calculates the resulting Brillouin gain. If that exceeds roughly 90 dB, substantial Brillouin scattering is to be expected. The mentioned threshold value somewhat depends on the circumstances, e.g. whether continuous-wave light or light pulses are considered. That simple approach may not be highly accurate, but note that for a high accuracy of full numerical simulations one would require a detailed knowledge of the involved optical inputs, which is often not available. For example, such details are usually not known for common types of seed lasers and would be rather difficult to measure. Further uncertainties may occur e.g. concerning the details of optical fibers such as their detailed Brillouin scattering and how that may change over the length of fiber.

Stimulated Raman scattering

The nonlinear c(3) response of a transparent optical medium to the optical intensity of light propagating through the medium is very fast, but not instantaneous. In particular, a delayed nonlinear response is caused by vibrations of the crystal (or glass) lattice. When these vibrations are associated with optical phonons, the effect is called Raman scattering, whereas acoustical phonons are associated with Brillouin scattering. When e.g. two laser beams with different wavelengths (and normally with the same polarization direction) propagate together through a Raman-active medium, the longer wavelength beam (called the Stokes wave) can experience optical amplification at the expense of the shorter wavelength beam. In addition, lattice vibrations are excited, leading to a temperature rise. The Raman gain for the longer wavelength beam can be exploited in Raman amplifiers and Raman lasers. That gain can be substantial if the Stokes shift corresponds to a frequency difference of several terahertz.

Figure 5 shows a numerical simulation for Raman scattering in an optical fiber, exploited for signal amplification of nanosecond pulses. It demonstrates that under suitable conditions the Raman conversion can be highly efficient.

Raman scattering can occur not only in solid materials, but also in liquids or gases. For example, molecular glasses have vibrational/rotational excitations, and the observed Stokes shifts are related to those.

The Raman effect occurs together with the Kerr effect, which results from the (nearly) instantaneous c(3) response of the electrons.

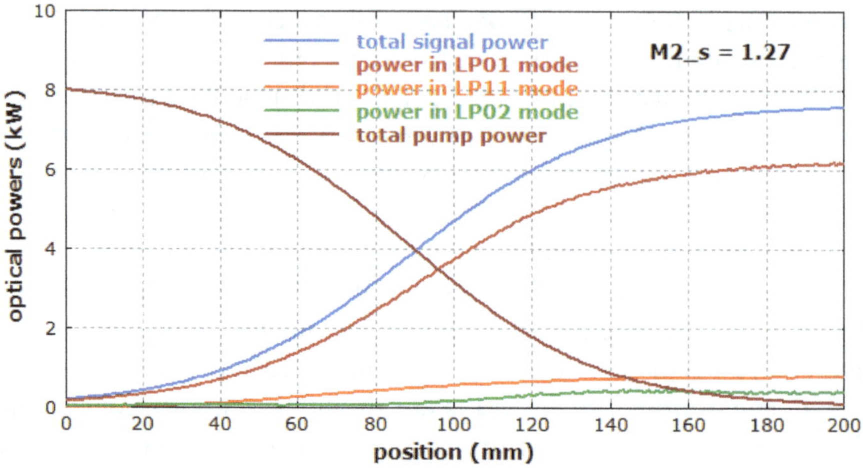

Fig. 5: Evolution of optical powers in a parabolic-index multimode fiber, simulated with the numerical beam propagation feature of the software RP Fiber Power as part of a case study. A signal wave is strongly amplified, while the pump wave is strongly depleted. Multiple modes are involved in the conversion process.

Interaction of Photons and Phonons

In the Raman scattering process, one pump photon is converted into one lower-energy signal photon, and the difference of photon energies is carried away by a phonon (a quantum of the lattice vibrations).

In principle, it is also possible that an already existing phonon interacts with a pump photon to generate one higher-energy photon, belonging to an anti-Stokes wave at a shorter wavelength. That process, however, is usually weak, particularly at low temperatures. Note, however, that strong anti-Stokes light can also arise from four wave mixing if that process is phase-matched.

Raman scattering can be spontaneous or stimulated. Spontaneous Raman scattering occurs when there is a pump wave but no signal input wave (i.e., no signal input photons). It can be considered as a quantum effect – amplification of zero-point oscillations of the signal field.

Raman scattering is also called inelastic scattering, because the involved loss of photon energy is somewhat reminiscent of losses of kinetic energy in collisions of mechanical objects.

Raman-active Media

Some typical Raman-active media are

- certain molecular gases, e.g. hydrogen (H2), methane (CH4) and carbon dioxide (CO2), used in, e.g., high pressure cells for Raman shifters

- solid-state media such as glass fibers or certain optical crystals, e.g. barium nitride = Ba(NO3)2, various tungstates such as KGd(WO4)2 = KGW and KY(WO4)2 = KYW, and synthetic diamond

Cascaded Raman Scattering

When the intensity of the generated Stokes wave becomes sufficiently high, that wave may again act as the pump for a further Raman process. Particularly in some Raman lasers, it is possible to observe several Stokes orders (cascaded Raman lasers).

Raman Scattering with Ultrashort Pulses

Raman scattering can also occur within the broad optical spectrum of, e.g., an ultrashort optical pulse, effectively shifting the spectral envelope of the pulse towards longer wavelengths (Raman self-frequency shift, also called soliton self-frequency shift when soliton pulses are involved).

In optical fiber devices such as fiber amplifiers for intense pulses, Raman scattering can be detrimental: it can transfer much of the pulse energy into a wavelength range where laser amplification does not occur. Figure 6 shows a simulation for an example case. That effect can limit the peak power achievable with such devices.

More on Unwanted Raman Scattering

Even in continuous-wave high-power fiber lasers and amplifiers, Raman scattering can be a problem. There are, however, various kinds of solutions to such problems, including chirped-pulse amplification and the use of special fiber designs which suppress Raman scattering by attenuating the Raman-shifted wavelength component.

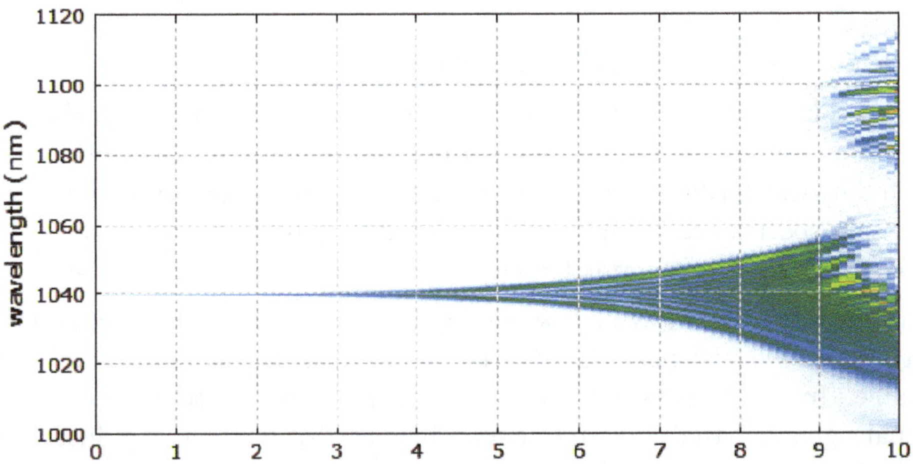

Fig. 6: Evolution of the pulse spectrum in a fiber amplifier. Near the right end, a significant part of the power is shifted into longer-wavelength components by stimulated Raman scattering. The simulation has been made with the software RP Fiber Power as part of a case study.

In bulk media, such as certain nonlinear crystal materials, unwanted stimulated Raman scattering can occur even via non-collinear phase matching, if the pump intensity is rather high and the beam width is large enough. This can occur e.g. in optical parametric generators operated with intense pump pulses.

Raman Spectroscopy

Raman scattering can be exploited for Raman spectroscopy. In particular, it allows one to investigate the vibrational modes of solid materials and vibrational/rotational states of molecules.

Mid-infrared and far-infrared transmission

Most optical fibers are silica fibers; these cannot be used for transmitting mid-infrared light (with wavelengths between 3 μm and 50 μm according to ISO 20473:2007), since silica is strongly absorbing for optical wavelengths above »2 μm due to vibrational resonances (multiphonon absorption). However, there are other glass materials from which optical fibers can be made which can transmit light at longer wavelengths. In other cases, crystalline media or hollow waveguides are used.

Chalcogenide Glasses

Already since the mid 1960's, various types of chalcogenide glasses are used, which are sulfides, selenides or tellurides e.g. of arsenic (As) or germanium (Ge) and have substantially lower vibration frequencies (thus also lower phonon energies) than silica due to the higher mass of chalcogenide ions (comparing with oxygen ions). Consequently, their infrared absorption starts at longer wavelengths:

- Sulphide glasses like arsenic trisulfide (As2S3) are suitable e.g. for wavelengths of 1 to 6 μm, but can exhibit strong absorption peaks within that wavelength range, e.g. at »4 μm.
- Selenide glass (As2Se3) can be used up to »9 μm.

Extrinsic propagation losses are usually strong; for example, hydrogen as an impurity causes absorption in the region around 4 μm related to the vibration of S–H bonds, and hydroxyl (OH") groups cause additional absorption lines. Therefore, the transmission spectra are often quite structured. The propagation losses are much higher e.g. than those of fluoride fibers.

Chalcogenide glasses have a low band gap energy and are thus not suitable for guiding shorter-wavelength light. (For that reason, such glasses also look colored, e.g. orange.) Infrared fibers based on chalcogenide glass are sometimes labeled as CIR = chalcogenide-based infrared fibers. They are fairly stable, e.g. insensitive to moisture, in contrast to some other mid-IR materials.

Germanate Glasses

Germanate glasses are chemically similar to silica glasses, just containing germanium (Ge) instead of silicon (Si). An increased refractive index for the fiber core can be achieved by doping with PbO.

The transmission range extends to roughly 3 μm or even 4 μm, to be compared with only H"2 μm for silica. The transmission losses are of the order of 0.2 dB/m, and it is possible to transmit laser beams with more than 10 W of optical power through multimode fibers.

Heavy Metal Fluoride Glasses

Some heavy metal fluoride glasses (HMFG) can be used for mid-infrared fibers. The most common one is ZBLAN, chemically ZrF4-BaF2-LaF3-AlF3-NaF. It has a loss minimum around 2.5 μm, where propagation losses of a few dB/m can be reached, and can be used even well above 3 μm, although with increasing propagation losses.

Unfortunately, these fluoride glasses are rather fragile, having a failure strain which is more than an order of magnitude lower than for silica, and the fragility can further increase due to the influence of moisture. There is also a tendency for surface crystallization. The small glass softening temperature range makes it relatively difficult to draw fibers from such materials.

Variants of such glasses exist (for example AlF3-ZrF4-BaF2-CaF2-YF3) which can exhibit an improved durability, but possibly at the cost of higher propagation losses.

There are also indium fluoride fibers with improved infrared transmission at longer wavelengths.

Polycrystalline Fibers

Various halide materials exhibit excellent infrared transmission, but have a crystalline nature; one cannot make glasses out of them. Some of them are suitable for fiber fabrication by hot extrusion; one can press them through a die at a temperature near the melting temperature and obtains a polycrystalline structure with a grain size of the order of 10 μm. A fiber cladding can be produced in a second step with the rod-in-tube technique, or with a refined extrusion process. Silver halide fibers (the ternary alloy AgBrCl) are nowadays most common and allow transmission ranges from roughly 3 μm to 18 μm.

In the optimum wavelength region around 8–14 μm, the propagation losses can be well below 1 dB/m. However, the propagation losses tend to increase with time due to growth of the microcrystals, possibly also due to exposure to visible light (photodarkening), which leads to the formation of silver particles. Also, excessive bending can permanently increase the propagation losses.

Thalide fibers have also been developed but are not widely used because of the high toxicity of thallium (Tl) and their inferior mechanical properties.

Sapphire Fibers

Optical fibers can also be made of the monocrystalline sapphire ('! single-crystal fibers), which is aluminum oxide (Al2O3). Common fabrication methods are edge-defined, film-fed growth and laser-heated pedestal growth, possibly followed by an annealing process for reducing the propagation losses.

The transmission range extends up to roughly 3.2 μm, and the low propagation losses and tight robustness of that material allows the use with fairly high optical average powers of more than 10 W. For example, sapphire fibers are used for transmitting the output of Er:YAG lasers at 2.94 μm.

Hollow Waveguides

A completely different approach is to use hollow waveguides, made e.g. of glass or sapphire, where most of the optical power propagates within an air core (typically a few hundred micrometers in diameter), and the light field has only little overlap with the solid materials. Therefore, substantial absorption in the used materials can be tolerated without obtaining high propagation losses for the guided light, and a very wide transmission wavelength range can be achieved. Also, end reflections are very weak. Typically, one has single-mode or few-mode propagation, which implies a high beam quality at the output, but also a high beam quality requirement for the input. One can in principle simply use a hollow glass piece, but the performance can be optimized with additional layers deposited on the inner surface, e.g. a thin metallic layer reflecting the light, further assisted with a dielectric layer (e.g. AgI on Ag).

Polycrystalline Fibers

Various halide materials exhibit excellent infrared transmission, but have a crystalline nature; one cannot make glasses out of them. Some of them are suitable for fiber fabrication by hot extrusion; one can press them through a die at a temperature near the melting temperature and obtains a polycrystalline structure with a grain size of the order of 10 μm. A fiber cladding can be produced in a second step with the rod-in-tube technique, or with a refined extrusion process. Silver halide fibers (the ternary alloy AgBrCl) are nowadays most common and allow transmission ranges from roughly 3 μm to 18 μm. In the optimum wavelength region around 8–14 μm, the propagation losses can be well below 1 dB/m. However, the propagation losses tend to increase with time due to growth of the microcrystals, possibly also due to exposure to visible light (photodarkening), which leads to the formation of silver particles. Also, excessive bending can permanently increase the propagation losses.

Thalide fibers have also been developed but are not widely used because of the high toxicity of thallium (Tl) and their inferior mechanical properties.

Sapphire Fibers

Optical fibers can also be made of the monocrystalline sapphire ('! single-crystal fibers), which is aluminum oxide (Al2O3). Common fabrication methods are edge-defined, film-fed growth and laser-heated pedestal growth, possibly followed by an annealing process for reducing the propagation losses.

The transmission range extends up to roughly 3.2 μm, and the low propagation losses and tight robustness of that material allows the use with fairly high optical average powers of more than 10 W. For example, sapphire fibers are used for transmitting the output of Er:YAG lasers at 2.94 μm.

Hollow Waveguides

A completely different approach is to use hollow waveguides, made e.g. of glass or sapphire, where most of the optical power propagates within an air core (typically a few hundred micrometers in diameter), and the light field has only little overlap with the solid materials. Therefore, substantial absorption in the used materials can be tolerated without obtaining high propagation losses for the guided light, and a very wide transmission wavelength range can be achieved. Also, end reflections are very weak. Typically, one has single-mode or few-mode propagation, which implies a high beam quality at the output, but also a high beam quality requirement for the input.

One can in principle simply use a hollow glass piece, but the performance can be optimized with additional layers deposited on the inner surface, e.g. a thin metallic layer reflecting the light, further assisted with a dielectric layer (e.g. AgI on Ag).

In order to obtain lower loss fibers it is necessary to produce glasses exhibiting longer infrared cutoff wavelengths. Potentially, much lower losses can be achieved if the transmission window of the material can be extended further into the infrared by utilizing constituent atoms of higher atomic mass and if it can be drawn into fiber exhibiting suitable strength and chemical durability. The reason for this possible loss reduction is due to Rayleigh scattering which displays a l-4 dependence and hence becomes much reduced as the wavelength is increased. For example, the scattering loss is reduced by a factor of 16 when the optical wavelength is doubled.

Thus it may be possible to obtain losses of the order of 0.01 dB km-1 at a wavelength of 2.55 μm, with even lower losses at wavelengths of between 3 and 5 μm. Candidate glass-forming systems for mid-infrared transmission are fluoride, fluoride–chloride, chalcogenide and oxide.

In particular, oxide glasses such as Al2O3 (i.e. sapphire) offer a near equivalent transmittance range to many of the fluoride glasses and have benefits of high melting points, chemical inertness, and the ability to be readily melted and grown in air. Chalcogenide glasses, which generally comprise one or more elements Ge, Si, As and Sb, are capable of optical transmission in both the mid-infrared and far-infrared regions. A typical chalcogenide fiber glass is therefore arsenide trisulfide (As2S3). However, research activities into far-infrared transmission using chalcogenide glasses, halide glasses, polycrystalline halide fibers (e.g. silver and thallium) and hollow glass waveguides are primarily concerned with radiometry, infrared imaging, optical wireless, optical sensing and optical power transmission rather than telecommunications

The loss spectrum for a single-crystal sapphire fiber which also transmits in the midinfrared is also shown in Figure 7. Although they have robust physical properties, including a Young's modulus six times greater as well as a thermal expansion some ten times higher than that of silica, these fibers lend themselves to optical power delivery applications, not specifically optical communications. Chalcogenide glasses which have their lowest losses over both the mid- and far-infrared ranges are very stable, durable and insensitive to moisture. Arsenic trisulfide fiber, being one of the simplest, has a spectral range from 0.7 to around 6 μm. Hence it has a cut off at long wavelength significantly before the chalcogenide fibers containing heavier elements such as Te, Ge and Se, an attenuation spectrum for the latter being incorporated in Figure 7. In general, chalcogenide glass fibers have proved to be useful in areas such as optical sensing, infrared imaging and for the production of fiber infrared lasers and amplifiers.

FIBER BEND LOSS

Bend losses are a frequently encountered problem in the context of waveguides, and in particular in fiber optics, since fibers can be easily bent. Bend losses mean that optical fibers exhibit additional propagation losses by coupling light from core modes (guided modes) to cladding modes when they are bent. Typically, these losses rise very quickly once a certain critical bend radius is reached. This critical radius can be very small (a few millimeters) for fibers with robust guiding characteristics (high numerical aperture), whereas it is much larger (often tens of centimeters) for single-mode fibers with large mode areas.

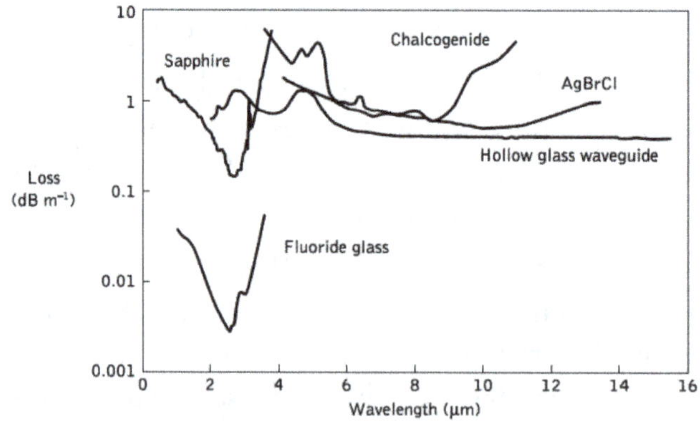

Fig. 7. Attenuation spectra for some common mid- and far-infrared fibers

Generally, bend losses increase strongly for longer wavelengths, although the wavelength dependence is often strongly oscillatory due to interference with light reflected at the cladding/coating boundary, and/or at the outer coating surface. The increasing bend losses at longer wavelengths often limit the usable wavelength range of a single-mode fiber. For example, a fiber with a single-mode cut-off wavelength of 800 nm, as is suitable for operation in the 1-μm region, may not be usable at 1500 nm, because they would exhibit excessive bend losses. Note that even without macroscopic bending of a fiber, bend losses can occur as a result of microbends, i.e., microscopic disturbances in the fiber, which can be caused by imperfect fabrication conditions.

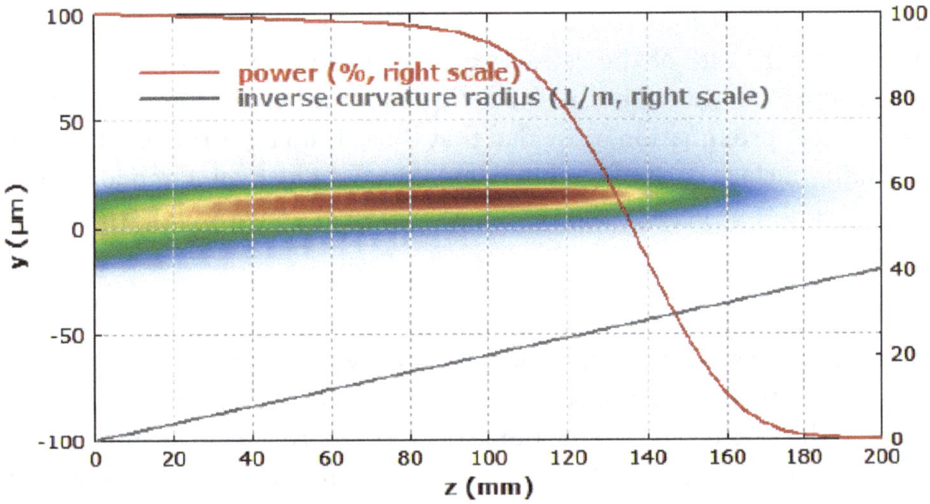

Fig. 8: Amplitude distribution in a large mode area fiber, which is bent more and more strongly towards the right side. The fiber mode becomes substantially smaller and then very lossy; the light is coupled out into cladding modes. The numerical simulation has been done with the RP Fiber Power software.

In multimode fibers, bend losses are usually strongly mode-dependent. The critical bend radius is typically larger for higher-order transverse modes. By properly adjusting the bend radius, it is possible to introduce significant losses for higher-order modes without affecting the lowest-order mode. This can be useful e.g. for the design of high-power fiber amplifiers and fiber lasers where a higher effective mode area can be achieved when using a fiber with multiple transverse modes.

The magnitude of bend losses has some dependence on the polarization. This can be exploited, for example, for obtaining stable single-polarization emission from a fiber laser.

Photonic crystal fibers can have very low bend losses even far beyond the single-mode cutoff wavelength. Therefore, they can be "endlessly single-mode", i.e., they exhibit usable single-mode characteristics over a very large wavelength range.

Note that bending not only introduces losses, but can also reduce the effective mode area. This is particularly true for large mode area step-index fibers. Also, bending induces birefringence.

Estimating Bend Losses

For estimating the magnitude of bend loss, the equivalent index method can be used. The basic idea behind this technique is to calculate the mode distributions for an effective index which contains a term accounting for the modified path lengths at different transverse positions. An elasto-optic correction term (taking into account local modifications of the refractive index by mechanical stress) leads to an effectively weaker "tilt" of the refractive index profile than when considering the geometrical effect alone. Such a method of calculating bend losses is convenient and usually a good approximation, provided that there is no light reflected e.g. from the outer cladding surface back to the fiber core. More sophisticated models can include such effects, and thus predict the full wavelength dependence, but are complicated to handle.

Bend-insensitive Fibers

In some application areas of fibers, and particularly in optical fiber communications, it is of interest to have a relatively bend-insensitive fibers. This is particularly important in the area of fiber to the home (FTTH), where it is problematic to demand that any tight bending is avoided in the fiber installation. The G.657 standard for telecom fibers defines the characteristics of such fibers, with the main categories A for use in access networks and B for short distances, but with even stronger bending tolerance – in category B3 allowing a bend radius down to only 5 mm. For comparison, the more general category G.652 allows a bend radius only down to 30 mm.

Bend-insensitive fibers can be made with different designs. It is common to have trench-assisted fiber designs, containing a ring (trench) with relatively low refractive index around the fiber core. Another possibility are certain designs of photonic crystal fibers, which may be called hole-assisted.

Note that there are trade-offs between bend insensitivity and other properties, such as low propagation losses in the straight form. However, lowest propagation losses are not essential e.g. for FTTH, where only a relatively short transmission distances need to be realized.

Bend Losses in Photonic Integrated Circuits

Bend losses are important not only in fiber optics, but also in the context of photonic integrated circuits. Compact circuits designs often require strong bending of waveguides on such chips, with desired values of the bend radius of a few microns only, but limited by the issue of bend losses. For strong bending and thus particularly compact designs, waveguides with a relatively high numerical aperture, i.e., with a large refractive index contrast (i.e., a high numerical aperture), are needed, e.g. based on silicon-on-insulator (SOI) technology. Those, however, are tentatively more sensitive to propagation losses by scattering at non-perfect boundaries between core and cladding.

DISPERSION

Dispersion of the transmitted optical signal causes distortion for both digital and analog transmission along optical fibers. When considering the major implementation of optical fiber transmission which involves some form of digital modulation, then dispersion mechanisms within the fiber cause broadening of the transmitted light pulses as they travel along the channel.

The phenomenon is illustrated in Fig. 9, where it may be observed that each pulse broadens and overlaps with its neighbors, eventually becoming indistinguishable at the receiver input. The effect is known as intersymbol interference (ISI).

Thus an increasing number of errors may be encountered on the digital optical channel as the ISI becomes more pronounced. The error rate is also a function of the signal attenuation on the link and the subsequent signal-to-noise ratio (SNR) at the receiver. For no overlapping of light pulses down on an optical fiber link the digital bit rate BT must be less than the reciprocal of the broadened (through dispersion) pulse duration (2t).

Hence:

$$B_r \leq \frac{1}{2\tau}$$

The conversion of bit rate to bandwidth in hertz depends on the digital coding format used. For metallic conductors when a nonreturn-to-zero code is employed, the binary 1 level is held for the whole bit period t. In this case there are two bit periods in one wavelength (i.e. 2 bits per second per hertz), as illustrated in Fig. 10(a). Hence the maximum bandwidth B is one-half the maximum data rate or:

Br(max) = 2B

Fig. 9. An illustration using the digital bit pattern 1011 of the broadening of light pulses as they are transmitted along a fiber: (a) fiber input; (b) fiber output at a distance L1; (c) fiber output at a distance L2 > L1

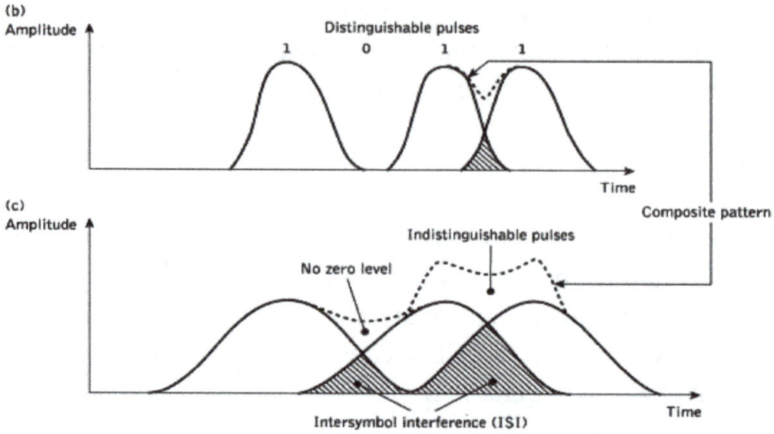

Fig. 10. Schematic illustration of the relationships of the bit rate to wavelength for digital codes: (a) nonreturn-to-zero (NRZ); (b) return-to-zero (RZ)

However, when a return-to-zero code is considered, as shown in Figure 10(b), the binary 1 level is held for only part (usually half) of the bit period. For this signaling scheme the data rate is equal to the bandwidth in hertz (i.e. 1 bit per second per hertz) and thus BT = B.

The bandwidth B for metallic conductors is also usually defined by the electrical 3 Db points (i.e. the frequencies at which the electric power has dropped to one-half of its constant maximum value). However, when the 3 dB optical bandwidth of a fiber is considered it is significantly larger than the corresponding 3 dB electrical bandwidth. Hence, when the limitations in the bandwidth of a fiber due to dispersion are stated (i.e. optical bandwidth Bopt), it is usually with regard to a return to zero code where the bandwidth in hertz is considered equal to the digital bit rate. Within the context of dispersion the bandwidths expressed in this chapter will follow this general criterion unless otherwise stated.

When electro-optic devices and optical fiber systems are considered it is more usual to state the electrical 3 dB bandwidth, this being the more useful measurement when interfacing an optical fiber link to electrical terminal equipment.

CHROMATIC DISPERSION

The chromatic dispersion of an optical material is the phenomenon that the phase velocity and group velocity of light propagating in a transparent medium depend on the optical frequency. That dependency results mostly from the interaction of light with electrons of the medium, and is related to absorption in some spectral regions; see the article on Kramers–Kronig relations. A quantitative measure is the group velocity dispersion.

The attribute "chromatic" is used to distinguish that type of dispersion from other types, which are relevant particularly for optical fibers: intermodal dispersion and polarization mode dispersion.

Mathematical Description of Chromatic Dispersion

An early measure for the magnitude of chromatic dispersion was the Abbe number VD, introduced by Ernst Abbe:

$$v_D = \frac{n_D - 1}{n_F - n_C}$$

The denominator is also called the principal dispersion. The Abbe number depends on the refractive indices at only three different wavelengths:

- 486.1 nm (blue Fraunhofer F line from hydrogen)
- 589.2 nm (orange Fraunhofer D line from sodium)
- 656.3 nm (red Fraunhofer C line from hydrogen)

Large values of the Abbe number indicate low chromatic dispersion and vice versa. Such values can be used for the design of achromatic optical elements, but can of course give only a quite rough indication.

The modern way of quantifying chromatic dispersion is based on a Taylor expansion of the wavenumber k (change in spectral phase per unit length) as a function of the angular frequency w. The expansion is made around some center frequency w0, e.g. the mean angular frequency of some light pulses:

$$k(\omega) = k_0 + \frac{\partial k}{\partial \omega}(\omega - \omega_0) + \frac{1}{2}\frac{\partial^2 k}{\partial \omega^2}(\omega - \omega_0)^2 + \frac{1}{6}\frac{\partial^3 k}{\partial \omega^3}(\omega - \omega_0)^3 + \ldots$$

where the terms corresponding to the different orders have the following meaning:

- The zero-order term describes a common phase shift.
- The first-order term contains the inverse group velocity (i.e., the group delay per unit length) and describes an overall time delay without an effect on the pulse shape:

$$k' \equiv \frac{\partial k}{\partial \omega} = \frac{1}{v_g}$$

- The second-order (quadratic) term contains the second-order dispersion or group delay dispersion (GDD) per unit length:

$$k'' \equiv \frac{\partial^2 k}{\partial \omega^2}$$

- The third-order (cubic) term contains the third-order dispersion (TOD) per unit length:

$$k''' \equiv \frac{\partial^3 k}{\partial \omega^3}$$

Second-order dispersion is often specified in units of s2/m. It is the derivative of the inverse group velocity with respect to angular frequency:

$$k' \equiv \frac{\partial k}{\partial \omega} = \frac{1}{v_g} \Rightarrow k'' \equiv \frac{\partial k}{\partial \omega}\left(\frac{1}{v_g}\right)$$

As an example, the group delay dispersion of silica is +36 fs2/mm at 800 nm, or -22 fs2/mm at 1500 nm. Zero group delay dispersion is reached close to 1270 nm.

Dispersion of third and higher order is called higher-order dispersion. When dealing with very broad optical spectra, one sometimes has to consider dispersion up to the fourth or even fifth and sixth order. Ultimately, the concept of Taylor expansion loses its value in this regime, where many dispersion orders have to be considered. It is therefore often more convenient e.g. in numerical modeling to work directly with a table of frequency-dependent phase changes.

The dispersion of various orders for a medium can most conveniently be calculated if the refractive index is specified with a kind of Sellmeier formula. Tabulated index data are less suitable, since the numerical differentiation is sensitive to noise.

For light propagating in waveguides such as optical fibers, one considers the phase constant b instead of the wavenumber k. The second-order dispersion, for example, is then given as the second derivative of b with respect to angular frequency. This is often called b'' or b2.

Normal and Anomalous Dispersion

One distinguishes normal dispersion (for k'' > 0) and anomalous dispersion (for k'' < 0). Normal dispersion, where the group velocity decreases with increasing optical frequency, occurs for most transparent media in the visible spectral region. Anomalous dispersion sometimes occurs at longer wavelengths, e.g. in silica (the basis of most optical fibers) for wavelengths longer than the zero-dispersion wavelength of »1.3 μm.

Great care is recommended when the sign of dispersion is specified. The ultrafast optics community identifies that sign with the sign of k''. The opposite sign is usually used in optical fiber communications, where the dispersion is often specified with the dispersion parameter in units of picoseconds per nanometer and kilometer (ps/(nm km)). The different signs result from using a frequency derivative in one case and a wavelength derivative in the other. Note also that the conversion factor depends on the wavelength.

$$D_\lambda = -\frac{2\pi c}{\lambda^2} - \frac{\partial^2 k}{\partial \omega^2}$$

Between wavelength regions with normal and anomalous dispersion, there is a zero dispersion wavelength. The region around this wavelength can be special in some respects, not only concerning weak dispersive pulse broadening.

EFFECTS OF CHROMATIC DISPERSION

Wavelength-dependent Refraction and Diffraction

A frequently encountered phenomenon is that dispersion causes wavelength-dependent refraction, which is responsible, e.g., for the occurrence of rainbows. Similarly, wavelength-dependent diffraction at a diffraction grating allows the spatial separation of different frequency components of light. That is exploited in grating spectrometers, for example.

Dispersive Pulse Broadening and Chirping

Chromatic dispersion has an important impact on the propagation of light pulses, because a pulse always has a finite spectral width (bandwidth), so that dispersion can cause frequency-dependent phase changes. Consequently, its frequency components propagate with different group velocities. Normal dispersion, for example, leads to a lower group velocity of higher-frequency components, and thus to a positive chirp (instantaneous frequency rising over time), whereas anomalous dispersion creates negative chirps.

Note that the spectral shape (i.e., the frequency dependence of the power spectral density of the optical field) is never changed by chromatic dispersion, since only phase changes are caused. However, if chromatic dispersion occurs together with optical nonlinearities, e.g. in an optical fiber, it can indirectly affect the spectral shape, since the phase changes modify the temporal shape and thus the effect of nonlinearities. Dispersion of the opposite sign can be used later to recompress the pulse (dispersive pulse compression). This is important, e.g., in the context of chirped-pulse amplification. Depending on the sign and amount of required dispersion, and also on other factors such as the optical peak power, different devices are used as dispersive compressors. Examples are prism pairs, pairs of diffraction gratings, chirped mirrors or other dispersive mirrors, chirped Bragg gratings, and dispersive glass fibers.

Higher-order dispersion causes more complicated changes of the pulse shape. A challenge in dispersion compensation is often to compensate not only the lowest (second) order of dispersion, but also higher orders for approaching the transform limit.

Soliton Effects

Although dispersion alone leads to pulse broadening, dispersion combined with the Kerr non-linearity can give rise to the formation of solitons and in this way assist the generation of extremely short pulses of light with a soliton mode-locked laser. However, this usually requires the careful management of dispersion over a wide wavelength range, i.e., taking care also of higher-order terms.

Dispersion in Nonlinear Optics

In nonlinear optics, particularly concerning nonlinear frequency conversion, chromatic dispersion has manifold influences, which can be grouped into three categories:

- It determines the options for phase matching of parametric nonlinear processes.
- It limits the effective interaction length for short pulses via the group velocity mismatch (temporal walk-off, limited phase-matching bandwidth).
- It can also limit the nonlinear interaction length via dispersive pulse broadening: eventually, the pulses become so long that their peak power is too low for efficient nonlinear interactions. However, the group velocity mismatch normally introduces the more stringent limitation.

CHROMATIC DISPERSION OF OPTICAL COMPONENTS

Dispersion can also be defined for optical components rather than media. In that case, one performs the Taylor expansion as shown above for the total phase delay of the component (rather than for the wavenumber, i.e., the phase delay per unit length), and obtains the total group delay dispersion (in units of seconds squared) rather than the dispersion per unit length. The chromatic dispersion of an optical component may simply result from the dispersion of its parts, but in some cases large amounts of dispersion can arise from interference effects. For example, a Gires-Tournois interferometer generates dispersion by interference effects in an optical resonator, and not primarily via material dispersion. The same happens in other types of interferometers. The amount of dispersion arising from such effects can be huge when large differences in propagation lengths are involved.

Chromatic dispersion can also result from geometric effects. An important example is that of a prism pair as is often used for dispersion compensation in mode-locked lasers. Here, chromatic dispersion arises from wavelength-dependent path lengths, caused by wavelength-dependent refraction at prism surfaces. Similar effects occur in laser resonators containing Brewster-angled optical components, and in pairs of diffraction gratings as used for dispersive pulse compression.

WAVEGUIDE DISPERSION

The discussion above is based on the assumption of plane waves. In practice, significant deviations from this situation can occur, in particular in the context of waveguides. Here, the quantity of interest is usually not the magnitude of the wave vector (k vector) (which anyway is no longer well defined), but rather the value b (the imaginary part of the propagation constant), which specifies the phase change per unit length in the propagation direction. As b is influenced by the waveguide (particularly for mode diameters of only a few wavelengths or even less), the dispersion is also affected (waveguide dispersion). This is important e.g. in optical fibers, and particularly in photonic crystal fibers with very small effective mode areas. In some cases, waveguide dispersion makes the overall dispersion anomalous even in the visible wavelength region, where the material dispersion of silica alone is clearly in the normal dispersion regime. For telecom applications, fiber designs are often made for tailored dispersion properties, resulting in, e.g., dispersion-shifted fibers.

MEASUREMENT OF CHROMATIC DISPERSION

There are several techniques for measuring chromatic dispersion:

- The pulse delay technique (for fibers) is based on measuring the difference in propagation time (group delay) for pulses with different center wavelengths. This is typically done using hundreds of meters (or even some kilometers) of a fiber. The dispersion is obtained by differentiation of these data.
- The phase shift technique or "difference method" (also for fibers): a light beam with a sinusoidally modulated intensity is sent through a fiber, and the phases of the oscillations of input and output power are compared. The group delay can be calculated from that phase, and the dispersion can be measured by performing the measurement at different wavelengths.
- Dispersion in the resonator of a wavelength-tunable passively mode-locked laser can be measured by monitoring changes in the pulse repetition frequency when the laser wavelength is changed, as this reveals the wavelength-dependent group delay. Different types of interferometry (e.g. white-light interferometry or spectral phase interferometry) can be used to measure the phase delay caused by a dispersive component. The dispersion properties can be obtained from this phase by numerical differentiation. The method is normally used for dispersion measurements on dispersive laser mirrors and sometimes for fibers.

INTERMODAL DISPERSION

Intermodal dispersion (also called modal dispersion) is the phenomenon that the group velocity of light propagating in a multimode fiber (or other waveguide) depends not only on the optical frequency but also on the propagation mode involved.

Figure 11 shows a numerical simulation, where a 200-fs ultrashort pulse is launched into a 50 cm long multimode fiber such that multiple modes are excited.

After the fiber, the corresponding contributions appear at different times due to different group velocities of the modes. The fundamental mode comes first, as it is the fastest.

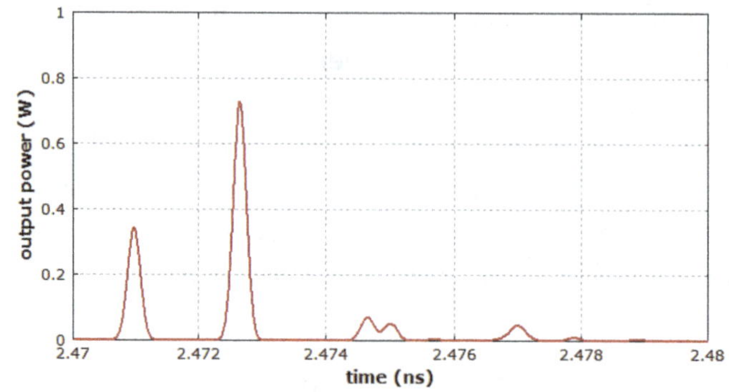

Fig. 11: Output power versus time for a 200-fs input pulse injected into a 50 cm long multimode fiber. The numerical simulation has been done with the RP Fiber Power software.

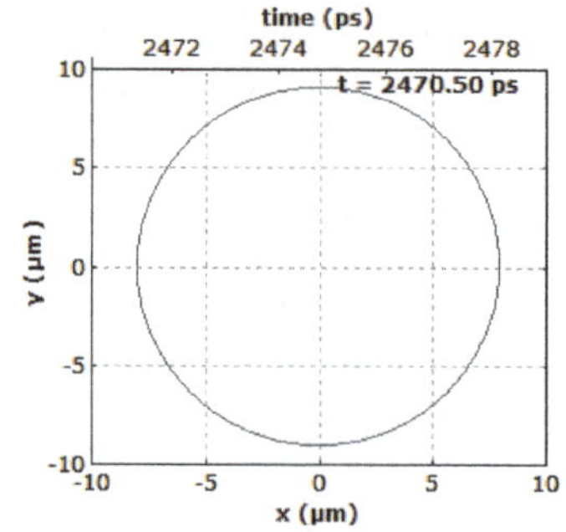

Fig 12: Time-dependent output beam profile for the same situation as in Fig. 11.

The strength of intermodal dispersion can be quantified as the differential mode delay (DMD). It depends strongly on the refractive index profile of the fiber in and around the fiber core. For example, for a step-index profile the higher-order modes have lower group velocities, and this can lead to differential group delays of the order of 10 ps/m = 10 ns/km. It is then hardly possible to realize data rates of multiple Gbit/s in an fiber-optic link with a kilometer length. In systems for optical fiber communications based on multimode fibers, intermodal dispersion can severely limit the achievable data transmission rate (bit rate). In order to avoid strong signal distortion, it is usually necessary to keep the pulses long enough to maintain a reasonable temporal overlap of components from different modes, and this unavoidably sets a limit on the data rate. The natural way of eliminating intermodal dispersion is to use fiber links based on single-mode fibers: if there is only one propagation mode available (disregarding possible polarization mode dispersion and cladding modes), there cannot be difference between propagation times. However, intermodal dispersion can also be minimized by using multimode fibers with a parabolic refractive index profile, where intermodal dispersion is minimized.

CHAPTER

OPTICAL FIBERS AND CABLES 4

INTRODUCTION

A fiber-optic cable is composed of very thin strands of glass or plastic known as optical fibers; one cable can have as few as two strands or as many as several hundreds of them. These optical fiber cables carry information in the form of data between two places using optical or light-based technology. Once the light beams travel down the optical fiber cable (OFC), they would emerge at the other end. A photoelectric cell will be required to turn the pulses of light back into electrical information the computer can understand.

While traveling down fiber optic cable, light bounces repeatedly off the walls. The beam of light does not leak out of the edges because it hits the glass at really shallow angles. And then it reflects back again as if the glass were really a mirror. This is called total internal reflection. The other factor that keeps it in the pipe is the cable structure.

Fiber cable offers many advantages, the prime ones being higher bandwidth and reach. Optical fibre cables (OFC) are now preferred over old copper telecom cables as they provide high-speed broadband services. Optical fibre loses 3% of the signal over 100 meters of distance, while copper wires lose 94%. Additionally, optical fibres are more long-lasting as compared to copper wires, which are much more fragile. Copper wire can be tapped very easily, while optical fibres do not radiate signals that can be tapped. Optical fibre offers much lower latency (the amount of time required to perform data transmission) compared to copper wires.

The telecom industry is among the primary applications of optical fiber communication. The increasing need for 24×7 high-speed connectivity and an increase in data traffic due to services such as voice, messaging, emails, downloads, and video streaming is made possible only with optical fibre communication.

In several developed countries, initiatives like tax incentives, substantial project subsidies, and financial support for R&D have assisted domestic players to proliferate.

The Indian manufacturers have the capability and capacity, and they also need a similar kind of support from the government.

STL Tech is one of the most trusted end-to-end optical fibre manufacturers in India and the only company in the world with a fully integrated silicon-to-software pipeline. We specialize in everything from manufacturing optical fiber cables of various types and for various applications, to designing and installing integrated hyper-scale fiber optic networks. STL's innovative optical fibre products are utilized in a plethora of large-scale applications such as long-haul, metropolitan, FTTx, CATV, smart city, intelligent traffic networks, access, defence-security, and premise. We also specialize in various optical fiber cable configurations such as ribbon, loose tube, and tight buffered for all types of aerial and underground deployments.

Since 2019, STL has invested over INR 1500 crores in expanding its optical fibre manufacturing capacity from 30,000 fiber km to 50,000 fiber km. Our global optical fibre working network currently stands at 33 million fibre km, catering to India and other international geographies like the UK and the US. STL's worldwide footprint encapsualtes next-gen optical preform, fibre, cable, and interconnect subsystem manufacturing facilities in countries like India, China, Brazil, and Italy.

STL partners with telecom operators, government organisations, internet service providers, infrastructure companies, public utility providers, and more in order to deliver and install high-end, high-performance optical fibre networks of the future. In India, STL has partnered with various state governments to maintain the country's nationwide optic fiber network program – BharatNet.

Moreover, STL's Centre of Excellence offers state-of-the-art core research and development in fibre technology and has yielded over 216 patents in the industry.

OPTICAL FIBERS

Optical fibers, fibres (British spelling), passive fibers, active fibers, rare-earth-doped fibers, double-clad fibers, triple-clad fibers, step-index fibers, graded-index fibers, polarization-maintaining fibers, silica fibers, fluoride fibers, mid-infrared fibers, single-crystal fibers, plastic optical fibers, dispersion-shifted fibers, photonic crystal fibers, photonic bandgap fibers, hollow-core fibers, nanofibers, single-mode fibers, single-polarization fibers, few-mode fibers, multimode fibers, large-core fibers, large mode area fibers, tapered fibers, telecom fibers, specialty fibers

Optical fibers are the core components of fiber optics and play a particularly important role in photonics in general. They are a kind of optical waveguides, which are in most cases made of some kind of glass, can potentially be very long (hundreds of kilometers), and are – in contrast to other waveguides – fairly flexible. Although glasses are generally quite fragile, particularly thin silica fibers protected with a polymer coating can be bent substantially – for example around your finger – without breaking.

The most commonly used glass for optical fibers is fused silica (quartz glass, amorphous silicon dioxide = SiO_2), either in pure form or with some dopants. Silica is so widely used because of its outstanding properties, in particular its potential for extremely low propagation losses (realized with ultrapure material) and its amazingly high mechanical strength against pulling and even bending (provided that the surfaces are well prepared). However, for some special applications one requires specialty fibers made from other types of glasses, e.g. fluoride fibers or chalcogenide fibers (see below).

Most optical fibers used in laser technology have a fiber core with a refractive index which is somewhat higher than that of the surrounding medium (called the *fiber cladding*). The simplest case is that of a step-index fiber, where the refractive index is constant within the core and within the cladding. The index contrast between core and cladding determines the numerical aperture of the fiber (Fig. 1), and is typically small, so that optical fibers are weakly guiding. Light launched into the core is guided along the core, i.e., it propagates mainly in the core region, although the intensity distribution may extend somewhat beyond the core. Due to the guidance and the low propagation losses, the optical intensity can be maintained over long lengths of fiber. Although most fibers have only a single core, there are also so-called multi-core fibers containing a few cores or even a large number of those.

A less frequently used principle of guiding light is based on a photonic bandgap (*photonic bandgap fibers*). For example, this can be realized with concentric rings of different refractive index, forming a kind of two-dimensional Bragg mirror.

The term *specialty fibers* is used for many different kinds of optical fibers with special properties, and is thus not very specific. Such fibers may be based on unusual glasses or just have unusual designs.

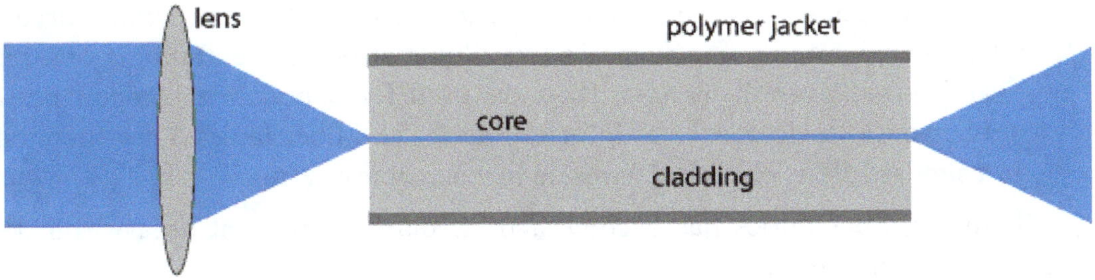

Fig. 1: Simple setup for launching light into an optical fiber (not to scale). A collimated laser beam is focused into the fiber core. The light propagates along the core and leaves the other fiber end as a divergent beam. The fiber core and cladding are made of glass. A polymer jacket protects the glass fiber.

Fig. 2: A fiber in which supercontinuum generation occurs. Some of the light is scattered out of the fiber, so that it can be seen.

APPLICATIONS OF OPTICAL FIBERS

There are many important applications of fiber optics (largely of glass fibers). Some of the most important ones are:

- Optical fiber communications utilize optical fibers mostly for long-range data transmission, but sometimes also for short distances. Huge amounts of data can be quickly sent through a single fiber, which is also immune to external influences such as electric and magnetic fields.
- Active fiber-optic devices contain some rare-earth-doped fiber. Fiber lasers can generate laser light at various wavelengths, and fiber amplifiers can be used e.g. for boosting the optical power or amplifying some weak telecom signals.
- Fiber-optic sensors can be used e.g. for distributed temperature and strain measurements in buildings, oil pipelines, and wings of airplanes.
- Passive optical fibers are useful for transporting light from some source to another point, e.g. for purposes like illumination, diode pumping of lasers and power over fiber. Also, they are used for connecting components in fiber-optic devices, such as interferometers and fiber lasers. They then play a similar role as electrical wires do in electronic devices.

Therefore, fiber optics has become a particularly important area within the technology of photonics.

FIBER MODES – SINGLE-MODE VERSUS MULTIMODE FIBERS

An optical fiber can support one or several (sometimes even many) *guided modes*, the intensity distributions of which are located at or immediately around the fiber core, although some of the intensity may propagate within the fiber cladding. In addition, there is a multitude of cladding modes, which are not restricted to the core region. The optical power in cladding modes is usually lost after some moderate distance of propagation, but can in some cases propagate over longer distances.

Outside the cladding, there is typically a thin protective polymer coating with one or two layers, which gives the fiber improved mechanical strength and protection against moisture. Such coatings may consist of acrylate, silicone or polyimide, for example. (At the fiber ends, the coating often has to be stripped off.) Optical properties of such coatings are sometimes also relevant, e.g. concerning the attenuation of cladding modes which they cause.

In addition to the polymer coating, some fibers are also equipped with a much thicker buffer (secondary coating), e.g. with an outer diameter of 900 μm. Such buffered fibers are integrated into indoor fiber cables, for example. Another possibility is the integration of fibers in a loose tube, which is also filled with a viscous gel. That approach is common for outdoor cables.

An important distinction is that between single-mode and multimode fibers:

- Single-mode fibers usually have a relatively small core (with a diameter of only a few micrometers) and can guide only a single spatial mode (disregarding the fact that there are two different polarization directions), the profile of which in most cases has roughly a Gaussian shape. Changing the launch conditions only affects the power launched into the guided mode, whereas the spatial distribution of the light exiting the fiber is fixed. Efficiently launching light into a single-mode fiber requires a laser source with high beam quality and precise alignment of the focusing optics in order to achieve mode matching. The mode radius of a single-mode fiber is often of the order of 5 μm, but there are also large mode area fibers with single-mode guidance. In the latter case, the alignment tolerances are lower in terms of position but higher in terms of angle (which may be less problematic).
- Multimode fibers have a larger core and/or a larger index difference between core and cladding, so that they support multiple modes (possibly hundreds of more) with different intensity distributions (Figure 3). In this case, the spatial profile of light exiting the fiber core depends on the launch conditions, which determine the distribution of power among the spatial modes.

Long-range optical fiber communication systems usually use single-mode glass fibers, because the different group velocities of different modes would distort the signal at high data rates ('! intermodal dispersion). For shorter distances, however, multimode fibers are more convenient as the demands on light sources and component alignment are lower. Therefore, local area networks (LANs), except those for highest bandwidth, normally use multimode fiber.

Single-mode fibers are also normally used for fiber lasers and amplifiers. Multimode fibers are often used, e.g., for the transport of light from a laser source to the place where it is needed, particularly when the light source has a poor beam quality and/or the high optical power requires a large mode area.

Different modes of an optical fiber can be coupled via various effects, e.g. by bending or often by irregularities in the refractive index profile. These may be unwanted or purposely introduced, e.g. as fiber Bragg gratings. Waveguide theory shows that an important factor for the coupling between different fiber modes is the difference in their wavenumbers, which for efficient coupling has to match the spatial frequency of a coupling disturbance.

FIBER PROPERTIES

Main Parameters

The design of a step-index fiber can be characterized with only two parameters, e.g. the core radius *a* and the refractive index difference *Än* between core and cladding. Typical values of the core radius are a few microns for single-mode fibers and tens of microns or more for multimode fibers.

Instead of the refractive index difference, one usually uses the numerical aperture, defined as

$$NA = n \sin \theta_{max}$$

which is the sine of the maximum acceptable angle of an incident beam with respect to the fiber axis (considering the launch from air into the core in a ray-optic picture). (*n* is the refractive index of the surrounding medium - close to 1 in case of air.) The NA also quantifies the strength of guidance. Typical values are of the order of 0.1 for single-mode fibers, even though actual values vary in a relatively large range. For example, large mode area single-mode fibers can have low numerical apertures of 0.05 or even lower, whereas some rare-earth-doped fibers have values of 0.3 and higher for a high gain efficiency. NA values around 0.3 are also typical for multimode fibers. The sensitivity of a fiber to bend losses strongly diminishes with increasing NA, which causes strong confinement of the mode field to the core.

Another frequently used parameter is the *V* number

$$V = \frac{2\pi}{\lambda} aNA = \frac{2\pi}{\lambda} a\sqrt{n_{core}^2 - n_{cladding}^2}$$

which is a kind of normalized frequency. Single-mode guidance is achieved when the *V* number is below ≈2.405. Multimode fibers can have huge *V* values. The number of guided modes then scales with V^2.

As a numerical example, consider a typical step-index silica fiber for single-mode operation in the 1.5-μm spectral region, with a cut-off wavelength of 1.3 μm and a numerical aperture of 0.1. The refractive index of the pure silica cladding at 1.5 μm is ≈1.444. The core index is ≈1.4475, i.e., the index difference is ≈0.0035. The core diameter is 10 μm, and the *V* number is 2.1.

Refractive Index Profiles

The refractive index profile of optical fibers often deviates substantially from that of a step-index profile (with constant refractive index within the core):

- Due to preferential evaporation of the dopant during the collapse of the preform (assuming that the preform is made with chemical vapor deposition, see below), there is often a pronounced index dip at the center. For single-mode fibers with small mode areas, this does not need to have a strong impact on the mode field distribution, which in most cases closely resembles a Gaussian shape.
- Some fibers are made with graded index profiles (*graded-index fibers*) where the refractive index is gradually reduced away from the center, e.g. with a parabolic shape. Parabolic index profiles are useful e.g. for multimode telecom fibers because they minimize intermodal dispersion (see below).
- There are also "W profiles", where the core is surrounded by a region with a refractive index lower than that of the cladding (*depressed cladding*). In principle, there can even be additional index steps, or combinations with smooth refractive index variations.
- Triangular, trapezoidal and Gaussian index profiles are used for dispersion-shifted fibers.
- Index profiles are not always radially symmetric. For example, an elliptical core shape can provide increased birefringence ('! *polarization-maintaining fibers*) or even single-polarization guidance ('! *single-polarization fibers*) (see below).

Note that the definitions of the numerical aperture and consequently of the *V* number become somewhat ambiguous for non-rectangular index profiles.

In addition, there are so-called photonic crystal fibers (see below), where the refractive index profile is strongly structured.

Propagation Losses

The power losses (propagation losses) for light propagating in an optical fiber can be extremely small, particularly for single-mode silica fibers as used in telecommunications. The resulting attenuation is typically dominated by Rayleigh scattering for short wavelengths (and absorption if one gets closer to the band gap) and by multiphonon absorption at long wavelengths. Rayleigh scattering results from refractive index fluctuations, which are to some extent unavoidable in a glass, but may be strongly increased due to concentration fluctuations in fibers with high numerical aperture and by other imperfections. Other loss contributions come from inelastic scattering (spontaneous Brillouin scattering and Raman scattering), from absorbing impurities, and from fluctuations of the core diameter.

One may further reduce Rayleigh scattering by optimizing the glass composition (e.g. low with germania content and with dopants reducing the viscosity and therefore the fictitious glass temperature), but also with particularly slow fiber drawing to allow for some degree of annealing.

For silica fibers, the loss minimum occurs around 1.5–1.6 μm and can be a bit below 0.15 dB/km (≈3.4% per km), which is close to the theoretical limit based on Rayleigh scattering in fused silica. There is often a loss peak around 1.4 μm, which can be largely eliminated, however, by carefully optimizing the chemical composition of the core so as to reduce the OH content (i.e. the concentration of hydroxyl bonds). Interestingly, fibers with high OH content can exhibit lower losses for ultraviolet light, whereas they exhibit pronounced loss peaks in the infrared spectral region.

Multimode fibers, and in general fibers with high numerical aperture, tend to have significantly higher propagation losses, essentially because the higher doping level of the core increases the scattering losses. Rare-earth-doped fibers also have much higher losses, but as more than some tens of meters of such a fiber are rarely used, this usually does not matter for their applications.

POLARIZATION PROPERTIES

Despite the typically cylindrical symmetry, fibers usually exhibit some amount of random birefringence which can cause the polarization state of light to evolve in an uncontrolled way (*polarization of light*). There are special polarization-maintaining fibers with a strong built-in birefringence to solve this problem. In addition, there are single-polarization fibers, which guide only light with one polarization direction. There are also various types of fiber polarization controllers, which allow one to adjust the state of polarization in a fiber.

Chromatic Dispersion Properties

As a result of the waveguide properties, the chromatic dispersion of an optical fiber can deviate significantly from its material dispersion, particularly when the mode area is small (*waveguide dispersion*). This makes it possible to obtain unusual dispersion properties by engineering the waveguide properties. For example, dispersion-shifted fibers can have near zero dispersion in the 1.5-μm spectral region, and there are dispersion-flattened fibers with small dispersion over a large wavelength range or dispersion-decreasing fibers. A particularly high design freedom exists for photonic crystal fibers.

Other Kinds of Dispersion

The birefringence makes the group delay polarization-dependent; this is often called *polarization mode dispersion*. (That effect can be largely suppressed by using spun fibers.)

For multimode fibers, there is also intermodal dispersion, i.e., a dependence of the group velocity on the fiber mode, which may be minimized by choosing a suitable refractive index profile but is typically larger than the dispersion of single-mode fibers.

FIBER FABRICATION

Most optical fibers are fabricated by pulling from a so-called *preform*, which is a glass rod, e.g. with a diameter of a few centimeters and 1 m length. Along its axis, the preform contains a region with increased refractive index, which will form the core. When the preform is heated close to the melting point in a furnace (oven), a thin fiber with a diameter of typically 125 μm and a length of many kilometers can be pulled from the bottom of the preform. Before the fiber is wound up, it usually obtains a polymer coating for mechanical and chemical protection.

The core of a fiber can be doped with laser-active ions, normally rare earth ions of erbium, neodymium, ytterbium, or thulium. When these ions are excited with suitable pump light, optical amplification occurs, which can be used in fiber lasers or amplifiers.

MULTIMODE FIBERS

Multimode fibers are optical fibers which support multiple (often many) transverse guided modes for a given optical frequency and polarization.

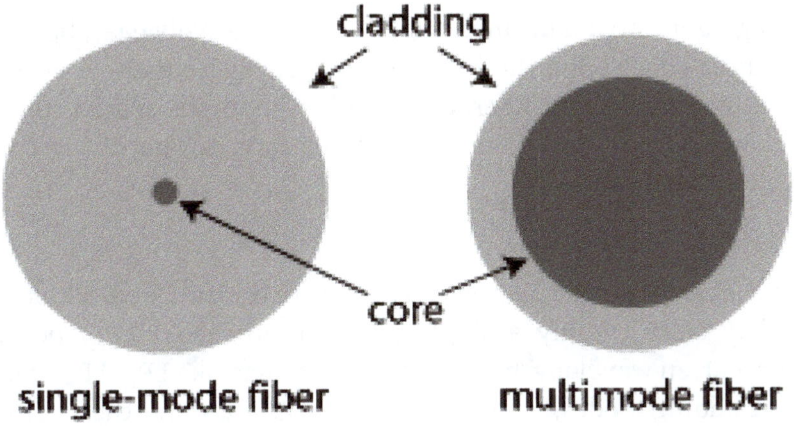

Fig. 3: A single-mode fiber (left) has a core which is very small compared with the cladding, whereas a multimode fiber (right) can have a large core.

The number of guided modes is determined by the wavelength (or optical frequency) and the refractive index profile. For step-index fibers, the relevant quantities are the core radius and the numerical aperture; the latter depends on the refractive index contrast between fiber core and cladding. In combination, these quantities determine the *V* number of step-index fibers. For large *V* values, the number of modes is roughly $V^2/2$, when counting both polarization directions. Particularly for fibers with a relatively large core (right-hand side in Fig. 3), the number of supported modes can be very high.

Highly multimode fibers can guide light with poor beam quality and high optical power (e.g. generated with a high-power diode bar), but for preserving the beam quality of a light source with higher radiance it can be better to use a fiber with smaller core and moderate numerical aperture, even though efficient launching of light into the fiber may then be more difficult.

Compared with standard single-mode fibers, multimode fibers usually have significantly larger core areas, but also generally a higher numerical aperture of e.g. 0.2–0.3. The latter leads to robust guidance, even under conditions of tight bending, but also to higher propagation losses without bending, as irregularities at the core–cladding interface can scatter light more effectively. The refractive index profile is usually rectangular (*tep-index fibers*), but sometimes parabolic.

A basic specification of a multimode fiber contains the core diameter and the outer diameter of a multimode fiber. Common types for fiber-optic communications (see below) are 50/125 μm and 62.5/125 μm fibers, having a core diameter of 50 μm or 62.5 μm, respectively, and a cladding diameter of 125 μm. Such fibers support hundreds of guided modes. There are also large-core fibers with even substantially larger core diameters of hundreds of micrometers.

Launching light into a multimode fiber is comparatively easy, because there are larger tolerances concerning the location and propagation angle of incident light, compared with a single-mode fiber. On the other hand, the spatial coherence of the fiber output is reduced, and the output field pattern can hardly be controlled, for reasons explained below.

SINGLE-MODE FIBERS

Single-mode fibers (also called *monomode fibers*) are optical fibers which are designed such that they support only a single propagation mode (LP_{01}) per polarization direction for a given wavelength. Higher-order modes like LP_{11}, LP_{20} etc. then do not exist – only cladding modes, which are not localized around the fiber core. Single-mode fibers usually have a relatively small core (with a diameter of only a few micrometers) and a small refractive index difference between core and cladding; the mode radius is typically a few microns. However, there are also large mode area fibers with a relatively large mode area.

A peculiar property of single-mode fibers is that the transverse intensity profile at the fiber output has a fixed shape, which is independent of the launch conditions and the spatial properties of the injected light, assuming that no cladding modes can carry substantial power to the fiber end. The launch conditions only influence the efficiency with which light can be coupled into the guided mode.

Intermodal dispersion can of course not occur in single-mode fibers. This is an important advantage for the application in optical fiber communications at high data rates (multiple Gbit/s), particularly for long distances. Essentially for that reason, and partly because of their tentatively lower propagation losses, single-mode fibers are exclusively used for long-haul data transmission, and nearly always for outdoor applications even over shorter distances. For short-distance indoor use, multimode fibers are more common, mostly because that allows the use of cheaper multimode data transmitters based on light-emitting diodes instead of laser diodes.

As a *standard single-mode fiber* for use in optical fiber communications in the 1.3-μm or 1.5-μm wavelength region, the SMF-28 of Corning (or the enhanced version SMF-28e) is common. This has a core diameter of 8.2 μm and a numerical aperture of 0.14. The mode field diameter is ≈9.2 μm at 1310 nm, or 10.4 μm at 1550 nm. The single-mode cut-off is at ≈1260 nm.

Conditions for Single-mode Guidance

For step-index fibers, the condition for single-mode guidance can be formulated using the V number (normalized frequency), which can be calculated from the wavelength, the core radius, and the numerical aperture (NA): the V number must be below ≈2.405. This requires that the core radius is small, particularly for fibers with high NA.

Typically, a fiber has single-mode characteristics only over a limited wavelength range with a width of a few hundred nanometers. The limit towards smaller wavelengths is given by the *single-mode cut-off wavelength*, beyond which the fiber supports multiple modes. This transition is very sharp and can easily be seen e.g. when tuning the launched wavelength around the cut-off wavelength: the shape of the transmitted beam varies rapidly in the multimode regime but remains constant in the single-mode regime. The long-wavelength limit of the useful single-mode region is usually given by excessive bend losses, by absorption of the material or (for certain fiber designs, e.g. with index-depressed cladding) by leakage into the cladding.

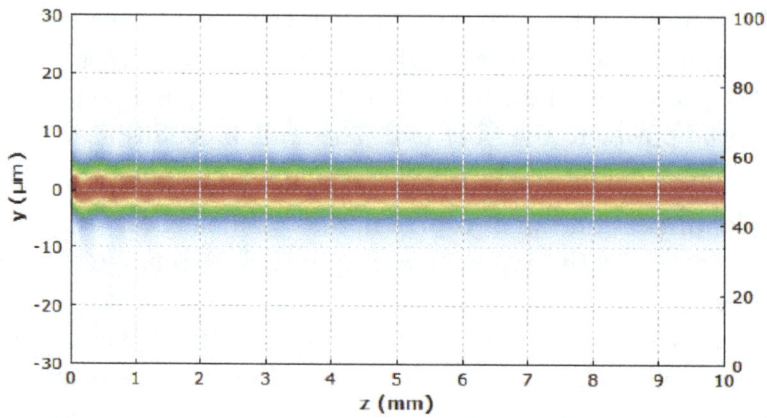

Fig. 4: Light propagation at 1.5 μm wavelength in a single-mode fiber with imperfect launch conditions. After some length, only light in the guided mode remains. The numerical simulation has been done with the software RP Fiber Power.

Figure 4 shows how light evolves when it is launched into the left end of a single-mode fiber with a slightly misaligned laser beam. Initially, some of the light gets into cladding modes, which however are attenuated substantially. After some length, mostly light in the guided mode remains.

For a shorter wavelength (Fig. 5), the fiber gets into the slightly multimode regime: in addition to the LP_{01} mode, there are now two LP_{11} modes. Here, one obtains a periodic oscillation of the field profile due to mode interference.

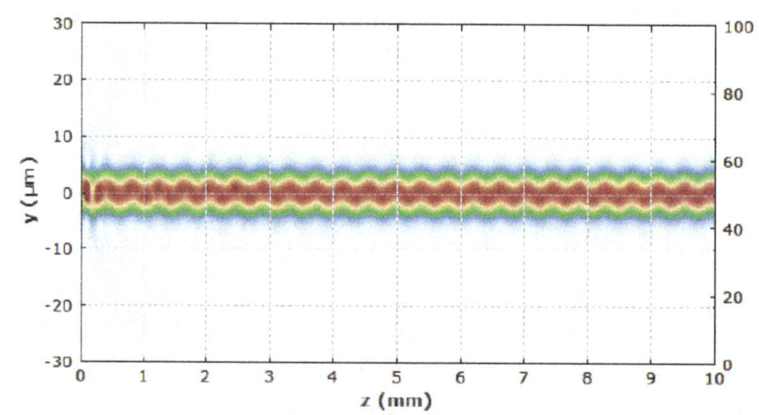

Fig 5: Light propagation in the same fiber as above, but for a shorter wavelength of 1 μm.

For a still shorter wavelength of 0.5 μm (see Fig. 6), the propagation gets more complicated, since the fiber supports more modes.

Note that mode interferences can be affected by any effects acting on the propagation constants of the modes. For example, even slight bending would affect the phase relationship of the modes. Also, the beam profiles are strongly wavelength-dependent. All these effects are not observed in single-mode fibers.

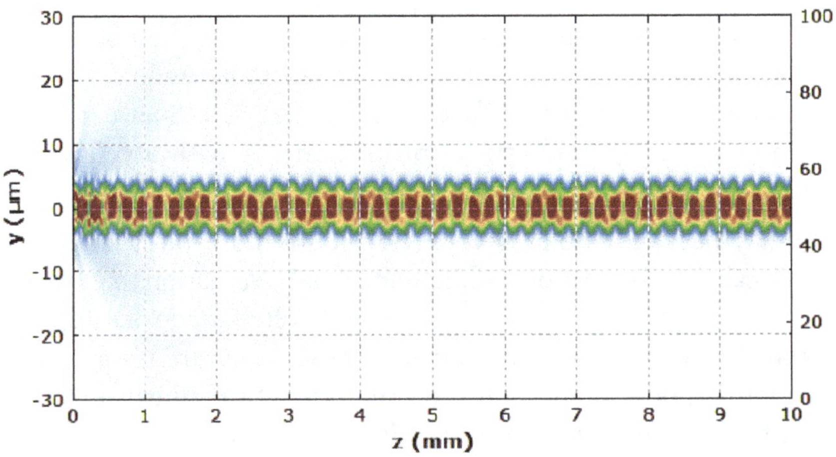

Fig. 6: Light propagation in the same fiber as above, but for a still shorter wavelength of 0.5 µm.

Photonic crystal fibers can have substantially broader or narrower wavelength regions with single-mode propagation. For example, there are *endlessly single-mode fibers* with a very wide single-mode wavelength region. On the other hand, there are also photonic bandgap fibers, based on an entirely different operation principle, with narrow single-mode regions and no guidance at other wavelengths. Besides, there are hollow-core fibers, which can offer single-mode guidance in combination with very high power levels. This is because in such fibers the light has very little spatial overlap with the glass, in contrast to the situation in a standard solid-glass single-mode fiber, where the average power level is usually limited to a couple of watts.

FIBER CABLES

An optical fiber cable (or *fiber-optic cable*) is a more or less flexible cable which contains one or several (sometimes even many hundreds of) optical fibers. Due to protective layers (made of polymers and sometimes also metals) which prevent excessive bending as well as damage due to externally applied stress, fiber cables can be much more robust than bare fibers. Also, fiber cables protect glass fibers against moisture.

A wide variety of constructions is used for fiber cables, depending on the requirements of the application area:

- There are simple *coated fibers* (called *bare fibers* despite the coating), where a single glass fiber is surrounded with a polymer coating with a diameter which is often 250 µm, i.e., twice the fiber diameter. This low level of protection is often sufficient for fibers within some optical setup.

- lternatively, one may use *tight buffered fibers*, where a thicker polymer buffer (often with an outer diameter of 900 µm) protects the fiber. The buffer can protect the fiber against excessive bending, but not much against stretching. Such buffered fibers can be further integrated into fiber cables, particularly for indoor use, with additional protection e.g. with a yellow (well visible) thick polymer layer.
- One may package one or several tight buffered cables, for example, into a larger cable structure providing additional protection, e.g. with an aramid yarn, a ripcord and an outer jacket (sheath, often made of PVC). There may be a *central strength member* around which the fibers are placed. (When pulling such a fiber, one should pull it on the strength member, not at the jacket.) Some fibers of this type are used as *distribution cables* within buildings. Others, called *breakout cables* or *fanout cables*, contain multiple smaller cables which are stronger than the single fibers in distribution cables; they are usually larger and more expensive than distribution cables. Some fiber cables even contain hundreds of glass fibers.
- Two *simplex fiber cables* can be combined to obtain a *duplex cable* – for example, for transmitting data signals in two directions through separate fibers.
- *Ribbon cables* contain multiple fibers (typically 12) which are arranged side by side, resulting in a flat ribbon. Multiple ribbons can be placed on top of teach other. That geometry allows for a high packing density.
- *Fiber bundle cables* can contain a large number of relatively small fibers, for example for illumination or imaging applications. Such bundles may be ordered ("coherent") or unordered. See also the article on fiber bundles.
- There are also fiber cables of the *loose-tube type*, where one or more coated fibers are loosely lying in a semi-rigid plastic tube (in air or in a moisture-blocking gel). There can be additional layers around the loose tube (or several loose tubes), such as a water blocking wrap and an outer jacket (e.g. made of polyethylene). The construction may also contain a metal jacket for increased mechanical strength. Loose-tube cables can tolerate some amount of stretching, which would usually break glass fibers, because the loose tubes can contain some excess length of fiber. Generally, there is little external stress influence, which can also be helpful in terms of polarization mode dispersion. The protection is often sufficient for outdoor use. However, loose-tube cables are less convenient concerning connectorization and sometimes not suitable for installation in vertical ducts.
- Armored fiber-optic cables have an additional outer armor layer for further increased protection. They are suitable for being directly buried in the ground, or laid out aerially, or in outdoor ducts.

- Submarine cables are equipped for use in fresh water or salt water. They must be protected against severe mechanical stress e.g. from boat anchors or fishing equipment.
- Some fiber cables contain specialty fibers. For example, there are hollow-core photonic crystal fibers for the delivery of intense ultrashort light pulses.

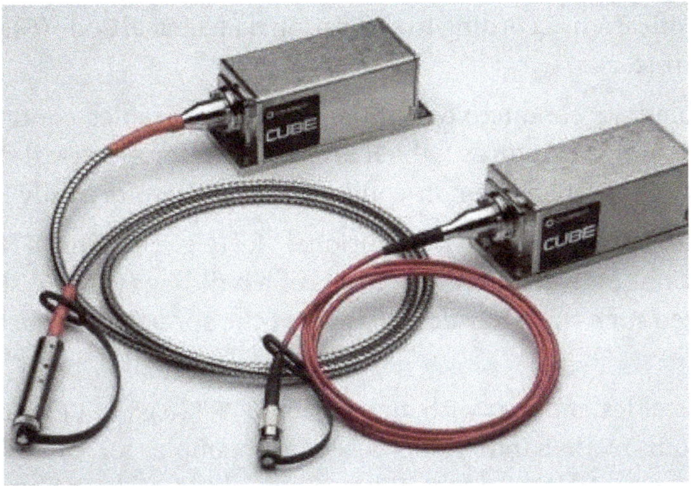

Fig. 7: Photograph of fiber-coupled diode lasers from Coherent, having different kinds of fiber-optic cables. The image was kindly provided by Coherent.

For indoor cables, fire safety is an important aspect, whereas for outdoor cables, moisture resistance and temperature tolerance are factors of high interest. Special aerial/self-supporting cables, which can be exposed to wind and sunlight in free air (hanging on poles), must have a particularly high pulling strength and overall robustness, using some steel or aramid yarn. Outdoor cables could easily tolerate indoor conditions, but are often not allowed for indoor use (or only with certain restrictions) due to issues with fire protection.

The detailed construction for a fiber cable can be relatively sophisticated, optimized for an appropriate trade-off between strength, diameter, weight, fire resistance, cost and other possibly relevant properties. For example, the mechanical details can have an influence on polarization mode dispersion.

The terminations (ends) of fiber cables are often equipped with fiber-optic connectors, which allow relatively simple plugging, as is possible with electrical cables. However, fiber-optic connections are usually more sensitive than electrical connections, and more sophisticated procedures and equipment are needed to prepare or clean them.

Various types of *fiber-optic patch cables* (*patch cords*) are available, which are terminated with standardized fiber connectors.

Long-distance cables are made by connecting multiple cables, each one being a few kilometers long. Stable connections can be made with splicing, in particular with fusion splicing.

NEC CLASSIFICATION OF OPTICAL FIBER CABLES

There exist many variations of fiber cables, to which some classifications are applied. A common classification according to the National Electrical Code (NEC) in the United States is as follows:

- OFC cables are conductive (C) fiber cables, i.e., cables containing metals in strengthening structures (which are not used for electrical connections). In contrast, OFN cables are non-conductive and therefore fully insulating.
- Riser cables, marked with an additional "R" (e.g., OFCR or OFNR), are used for connections between different levels of a building (vertical shafts). They are optimized such that they do not support the spread of fire between different floors.
- Plenum cables, marked with an additional "P" (e.g., OFCP or OFNP), are used in plenum or air-handling spaces – for example, for cables running along ceilings or ventilation ducks. When exposed to fire, they should not produce too much smoke and should be flame-retardant.
- General purpose cables are marked with a "G", e.g. OFCG or OFNG.

The above mentioned codes do *not* specify optical properties of the contained fibers; these can be single-mode or multimode fibers, for example, of any type.

Other Elements of Fiber Cable Specifications

Various elements are often encountered in the nomenclature and specifications of fiber cables or the fibers therein:

- For telecom fibers, ITU has developed some standardized characteristics like "G.651.1"; their mentioning implies that such specifications should be fulfilled.
- Fiber Distributed Data Interface (FDDI) is a standard concerning LAN fiber cables, which however with its quite limited transmission speed of 100 Mbit/s has become obsolete due to the improved availability of advanced copper cables.
- There can be additional classifications like OM1, OM2, OM3, OM4 and OM5 for graded-index fibers, which concern the residual level of intermodal dispersion. That limits the transmission bandwidth (or the bandwidth–distance product) of the cable. High performance is achieved e.g. with OM4 50/125-μm laser-optimized fibers, having a very precisely controlled refractive index profile.

- Sometimes, the maximum achievable bandwidth–distance product is specified (in units of MHz·km).
- Specs like "50/125 microns" indicate the core and cladding diameter. Large values of the core diameter – tens of microns – generally indicate multimode fibers.
- The propagation losses are typically specified in units of dB/km (decibels per kilometer). Those apply without significant bending of the cable, so that only some level of microbend losses occurs, apart from absorption and scattering losses.

APPLICATIONS OF FIBER CABLES

Optical fiber cables are often used in optical fiber communications. Large and strong cables are used for onshore and submarine data transmission, often bridging distances of thousands of kilometers (using fiber amplifiers). Smaller fiber patch cables can be used e.g. to connect the components of fiber-optic data links within buildings. The key advantages of fiber-optic cables over electrical data cables are the enormous transmission bandwidth and the low losses (particularly in the 1.5-μm wavelength region), but one may also profit from the immunity against electromagnetic interference and from the possibility to have fully insulating cables.

Nowadays, many fiber cables for data transmission have been laid down which are not or not yet used. This is partly because burying a fiber is much cheaper when a channel in the ground is already opened e.g. for laying electrical cables. One may prefer the risk of never used fiber cables over the other risk that additional ground work will be required later on.

There are also many "dark fibers" within fiber cables, i.e., fibers which are not (yet) used. They may be held in reserve for future use.

In laboratory and industrial setups, fiber cables are often convenient for transporting light from a source to an application – for example, from a high-power fiber laser to a welding robot in a car factory, or from an optical sampling head to a measurement instrument. For short-distance transmission of low optical powers, fiber patch cables are often used.

PREPARATION OF OPTICAL FIBERS

In a fiber optic cable, many individual optical fibers are bound together around a central steel cable or high-strength plastic carrier for support. This core is then covered with protective layers of materials such as aluminum, Kevlar, and polyethylene (the cladding). Because the core and the cladding are constructed of slightly differing materials, light travels through them at different speeds. As a light wave traveling in the fiber core reaches the boundary between the core and cladding, these compositional differences between the two cause the light wave to bend back into the core. Thus, as a pulse of light travels through an optical fiber, it is constantly bouncing away from the cladding. A pulse moves through the optical fiber at the speed of light—186,290 miles per second (299,340 kilometers per second) in a vacuum, somewhat slower in practice—losing energy only because of impurities in the glass and because of energy absorption by irregularities in the glass structure.

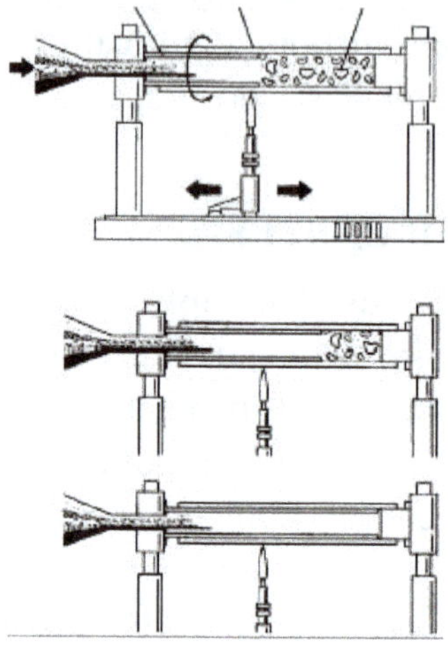

Fig. 8. To make an optical fiber, layers of silicon dioxide are first deposited on the inside surface of a hollow substrate rod. This is done using Modified Chemical Vapor Deposition, in which a gaseous stream of pure oxygen combined with various chemical vapors is applied to the rod. As the gas contacts the hot surface of the rod, a glassy soot several layers thick forms inside the rod. After the soot is built up to the desired thickness, the substrate rod is moved through other heating steps to drive out any moisture and bubbles trapped in the soot layers. During heating, the substrate rod and internal soot layers solidify to form the boule or preform of highly pure silicon dioxide.

Energy losses (attenuation) in an optical fiber are measured in terms of loss (in decibels, a unit of energy) per distance of fiber. Typically, an optical fiber has losses as low as 0.2 decibels per kilometer, meaning that after a certain distance the signal becomes weak and must be strengthened, or *repeated.* With current datalink technology, laser signal repeaters are necessary about every 30 kilometers (18.5 miles) in a long-distance cable. However, on-going research in optical material purity is aimed at extending the distance between repeaters of an optical fiber up to 100 kilometers (62 miles).

There are two types of optical fibers. In a single-mode fiber, the core is smaller, typically 10 micrometers (a micrometer is one-millionth of a meter) in diameter, and the cladding is 100 micrometers in diameter. A single-mode fiber is used to carry just one light wave over very long distances. Bundles of single-mode optical fibers are used in long-distance telephone lines and undersea cables. Multimode optical fibers, which have a core diameter of 50 micrometers and a cladding diameter of 125 micrometers, can carry hundreds of separate light wave signals over shorter distances. This type of fiber is used in urban systems where many signals must be carried to central switching stations for distribution.

PROCESS OF MANUFACTURING

There are two main steps in the process of converting raw materials into optical fiber ready to be shipped.

- Manufacturing of the pure glass preform
- Drawing of the preform

The first step in manufacturing glass optical fibers is to make a solid glass rod, known as a preform. Ultra-pure chemicals - primarily silicon tetrachloride (SiCl4) and germanium tetrachloride (GeCl4) - are converted into glass during preform manufacturing. These chemicals are used in varying proportions to fabricate the core regions for the different types of preforms. The basic chemical reaction of manufacturing optical glass is:

- $SiCl_4$ (gas) + O_2 > SiO_2 (solid) + $2Cl_2$ (in the presence of heat)
- $GeCl_4$ (gas) + O_2 > GeO_2 (solid) + $2Cl_2$ (in the presence of heat)

The core composition of all standard communication fibers consists primarily of silica, with varying amounts of germania added to increase the fiber's refractive index to the desired level. Single-mode fibers typically have only small amounts of germania and have a uniform composition within the core. Multimode fibers typically have a much higher refractive index, and therefore much higher germania content. Also, the core composition and the refractive index of graded-index multimode fibers changes across the core of the fiber to give the refractive index a parabolic shape.

Modified Chemical Vapor Deposition(MCVD)

First, a cylindrical preform is made by depositing layers of specially formulated silicon dioxide on the inside surface of a hollow substrate rod. The layers are deposited by applying a gaseous stream of pure oxygen to the substrate rod. Various chemical vapors, such as silicon tetrachloride ($SiCl_4$), germanium tetrachloride ($GeCl_4$), and phosphorous oxychloride ($POCl_3$), are added to the stream of oxygen. As the oxygen contacts the hot surface of the rod-a flame underneath the rod keeps the walls of the rod very hot-silicon dioxide of high purity is formed. The result is a glassy soot, several layers thick, deposited inside the rod. This soot will become the core. The properties of these layers of soot can be altered depending on the types of chemical vapors used.

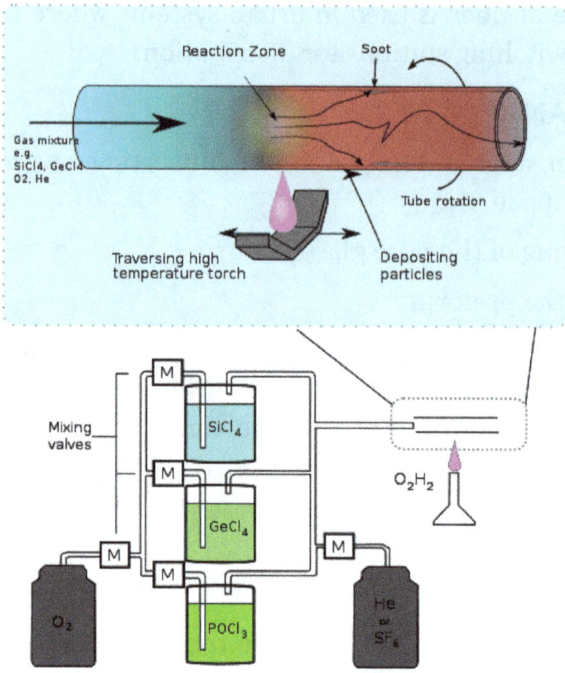

Fig 9. Illustration of MCVD (inside) process

After sufficient layers are built up, the tube is collapsed into a solid glass rod referred to as a preform. It is now a scale model of the desired fiber, but much shorter and thicker. The preform is then taken to the drawing tower, where it is pulled into a length of fiber up to 10 kilometers long.

OUTSIDE VAPOR DEPOSITION(OVD)

One of many variations of vapour deposition technique for fabricating optical fiber. Here an inert rod is layered with core and cladding glass deposits built up on the outside. Once enough layers are in place, the rod is removed and the layers consolidated into a solid preform which can be drawn into fiber. Silicon chloride, $SiCl_4$ and germanium chloride, $GeCl_4$ are oxidised to form silica and germania particles for the deposition.

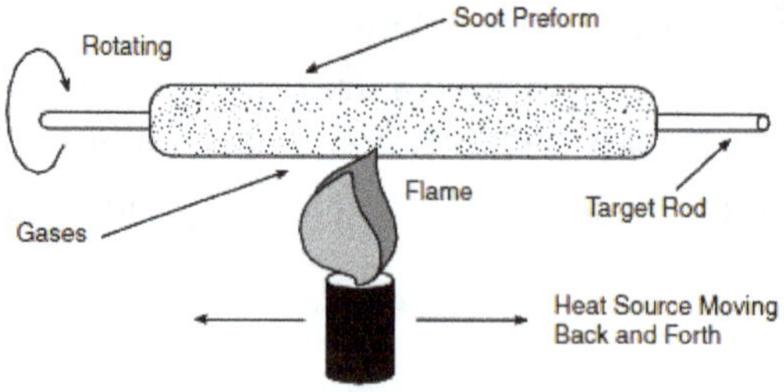

Outside vapor deposition (OVD).

Fig. 10. Outside vapor deposition

Vapour Phase Axial Deposition (VPAD)

In the diagram (Fig. 11), we see how the preform is made. A seed rod is slowly rotated and pulled upward. As the seed rod is pulled, two burners deposit fine glass soot. The lower burner in this diagram is depositing the core glass material, and above it is a burner depositing the cladding glass. The rate at which the seed rod is pulled is carefully controlled by servo mechanisms. After deposition the glass soot rod is dehydrated and sintered into a solid preform in a furnace.

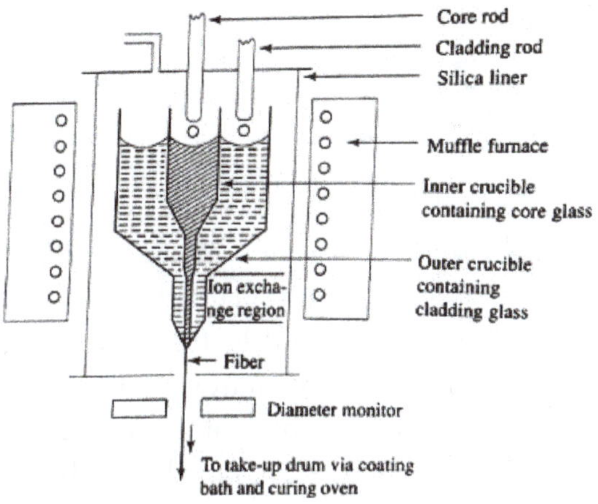

Fig. 11. Vapour Phase Axial Deposition

LIQUID-PHASE (OR MELTING) METHODS

This fiber fabrication method or Technique employs conventional glass-refining techniques for producing ultra-pure powders of the starting materials, which are oxides such as SiO_2, $Ge\ O_2$, $B_2\ O_3$, $Al_2\ O_3$, etc., which decompose into oxides during the melting process. An appropriate mixture of these materials is then melted in silica or platinum crucibles at temperatures varying from 1000 °C to 1300 °C. After the melt has been suitably processed, it is cooled and drawn into rods or tubes (of about 1 m length) of multicomponent glass.

The rod of core glass is then inserted into a tube of cladding glass to make a preform. The fiber is drawn from this preform using the apparatus shown in the figure below. Here, the preform is precision-fed into a cylindrical furnace capable of maintaining high temperature, normally called a drawing furnace. During its passage through the hot zone, its end is softened to the extent that a very thin fiber can be drawn from it. The outer diameter of the fiber is monitored through a feedback mechanism, which controls the feed rate of the preform and also the winding rate of the fiber. The bare fiber is then given a primary protective coating of the polymer by passing it through the coating bath. This coating is cured either by UV lamps or thermally. The finished fiber is then wound on a take-up drum.

A fiber 20-30 km long can be drawn from a preform of about 1 m in 2-3 h. Higher pulling rates are limited by the pulling process as well as the subsequent primary Coating operation. This method of preparing fibers tends to be a batch process and hence continuous production is not possible. Continuous manufacture is possible using another technique, which is called the *double crucible method.* The apparatus used is shown in Fig. 12.

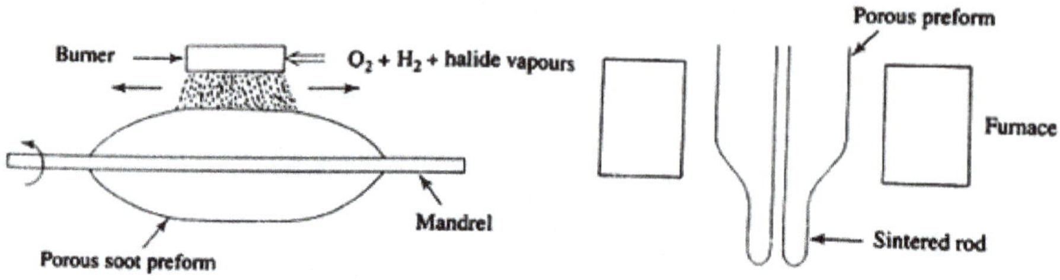

Fig. 12. Double crucible method for continuous production of fibers

It consists of two concentric platinum crucibles (also called a double crucible) mounted inside a vertical cylindrical muffle furnace whose temperature may be varied from 800 °C to 1200 °C. The starting material-core and cladding glasses, either directly in the powdered form or in the form of preformed rods are fed into the two crucibles separately. Both the crucibles have nozzles at their bases from which a clad fiber may be drawn from the melt in a manner similar to that shown in the figure of the Fiber-drawing apparatus. Index grading may be achieved by diffusion of dopant ions across the core–cladding interface, within the melt. Relatively inexpensive fibers of large core diameters and, therefore, large numerical apertures may be produced continuously by this method. An attenuation level of the order of 3 dB/km for sodium borosilicate glass fiber, which is prepared using this technique, has been reported.

Vapor-phase Deposition (VPD) Methods

In this fiber fabrication method or technique, the melting temperatures of silica-rich glasses are too high for liquid-phase melting techniques; therefore, vapor-phase deposition methods are used. Herein, the starting materials are halides of silica (e.g., $SiCl_4$) and the dopants, e.g., $GeCl_4$, $TiCl_4$, BBr_3, etc., which are purified to reduce the concentration of transition-metal impurities to below 10 ppb.

OUTSIDE VAPOUR-PHASE OXIDATION (OVPO) METHOD

This method uses flame hydrolysis to deposit the required glass composition onto a rotating mandrel (an alumina rod) as shown in the figure below. The mixture of vapors of the starting materials, e.g., $SiCl_4$, Ge Cl_4, BBr_3, etc. is blown through the oxygen-hydrogen flame. The soot produced by the oxidation of halide vapors is deposited on a cool Wmandrel. The flame is moved back and forth over the length of the mandrel so that a sufficient number of layers are deposited on it. The concentration of the dopant halides is either varied gradually, if index grading is required, or maintained constant if step-index fiber is required.

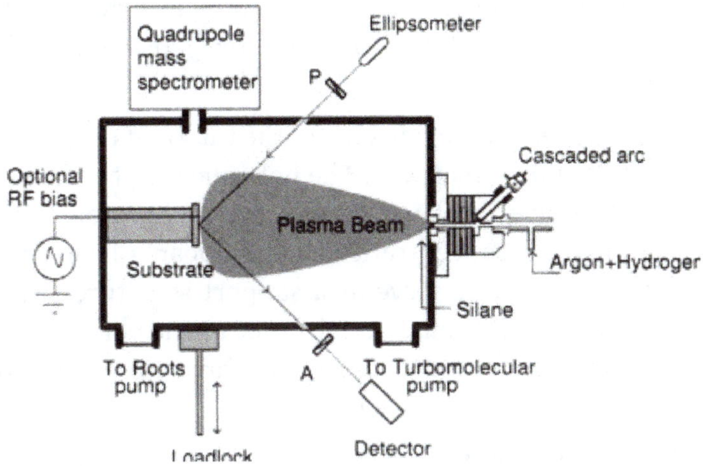

Fig. 13. Glass deposition by the OVPO method

After the deposition of the core and cladding layers is complete, the mandrel is removed. The hollow and porous preform left behind is then sintered in a furnace to form a solid transparent glass rod. This is then drawn into a fiber by the apparatus discussed earlier. With a single preform, 30-40 km of fiber can be easily prepared. The index profile can be controlled well using this method, as the flow rate of vapors can be adjusted after the deposition of each layer. An attenuation of less than 1 dB/km at an operating wavelength of 1 3 μm has been reported, but the fibers show an axial dip in the refractive index.

Vapour axial deposition (VAD) method

In this fiber fabrication method, core and cladding glasses are simultaneously deposited onto the end of a seed rod, which is rotated to maintain azimuthal homogeneity and also pulled up. The porous perform so deposited, while the seed rod is being pulled up, is heated to about 1100 °C in an electric furnace in an atmosphere of O_2 and thionyl chloride. Any water vapor in the preform is removed through the following reaction:

$$SOCl_2 + H_2O \text{ ® } SO_2 + 2HCl$$

The porous preform is then heated to about 1500 °C in a carbon furnace, where it is sintered into a transparent solid glass rod.

A good control over the index profile may be achieved with germania-doped cores. Graded-index fiber with an attenuation level less than 0.5 dB/km at 1.3 μm has been produced by this method.

Modified chemical vapour deposition (MCVD) method

The tube has a length of about 1 m and a diameter of about 15 mm. It is horizontally mounted and rotated on a glass-working lathe, with an arrangement (normally an oxygen-hydrogen flame) for heating the outer surface of the tube to about 1500°C. The reactants in the form of halide vapors and oxygen are passed at a controlled rate through the tube.

The halides are oxidized in the hot zone of the tube and the generated soot (glass particles) is deposited on the inner wall. The hot zone (i.e., the flame) is moved back and forth allowing the layer-by-layer deposition of the soot. Index grading may be achieved by varying the concentration of the dopants layer by layer. The tube can form a cladding material or serve as a support structure only for the porous preform. After the deposition is complete, the tube or the porous preform is sintered at a higher temperature (1700-1900°C) to form a transparent glass rod. Fiber is then drawn from this rod in the usual manner.

At present this is a widely used method for fabricating fibers, as it allows the deposition to occur in a clean environment, with reduced OH impurity. This method is suitable for preparing a variety of glass compositions for multimode or single-mode step-index (SI) or graded-index fibers. Typically, attenuation of the order of 0.2 dB/km at a wavelength of 1.55 μm has been reported for single-mode germania-doped silica fibers prepared by this method. Further, this technique is also suitable for preparing polarization-maintaining single-mode fibers.

PLASMA-ACTIVATED CHEMICAL VAPOUR DEPOSITION (PCVD) METHOD

The deposition rates of the MCVD process may be increased if microwave-frequency plasma is created in the reaction zone. This process is called plasma-activated chemical vapor deposition and is illustrated in the figure shown below. Herein, a microwave cavity (operating at 2.45 GHz) surrounds the substrate tube.

The halide vapors of the silica-based compound or the dopants along with oxygen are introduced into the tube where they react in the microwave-excited plasma zone. The tube temperature is maintained at about 1900 °C using a stationary furnace. The reaction zone is moved back and forth along the tube enabling circularly symmetrical deposition of glass layers onto the inner wall of the substrate tube. High deposition efficiency and excellent control of index grading are possible with this method. Attenuation of the order of 0.3 dB/km at λ = 1.55 μm has been obtained for dispersion flattened single-mode fibers.

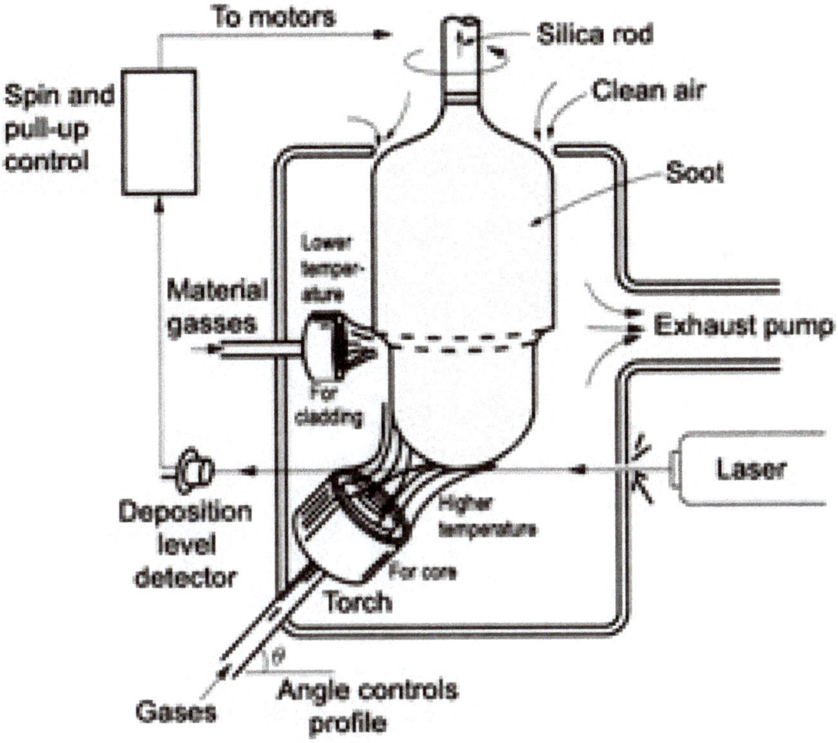

Fig. 14. Apparatus for the PCVD process

CHAPTER

OPTICAL FIBER CONNECTIONS: JOINTS, COUPLERS AND ISOLATORS 5

INTRODUCTION

There are many occasions when it is necessary to connect a fibre optic cable to another item. It may be that the fibre optic cable needs to be connected to another cable, or to an electronic interface device where the optical signal is converted to an electrical signal or to a light source. It is necessary that the fibre optic cable is correctly interfaced so that the minimum amount of light is lost. To achieve this it is necessary to use the correct form of fibre optic connector. In these cases fibre optic connectors are required.

While fibre optic connectors offer a very convenient method of connecting fibre optic cables, they should only be used where necessary. They introduce a loss at each connection. Typically the value is between 10 and 20 percent. Against this they make reconfiguring systems very much easier.

CONNECTOR BASICS

The fibre optic connector basically consists of a rigid cylindrical barrel surrounded by a sleeve. The barrel provides the mechanical means by which the connector is held in place wit the mating half. A variety of methods are used to ensure the connector is held in place, ranging from screw fit, to latch arrangements. The main requirement si that the end of the fibre optic cable is held accurately in place so that the maximum light transfer occurs.

As it is imperative that the optical fibre is held securely and accurately in place, connectors will normally be designed so that the fibre is glued in place, and in addition to this strain relief is also provided

Fibre ends may also be polished. For single mode fibre, the ends may be polished with a slight convex curvature so that the centres of the cables from the two connectors achieve physical contact. This approach reduces the back reflections, although the level of loss may be slightly higher.

FIBRE OPTIC CONNECTOR TYPES

Fibre optic connectors (fiber optic connectors) come in a variety of formats. These different fibre optic connectors may be used in slightly different applications or under different circumstances, as each type has its own capabilities.

When choosing a fibre optic connector, it is necessary to ensure that its properties meet the needs of the particular application in question. Some fibre optic connectors may be suitable for different optical fibres, and this needs to be taken into consideration.

There is a wide variety of different fiber optic connectors available. A selection of some is given below:

- **FC/PC:** This form of fibre optic connector is used for single-mode fiber optic cable. It provides very accurate positioning of the single-mode fiber optic cable with respect to transmitter (optical source) or the receiver (optical detector).
- **SC:** This form of connector is mainly used with single-mode fiber optic cables. The connector is simple low cost and reliable. The location and alignment is provided using a ceramic ferrule. It also has a locking tab to enable it to be mated and removed without fear of it accidentally falling loose.
- **Plastic fiber optic cable connectors:** As the name implies, these fibre optic cable connectors are only used with plastic fibre optic cabling.

Mechanical and fusion splices

The two types of fibre optic splices are used in different applications. The mechanical ones are used for applications where splices need to be made very quickly and where the expensive equipment for fusion splices may not be available. Some of the sleeves for mechanical fibre optic splices are advertised as allowing connection and disconnection. In this way a mechanical splice may be used in applications where the splice may be less permanent.

Fusion splices offer a lower level of loss and a high degree of permanence. However they require the use of the expensive fusion splicing equipment. In view of this they tend to be used more for the long high data rate lines that are installed that are unlikely to be changed once installed.

FIBER ALIGNMENT AND JOINT LOSS

A major consideration with all types of fiber–fiber connection is the optical loss encountered at the interface. Even when the two jointed fiber ends are smooth and perpendicular to the fiber axes, and the two fiber axes are perfectly aligned, a small proportion of the light may be reflected back into the transmitting fiber causing attenuation at the joint. This phenomenon, known as Fresnel reflection, is associated with the step changes in refractive index at the jointed interface (i.e. glass–air–glass).

In many applications of fiber optics, it is necessary to connect fiber ends (terminations) in some way such that light from one fiber can get into the other fiber without losing too much of its optical power. Examples are fiber lasers and systems for optical fiber communications.

There are different techniques for joining fiber ends:

- Permanent and stable connections with very low transition losses can be obtained by **fusion splicing**. Essentially, the **fiber** ends are fused together with a heat treatment.
- Semi-permanent connections can be made with **mechanical splices**, which are relatively simple alignment devices holding the fiber ends together. Typically, some **index matching gel** or an epoxy is used for reducing reflection losses. Compared with fusion splices, mechanical splices are faster to make and can be removed after use. The initial cost is much lower, but the cost per splice is higher. The splice losses and reflection loss are of the order of 0.5 dB, i.e., significantly higher than for fusion splices.
- **Fiber connectors** are convenient for connections which need to be released more often. Common connector types are named FC, SC and LC for single-mode applications and ST for multimode, but there are also dozens of other types, with special qualities such as duplex connections, particularly small size, built-in shutter for improved **laser safety**, etc. In most cases, the fiber is glued into the connector with some epoxy, and then the fiber end is **polished**. The transition losses of connectors are similar to those of mechanical splices, although the stability is generally lower.

Fig. 1: Microscope imagine of a fusion splice between a photonic crystal fiber (PCF, left side) and a conventional fiber (right side). The hole pattern of the PCF can be seen.

In any case, it is essential that the fiber ends are carefully prepared before joining them. In many cases, fiber ends with perpendicularly cut surfaces are needed. This has the advantage that the axis of the emitted light is in line with the fiber axis, and the alignment is comparatively simple. In other cases, the cleave angle has to have some deviation from 90°. Such angle cleaves have the advantage that light reflected at the air–glass interface will not get back into the fiber mode, provided that the cleave angle is sufficiently large.

It is also possible to use free-space optics for launching light from one fiber into the other. Typical configurations contain a single lens or two lenses. An advantage of that approach is that the fiber ends can be kept at a safe distance, making it less likely to damage them. Also, it is possible to efficiently couple light between fibers with different effective mode areas if the optics are designed accordingly. Furthermore, it is possible to insert other optical elements, such as an optical filter or a Faraday isolator. On the other hand, free-space optics have to be kept in stable alignment and are sensitive to dust. In addition, losses arising from the Fresnel reflections at the fiber ends cannot be avoided.

TOLERANCES FOR LOW-LOSS FIBER JOINTS

Multimode Fibers

The alignment tolerances for connections between fibers are relatively low for multimode fibers, particularly for those with a large core area. A high coupling efficiency is possible if the parameters of the two fibers differ substantially. It is sufficient that the second fiber (receiving the light) has values of the core diameter and numerical aperture which are equal or higher than those of the first fiber. If these parameters are higher for the second fiber, the light in that fiber will in general have nearly the same power but a lower brightness, whereas a substantial coupling loss can occur for light coming from the second fiber.

If there is a transverse mismatch or an angular mismatch of the two fiber cores, this can lead to a coupling loss which may be estimated with geometrical considerations, if the light uniformly fills the whole core area and angular range. Note that the coupling loss generally depends on how the optical power is distributed over the fiber modes. For this and other reasons, reliably estimating coupling losses for multimode fibers is hard to achieve.

The wave nature of light can lead to surprising behavior in multimode fiber coupling. For example, if the output fiber has a somewhat larger core than the input fiber and the same numerical aperture, one may expect to have zero coupling losses.

This is not the case, however, because the larger core changes the shapes of the mode profiles, so that the modes of the input fiber cannot fully be transformed into guided modes of the output fiber; some of the light gets into the cladding – but strongly depending on the considered mode of the input fiber, and often sensitively depending on the magnitude of core size mismatch (with complicated nonmonotonic dependencies). For such reasons, reliable estimations of coupling losses for multimode fibers require numerical software, not just some simple equations.

Single-mode Fibers

The situation is rather different for coupling of single-mode fibers. Here, the coupling loss can not depend on the direction of propagation. It can be low only if the following conditions are fulfilled:

- The modal shapes and particularly the effective mode areas are similar.
- The transverse and angular alignment of the modes is rather precise.

MECHANICAL FIBER SPLICES

In fiber optics, for example in the context of optical fiber communications, it is often necessary to connect the ends of optical fibers such that light from one fiber can get into the other fiber without losing much of its optical power.

One of the possibilities is *mechanical splicing*. Here, one uses a simple kind of alignment device which holds the fiber ends together such that the two fiber cores are well aligned to each other and there is no significant air gap in between the ends. Once such a splice is applied, the connection should be fairly stable, maintaining a high coupling efficiency even if the connecting fibers are moved.

The Splicing Process

The details of the splicing process depends on the equipment used; one should follow the instructions of the manufacturer. The process should normally not be too complicated. A typical process may be as described in the following.

In case that one works with fiber cables instead of simple fibers, the cable ends also need to be prepared appropriately. Typically, one has to nip through the jacket with a suitable tool, strip off the jacket, and cut off any kevlar to obtain the fiber alone over a length of a few centimeters.

The fiber ends need to be prepared such that one has clean and flat surfaces, which should be perpendicular to the fiber axis. First, one has to remove the buffer coating from the fibers over some length. Thereafter, one needs to clean and cleave the ends. Cleaving is normally sufficient for surface generation, although polishing the fiber ends could lead to even better results. However, accurate cleaving is key for obtaining high performance of the splice.

One then inserts one of the fibers into the splicing device. That may contain a transparent window, so that one can monitor the position of the fiber end before finally fixing it by tightening a locking nut. Thereafter, one carefully inserts the second fiber, until the two ends meet, and fixes that end again with its locking nut.

Sometimes, one applies a thin layer of some index matching gel or a transparent epoxy between the fiber ends in order to further reduce the reflection losses. That should lead to a small insertion loss (e.g. <0.1 dB) and also to a large return loss, i.e., to negligibly weak reflections at the joint. The exact width of the gap, filled with the index matching agent, is not critical in this case. However, this technique is not ideal for splices which are supposed to be removed later on. Also, one should be sure to have an index matching agent which stays stable over time.

Details may have to be adjusted, or other splicing equipment used, if the fibers have a non-standard diameter.

Depending on the used splicing equipment, the resulting connection may be permanent or removable. In the latter case, one may only need to clean the fiber ends (particularly if an index matching gel was used), but not to cleave them again.

Testing of Splices

After splicing, one may want to test the splice quality. For example, one can use a *visual fault locator*, which is essentially a visible laser source. When some light is launched into the fiber, a poor quality splice will exhibit a lot of stray light, which can be observed if the splicing device is transparent. In case of a good splice, not much stray light will be seen.

A more elegant method, which however requires expensive equipment, is measuring the insertion loss and the return loss with an optical time-domain reflectometer.

Comparison with Fusion Splicing

An alternative to mechanical splicing is fusion splicing of fibers. Compared with that, advantages of mechanical splicing are that no expensive equipment is needed, that the process can be tentatively faster, that it can more easily be performed in the field, and that the connection can usually be removed without re-cleaving the fibers. On the other hand, the splice quality is typically worse in terms of insertion loss and return loss. Also, the cost per splice (when ignoring the upfront cost of fusion splicing) is higher.

FIBER CONNECTORS

Fiber connectors are often used as the terminations of optical fiber cables in order to provide non-permanent connections between fiber-coupled devices (a kind of removable *fiber joints*). They are used in a similar manner as electrical connectors.

Typically, a connector assembly comprises an adapter and two connector plugs, where one fiber is inserted into every connector plug. The process of fitting a connector plug to a fiber is more delicate than for most electrical connections:

- First, a clean fiber end must be prepared, usually with a fiber cleaver. It must be made sure that the orientation of the fiber interface is correct, e.g.

perpendicular to the fiber axis or with some defined angle against it (e.g. 8°). This might be checked e.g. with a microscope.

- Then the fiber end needs to be carefully inserted into the connector plug. Its position must be precisely correct, and the fiber end should not be damaged during that process. Depending on the type of fiber connector, a detailed procedure must be followed, which normally includes the proper preparation and cleaning of the plug and some polishing of the fiber tip. Most connectors require a type of epoxy or other adhesive to fix the fiber.
- In sensitive cases, it may be necessary to measure the insertion loss and/or return loss, as it may otherwise be more cumbersome to locate a fault later on.

These termination assembly operations are often performed in factories, but with suitable equipment they can also be done in the field, preferably in a reasonably clean environment.

Although fiber connectors give some protection to the inserted fibers, that protection is quite incomplete, e.g. concerning moisture. Therefore, connectors may be put into protective enclosures.

Hermetically sealed enclosures protect most reliably against moisture and dirt, but can hinder the dissipating of heat (e.g. from connector losses in high-power operation). Dust caps are often used during times where a connector is not plugged in.

Most fiber connectors have a single fiber, but there are also specialized fiber connectors for fiber arrays.

Characteristics of Optical Fiber Connectors

The following characteristics can be relevant for the selection of fiber connectors:

Many properties of fiber connectors can be relevant, depending on the circumstances – this is why many different versions are in use.

- A connector should provide a reliable low-loss contact in the plugged-in state. One specifies a typical coupling loss in decibels. It is called the *attenuation* or *insertion loss*. Its measurement is defined by IEC standard 61753-1. Typical values of the insertion loss are of the order of 0.5 dB. When the same connector is plugged in and out repeatedly, the insertion loss may exhibit significant variations (non-perfect repeatability).
- For sensitive applications, the *reflection loss* (= *return loss*), i.e., the attenuation for light being reflected at the interface, can also be relevant; it should be as high as possible. Values above 20 dB (i.e., less than 1 % reflectivity) are easily achieved – with some connector versions (HRL = high return loss) even more than 50 dB.

- Note that the insertion losses are usually substantially higher than what one would expect from the return loss alone – particularly for fiber connectors using angled fiber ends, where reflected light does not get back into the fiber.

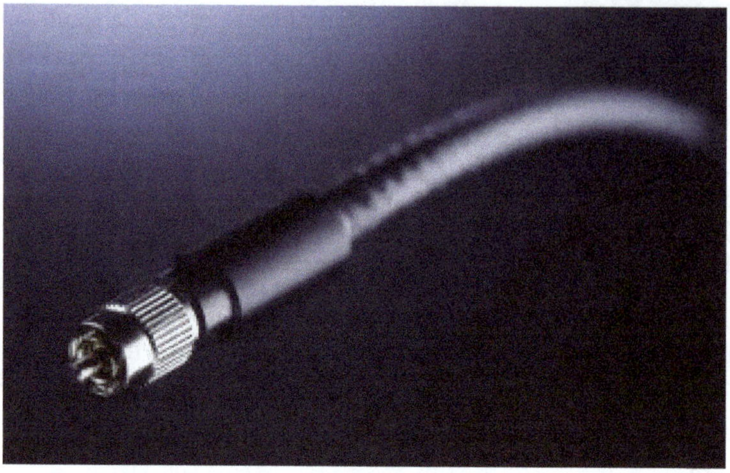

Fig. 2: A fiber connector at the end of a fiber cable.

- Note also that increased reflections can occur as a result of a tiny air gap between the connectorized fibers (e.g. due to an incorrect angle of the fiber tip): the Fresnel reflections at both interfaces can cancel each other only if the gap width is much smaller than one wavelength.
- In principle, one could also suppress reflections with anti-reflection coatings on the fiber ends, but this technique is not very common due to various practical issues, such as the fragility of coatings and the substantial effort to produce them.
- Many fiber connectors are rarely disconnected or reconnected in normal use, but for some application a higher number of reconnections (*mating cycles*) must be possible during the connector's lifetime. Thousands of mating cycles are not easily achieved, as the contact between the fiber ends needs to be very precise and clean.
- It may also matter how easily and reliably fibers can be connectorized, particularly if that needs to be done in the field. The sensitivity to environmental factors such as cleanliness and vibrations also varies, and the required equipment depends on the connector type.
- In some cases, the power handling capability is relevant.
- There are *multi-fiber connectors*, which can connect multiple fibers.
- The dimensions of connectors are relevant particularly if many connections need to be done to devices of moderate size.

TYPES OF OPTICAL FIBER CONNECTORS

The most common fiber connectors are those of ST, FC, SC and LC type, and there are various versions of those. A short overview is given in the following:

ST Connectors

ST connectors are most popular for fiber-optic networks based on multimode fibers, as are usually used within buildings. The fiber is placed in a relatively long cylindrical keyed ferrule (often made of ceramic), providing a tight contact (physical contact) of the fiber ends. ST connectors contain a spring-loaded bayonet mount.

FC Connectors

FC connectors ("fixed connection") are a popular type for single-mode fibers. They contain a floating ferrule with 2.5 mm diameter, into which the fiber is inserted (with the risk of scratching the fiber end). They provide a good mechanical isolation of the contact, allowing the use in high-vibration environments, but need to be handled quite carefully, with alignment of the ferrule's key when plugging it in. That key prevents rotation of the fiber when the plug is screwed into the receptacle.

This is important for angled fiber ends and for polarization-maintaining fibers. Unfortunately, there are two mutually incompatible versions with key widths of 2 mm (reduced version, "type R") and 2.14 mm ("type N"), respectively.

The standard version FC/PC involves a standard "physical contact" with a slightly rounded fiber surface; the fiber tip is polished after insertion into the ferrule.

The ferrule is spring-loaded, providing a well-defined contact force when the connector is plugged in. Higher-quality fiber tip polishing is used in FC/SPC and FC/UPC connectors (with "super" or "ultra" polish). This provides higher return loss and lower insertion loss.

FC/APC connectors use angled fiber ends in order to obtain a high return loss (even in the disconnected state), but normally at the cost of increased insertion loss. Of course, fiber ends with a carefully controlled tilt angle are more difficult to make, and the two angled fibers need to be inserted properly such the surfaces fit together.

FC connectors are often used for single-mode fibers including polarization-maintaining fibers, but are increasingly replaced with SC and LC connectors.

SC Connectors

SC connectors ("subscriber connectors") are general purpose connectors of push/pull snap-in type for single-mode fibers. The fiber lies in a ceramic ferrule, and its orientation is controlled with a key. The rectangular shape of the connector leads to a defined orientation concerning rotations (in contrast e.g. to FC connectors). The snap-in mechanism makes the handling less critical than with FC connectors.

SC connectors are increasingly used in network applications, particularly with single-mode fibers. They are also available in duplex versions.

LC Connectors

LC connectors ("Lucent connectors") are particularly small, belonging to the "small form factor" connectors (*SFF connectors*). It contains a ferrule with only 1.25 mm diameter. It is often used for single-mode fibers. Duplex versions are also available.

Applications of Fiber Connectors

As fiber optics are intensively used for optical fiber communications, many fiber connectors are also used in that context – for example, in central offices of telephone companies, at Internet backbones or in fiber to the home installations.

Laboratories also often use fiber cables with connectors, e.g. for transporting light to diagnostic instruments such as fiber-optic power meters and spectrometers. For complex setups such as mode-locked fiber lasers, for example, it is less common to use fiber connectors, partly because of the detrimental effects even of very small return losses. Mechanical splicing or fusion splicing is then often a better option.

Multimode fiber connectors are often installed in the field. Single-mode connectors, having tighter tolerances, are usually combined with their fibers in factories. In the field, one may fusion-splice such pig-tailed single-mode connectors to cables. (Single-mode connectors installed in the field tend to have higher losses.)

Expanded Beam Fiber Connectors

Expanded Beam connectors have distinct advantages and disadvantages compared to Physical Contact connectors. Amphenol Fiber Systems International (AFSI) is the only manufacturer that offers a complete suite of both Expanded Beam and Physical Contact connectors. Contact your AFSI sales representative to discuss which option is best suited for your application.

Fig. 3. Expanded Beam Fiber Connector

Expanded beam connectors are typically found in harsh environments such as mining, outside broadcasting, offshore and some factory automation applications.

They're also often used in medical applications that require thousands of mating cycles and are subject to microorganisms, fluids and other contaminants found in a clinical setting.

Due to their higher mating cycles, expanded beam connectors are commonly used in military and temporary applications such as federal emergency response scenarios. TFOCA (tactical fiber-optic cable assemblies) designed for military tactical field use often use expanded beam connectors due to easy mating and resistance to extreme temperatures, dust and sand.

Fiber Couplers

Fiber couplers belong to the basic components of many fiber-optic setups. Note that the term *fiber coupler* is used with two different meanings:

- It can be an optical fiber device with one or more input fibers and one or several output fibers. Light from an input fiber can appear at one or more outputs, with the power distribution potentially depending on the wavelength and polarization.
- It can also be a device for coupling (launching) light from free space into a fiber; see the article on fiber launch systems.

Fig. 4: A 2-by-2 fiber coupler.

Fiber couplers of the first type, coupling light from fibers to fibers. Such couplers can be fabricated in different ways:

- Two or more fibers can be thermally tapered and fused so that their cores come into intimate contact over some length of a few centimeters, for example. Such fused couplers can also be made with polarization-maintaining fibers, leading to polarization-maintaining couplers (PM couplers) or splitters.
- Some couplers use side-polished fibers, providing access to the fiber core.
- There are fiber-optic pump combiners and pump–signal combiners, which usually work with multimode pump fibers.
- There are planar lightwave circuits, containing things like branching waveguides, with fibers coupled to the inputs and outputs.

- Couplers can also be made from bulk optics, for example in the form of microlenses and beam splitters, which can be coupled to fibers ("fiber pig-tailed").

One may omit one of the input ports of a 2-by-2 fiber coupler, obtaining a *Y coupler*, also called *T coupler*. It may also be called a *tap coupler*, particularly if only a small fraction of power is obtained at one output and used e.g. for power monitoring. Couplers with many inputs or outputs are called *star couplers*; they may be used, e.g., as *fiber-optic splitters*, e.g. for distributing cable-TV signals.

Fiber couplers are usually *directional couplers*, which means that essentially no optical power sent into some input port can go back into one of the input ports. There is often a specification of *return loss*, which indicates how much weaker the back-reflected light is, compared with the input, and is usually quite large (many tens of decibels).

Limitations for Fiber Combiners

Coupling Loss

In many cases, all fibers involved are single-mode, i.e., they support only a single mode per polarization direction for a given wavelength. There are then certain physical restrictions on the performance of the coupler. In particular, it is not possible to combine two or more inputs of the same optical frequency into a single-polarization output without significant excess losses, except if the optical phases of the input beams are precisely adjusted and stabilized. That means that the two inputs to be combined would have to be mutually coherent.

However, such a restriction does not occur for different input wavelengths: there are couplers which can combine two inputs at different wavelengths into one output without exhibiting significant losses. Such *dichroic couplers* are used in fiber amplifiers to combine the signal input and the pump wave. Their insertion loss may be very small (e.g. far below 1 dB) for both inputs. Other wavelength-sensitive couplers are used as multiplexers (WDM couplers) in wavelength division multiplexing (WDM) telecom systems to combine several input channels with different wavelengths, or to separate channels.

Multimode fiber combiners allow the powers of two mutually incoherent beams to be combined without a power loss. However, this will cause some loss of brightness.

Bandwidth

Most types of couplers work only in a limited range of wavelength (a limited bandwidth), since the coupling strength is wavelength-dependent (and often also polarization-dependent). This is a typical property of those couplers where the coupling occurs over a certain length. Typical bandwidths of fused couplers are a few tens of nanometers. As mentioned above, they can be used as *dichroic couplers* or beam combiners. They are sometimes also called *WDM couplers* ('! wavelength division multiplexing).

Typical Applications

Some typical applications of fiber couplers are:

- In a cable TV system, the powerful signal from one transmitter is sent in to a *fiber-optic splitter*, which distributes the power over a large number of output fibers for different customers.
- Fiber couplers can be used in fiber interferometers, for example for optical coherence tomography (OCT). Specially designed broadband couplers are often required for such purposes.
- Within the resonator of a fiber laser, a dichroic fiber coupler can be used to inject pump light, and another fiber coupler can be used as the output coupler. This technique is used particularly in fiber ring lasers, having no resonator ends where light could be injected.
- In fiber amplifiers and lasers, dichroic couplers are often used for injecting pump light or eliminating residual pump light from the signal output.
- In high-power fiber lasers and amplifiers, multimode fiber couplers are often used for combining the radiation of several laser diodes and sending them into inner cladding of the active fiber (a double-clad fiber).

WAVELENGTH DIVISION MULTIPLEXING COUPLERS

Wavelength division multiplexing is a kind of *frequency division multiplexing* – a technique where optical signals with different wavelengths are combined, transmitted together, and separated again. It is mostly used for optical fiber communications to transmit data in several (or even many) channels with slightly different wavelengths. In this way, the transmission capacities of fiber-optic links can be increased strongly, so that most efficient use is made not only of the fibers themselves but also of the active components such as fiber amplifiers. Apart from telecom, wavelength division multiplexing is also used for, e.g., interrogating multiple fiber-optic sensors within a single fiber.

WDM in Telecom Systems

Theoretically, the full data transmission capacity of a fiber could be exploited with a single data channel of very high data rate, corresponding to a very large channel bandwidth. However, given the enormous available bandwidth (tens of terahertz) of the low-loss transmission window of silica single-mode fibers, this would lead to a data rate which is far higher than what can be handled by optoelectronic senders and receivers. Also, various types of dispersion in the transmission fiber would have very detrimental effects on such wide-bandwidth channels, so that the transmission distance would be strongly restricted. Wavelength division multiplexing solves these problems by keeping the transmission rates of each channel at reasonably low levels (e.g. 10 Gbit/s or 100 Gbit/s) and achieving a high total data rate by combining several or many channels.

Two different versions of WDM, defined by standards of the International Telecommunication Union (ITU), are distinguished:

Course Wavelength Division Multiplexing

Coarse wavelength division multiplexing (CWDM, ITU standard G.694.2 [11]) uses a relatively small number of channels, e.g. four or eight, and a large channel spacing of 20 nm. The nominal wavelengths range from 1310 nm to 1610 nm. The wavelength tolerance for the transmitters is fairly large, e.g. ±3 nm, so that unstabilized DFB lasers can be used. The single-channel bit rate is usually between 1 and 3.125 Gbit/s. The resulting total data rates are useful e.g. within metropolitan areas, as long as broadband technologies are not widespread in households ('! *fiber to the home*).

Dense Wavelength Division Multiplexing

Dense wavelength division multiplexing (DWDM, ITU standard G.694.1 [10]) is the extended method for very large data capacities, as required e.g. in the Internet backbone. It uses a large number of channels (e.g. 40, 80, or 160), and a correspondingly small channel spacing of 12.5, 25, 50 or 100 GHz. All optical channel frequencies refer to a reference frequency which has been fixed at 193.10 THz (1552.5 nm). The transmitters have to meet tight wavelength tolerances. Typically, they are temperature-stabilized DFB lasers. The single-channel bit rate can be between 1 and 100 Gbit/s, and in the future even higher.

Even much higher channel counts (>1000) have become practical by using novel methods of producing the channels. One such technique is spectral slicing of broadband femtosecond pulses [8]. While such a femtosecond laser is substantially more expensive than a DFB laser as usually used for one channel, the femtosecond laser can replace a large number of such lasers and at the same time substantially reduce the efforts required for signal multiplexing.

Common Fiber Amplifiers

Due to the wide amplification bandwidth of erbium-doped fiber amplifiers, all channels can often be amplified in a single device (except in cases where e.g. the full range of CWDM wavelengths is used). However, problems can arise from the variation of gain with wavelength or from interaction of the data channels (*crosstalk, channel interference*) e.g. via fiber nonlinearities. Enormous progress has been achieved with a combination of various techniques, such as the development of very broadband (double-band) fiber amplifiers, gain flattening filters, nonlinear data regeneration and the like.

The system parameters such as channel bandwidth, channel spacing, transmitted power levels, fiber and amplifier types, modulation formats, dispersion compensation schemes, etc., need to be well balanced to achieve optimum overall performance.

Upgrade of Existing Fiber Links

Even for existing fiber links with only one or a few channels per fiber, it can make sense to replace senders and receivers for operation with more channels, as this can be cheaper than replacing the whole system with a system with a higher transmission capacity. In fact, this approach often eliminates the need to install additional fibers, even though the demand on transmission capacities is increasing enormously.

Increased Flexibility

Apart from increasing the transmission capacity, wavelength division multiplexing also adds flexibility to complex communication systems. In particular, different data channels can be injected at different locations in a system, and other channels can be extracted. For such operations, *add-drop multiplexers* can be used, which allow one to add or drop data channels based on their wavelengths. Reconfigurable add-drop multiplexers make it possible to reconfigure the system flexibly so as to provide data connections between a large number of different stations.

The optimization of the key parameters such as number of optical channels, their frequency spacing, their optical bandwidth and the used modulation format is a complex task, involving various kinds of trade-offs.

OPTICAL ISOLATORS AND CIRCULATORS

Optical isolators are devices which transmit light in one direction but not in the opposite direction. More precisely, they exhibit a relatively low propagation loss in one direction, but a much higher propagation loss in the other direction.

Essential characteristics of optical isolators are the following:

- the insertion loss
- the degree of isolation, i.e., the power loss for light in the unwanted direction (usually expressed in decibels)
- the return loss (relevant e.g. in optical fiber communications)
- the operation wavelength and the optical bandwidth in which sufficiently high isolation is achieved
- requirements on the input polarization

- the maximum optical power (peak or average power) which can be transmitted

Many isolators are used with free-space beams, while others are coupled to waveguides, e.g. optical fibers.

A variant is an *optical circulator*, having at least three optical ports. Light injected into port 1 will exit at port 2, while input at port 2 will be sent to port 3, and input at port 3 to port 1.

Types of Optical Isolators

Faraday Isolators

The vast majority of optical isolators are Faraday isolators, based on the Faraday effect, i.e., a rotation of the polarization direction caused by a magnetic field. This is a kind of magnetically induced optical activity.

Essential characteristics of Faraday isolators are:

- Many Faraday isolators transmit only light with a suitable direction of linear polarization, but there are also polarization-independent versions.
- The insertion loss can be rather small, and a high degree of optical isolation (e.g. well above 30 dB) can be achieved.
- Faraday isolators can be made for a wide range of optical wavelengths. Although each device works well only within a limited bandwidth, that bandwidth can be substantial – sufficient for a wide range of applications.
- Such devices can be made for operation with very high optical powers by using a sufficiently large beam area. Power limitations arise due to thermal effects (particularly thermal lensing and depolarization) associated with parasitic absorption.
- Relatively compact low-power isolators can also be made. However, it is hard to develop ultra-compact devices for use in photonic integrated circuits. Here, the requirement of a strong magnetic field is a major problem.

Isolators Based on Waveplates

One realize a type of optical isolator using a waveplate instead of a Faraday rotator. While the basic purpose of optical isolation can be fulfilled with that concept, it comes with some limitations, such as the sensitivity to polarization changes in reflected light.

For some applications, this approach is superior to the one based on a Faraday isolator, for example because the setup can be substantially more compact.

Optical Isolators Based on Acoustic Effects

As mentioned above, conventional types of optical isolators based on the

Faraday effect are hard to implement in the context of photonic integrated circuits. Therefore, alternative methods are under development, where non-reciprocal light propagation is obtained based on completely different physical mechanisms, not involving magnetic fields.

In particular, it has been demonstrated that one can utilize rotating sound waves in a ring-shaped waveguide resonator, coupled to a straight waveguide.

Here, the rotation of the acoustic field causes a direction-dependent shift of resonance frequencies. If those optical residences are positioned such that the ring is resonant for one propagation direction, resident coupling of the ring to the straight waveguide can cause substantial propagation losses. For the opposite propagation direction, the induced loss can be much lower.

This operation principle allows one to realize rather compact isolators for use on photonic chips. The main limitations are that the usable optical bandwidth is small and that a substantial electric power is required to drive piezo actuators creating the sound wave. However, that power may be substantially reduced with further optimization.

CHAPTER

6 OPTICAL SOURCES 1: THE LASER

INTRODUCTION

"Laser" (rarely written as l.a.s.e.r.) is an acronym for "Light Amplification by Stimulated Emission of Radiation", coined in 1957 by the laser pioneer Gordon Gould. Although this original meaning denotes an principle of operation (exploiting stimulated emission from excited atoms or ions), the term is now mostly used for *devices* generating light based on the laser principle. More specifically, one usually means laser *oscillators*, but sometimes also includes devices with laser amplifiers, called master oscillator power amplifier (MOPA). An even wider interpretation includes nonlinear devices like optical parametric oscillators and Raman lasers, which also produce laser-like light beams and are usually pumped with a laser, but are strictly speaking not lasers themselves.

Laser technology is at the core of the wider area of photonics, essentially because laser light has a number of very special properties:

- It is usually emitted as a well directed laser beam which due to its high spatial coherence can propagate over long lengths without much divergence (often limited only by diffraction) and can be focused to very small spots, where a high intensity is achieved.
- It often has a very narrow optical bandwidth (high temporal coherence), whereas e.g. most lamps emit light with a very broad optical spectrum. However, there are also broadband lasers, particularly among ultrafast lasers.
- Laser light may be emitted continuously, or alternatively in the form of short or ultrashort pulses, with pulse durations from microseconds down to a few femtoseconds. The temporal concentration of pulse energy – in addition to the potential of strong spatial confinement in a beam focus – allows for even far higher intensities to be generated. Particularly extreme intensity values are used in high-intensity physics.

These properties, which make laser light very interesting for a range of applications, are to a large extent the consequences of the very high degree of spatial and/or temporal coherence of laser radiation. The articles on laser light and laser applications give more details.

The first laser was a pulsed lamp-pumped ruby laser (a kind of solid-state laser), demonstrated by Theodore Maiman in 1960. In the same year, the first gas laser (a helium–neon laser) and the first laser diode were made. Before this experimental work, Arthur Schawlow, Charles Hard Townes, Nikolay Basov and Alexander Prokhorov had published ground-breaking theoretical work on the operation principles of lasers, and a microwave amplifier and oscillator (maser) had been developed by Townes' group in 1953. The term "optical maser" (MASER = microwave amplification by stimulated amplification of radiation) was initially used, but later replaced with "laser".

In laser technology, a wide range of optical components such as laser crystals, laser mirrors, polarizers, Faraday isolators and tunable optical filters are used; see the article on laser optics.

HOW A LASER WORKS?

Basic Principle

A laser oscillator usually comprises an optical resonator (laser resonator, laser cavity) in which light can circulate (e.g. between two mirrors), and within this resonator a gain medium (e.g. a laser crystal), which serves to amplify the light. Without the gain medium, the circulating light would become weaker and weaker in each resonator round trip, because it experiences some losses, e.g. upon reflection at mirrors. However, the gain medium can amplify the circulating light, thus compensating the losses if the gain is high enough. The gain medium requires some external supply of energy – it needs to be "pumped", e.g. by injecting light (*optical pumping*) or an electric current (*electrical pumping semiconductor lasers*). The principle of laser amplification is stimulated emission.

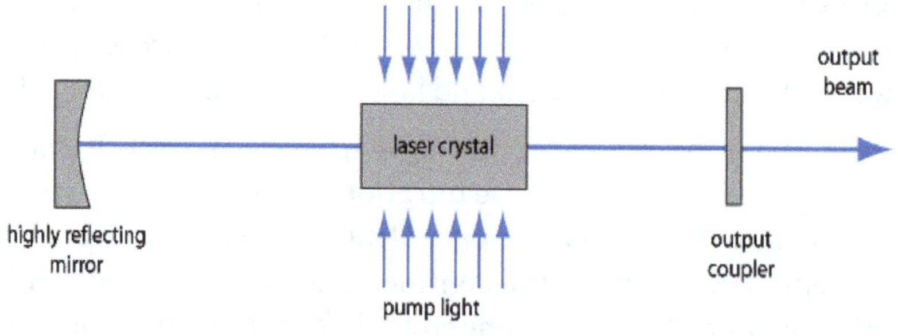

Fig. 1: Setup of a simple optically pumped solid-state laser. The laser resonator is made of a highly reflecting curved mirror and a partially transmissive flat mirror, the output coupler, which extracts some of the circulating laser light as the useful output. The gain medium is a laser crystal or rod, which is side-pumped, e.g. with light from laser diodes or a flash lamp.

A laser can not operate if the gain is smaller than the resonator losses; the device is then below the so-called laser threshold and only emits some weak luminescence light. Significant power output is achieved only for pump powers above the laser threshold, where the gain can reach (or temporarily exceed) the level of the resonator losses.

If the gain is larger than the losses, the power of the light in the laser resonator rises very rapidly, starting e.g. with low levels of light from fluorescence. Note that the resonator round-trip time is usually very small (e.g. a few nanoseconds, for compact laser types even much less), so that even a small net round-trip gain implies rapid exponential growth of the intracavity power. As high laser powers saturate the gain by extracting energy from the gain medium, the laser power will in the steady state reach a level so that the saturated gain just equals the resonator losses. Before reaching this steady state, a laser often undergoes relaxation oscillations (just one aspect of laser dynamics). The threshold pump power is the pump power where the small-signal gain is just sufficient for lasing.

Some fraction of the light power circulating in the resonator is usually transmitted by a partially transparent mirror, the so-called output coupler mirror. The resulting beam constitutes the useful output of the laser. The transmission of the output coupler mirror can be optimized for maximum output power (see also: slope efficiency). In most cases, one has only a single output coupler.

SPATIAL COHERENCE OF LASER RADIATION

A high degree of spatial coherence of the laser radiation can be achieved, essentially because the light emission is triggered (stimulated) by the intracavity radiation (i.e., the light circulating in the laser resonator) itself, rather than occurring spontaneously in an uncoordinated fashion. In the stimulated emission process, the laser-active ions are made to emit light in the direction of already existing light, and also with the same optical phase. In effect, the circulating laser light serves to strongly coordinate the emission of many atoms or ions. The resulting amplitude and phase profile of the laser beam is largely determined by the properties of the laser resonator, not usually by the laser gain medium.

As explained above, the spatial coherence is the physical basis for the potential of forming very directed laser beams with low divergence, and for focusing light to very small spots.

Temporal Coherence

Temporal coherence is a different issue, and it has completely different origins. Some laser gain media can emit light only in a narrow spectral range. However, even if that is not the case, a laser often (particularly in continuous-wave operation) emits light only at a precisely defined wavelength or frequency, because the conditions are such that a net zero round-trip gain is possible only for that wavelength, and other wavelengths exhibit a negative net round-trip gain. A laser may be tuned to the exact wanted wavelength (within the emission region of the gain medium) e.g. by using a tunable intracavity bandpass filter, such as a Lyot filter. Again, the mechanism of stimulated emission is of crucial importance: the laser-active ions can be made to emit at exactly the optical frequency of already existing light. The smaller the emission linewidth (i.e., the narrower the optical spectrum of emitted light), the higher the degree of temporal coherence.

In extreme cases, the linewidth of a laser can be forced to values below 1 Hz (with certain means of laser stabilization). This is many orders of magnitude below the mean frequency (hundreds of terahertz). Optical clocks involve such highly stabilized lasers.

Interestingly, even ultrashort pulses can exhibit a very high degree of temporal coherence, in that case involving coherence between subsequent pulses in a regular pulse train. This is related to the formation of a frequency comb as the optical spectrum. While the optical spectrum can then overall be very wide, each comb line can be extremely narrow and well defined in frequency.

GENERATION OF LIGHT PULSES

Some lasers are operated in a continuous fashion, whereas others generate pulses, which can be particularly intense. There are various (very different) methods for pulse generation with lasers, allowing the generation of pulses with durations of microseconds, nanoseconds, picoseconds, or even down a few femtoseconds (*ultrashort pulses* from *mode-locked lasers*).

Often, a laser medium can accumulate some amount of energy over some "pumping" time in order to then release it within a much shorter time.

The optical bandwidth (or linewidth) of a continuously operating laser may be very small when only a single resonator mode can oscillate. In other cases, particularly for mode-locked lasers, the bandwidth can be very large – in extreme cases, it can span about a full octave.

The center frequency of the laser radiation is typically near the frequency of maximum gain, but if the resonator losses are made frequency-dependent, the laser wavelength can be tuned within the range where sufficient gain is available. Some broadband gain media such as Ti:sapphire and Cr:ZnSe allow wavelength tuning over hundreds of nanometers.

LASER NOISE

Due to various influences, the output of lasers always contains some noise in properties such as the output power or phase. For pulsed lasers, additional quantities can come into play, for example the timing jitter. For more details, see the article on laser noise.

Types of Lasers

Laser technology is a rather diverse field, utilizing a wide range of very different kinds of laser gain media, optical elements and techniques. Common types of lasers are:

- *Semiconductor lasers* (mostly laser diodes), electrically (or sometimes optically) pumped, efficiently generating very high output powers (but typically with poor beam quality), or low powers with good spatial properties (e.g. for application in CD and DVD players), or pulses (e.g. for telecom applications) with very high pulse repetition rates. Special types include quantum cascade lasers (for mid-infrared light) and surface-emitting semiconductor lasers (VCSELs, VECSELs and PCSELs). Some of those are also suitable for pulse generation with high powers.
- *Solid-state lasers* based on ion-doped crystals or glasses (*doped insulator lasers*), pumped with discharge lamps or laser diodes, generating high output powers, or lower powers with very high beam quality, spectral purity and/or stability (e.g. for measurement purposes), or ultrashort pulses with picosecond or femtosecond durations. Common gain media are Nd:YAG, Nd:YVO_4, Nd:YLF, Nd:glass, Yb:YAG, Yb:glass, Ti:sapphire, Cr:YAG and Cr:LiSAF. A special type of ion-doped glass lasers are:
- *Fiber lasers*, based on optical glass fibers which are doped with some laser-active ions in the fiber core. Fiber lasers can achieve extremely high output powers (up to kilowatts) with high beam quality, allow for widely wavelength-tunable operation, narrow linewidth operation, etc.
- *Gas lasers* (e.g. helium–neon lasers, CO_2 lasers, argon ion lasers and excimer lasers), based on gases which are typically excited with electrical discharges. Frequently used gases include CO_2, argon, krypton, and gas mixtures such as helium–neon. Common excimers are ArF, KrF, XeF, and F_2. As far as gas molecules are involved in the laser process, such lasers are also called molecular lasers.

Not very common are chemical and nuclear pumped lasers, free electron lasers and X-ray lasers.

LASER SOURCES IN A WIDER SENSE

There are some light sources which are not strictly lasers, but are nevertheless often called *laser sources*:

- In some cases, the term is used for amplifying devices emitting light without an input (excluding seeded amplifiers). An example are X-ray lasers, which are usually *superluminescent sources*, based on spontaneous emission followed by single-pass amplification. There is then no laser resonator.
- A similar situation occurs for optical parametric generators, where the amplification, however, is not based on stimulated emission; it is parametric amplification based on optical nonlinearities.
- Raman lasers utilize amplification based on stimulated Raman scattering.

Light from such devices can have laser-like properties, such as strongly directional emission, high spatial and temporal coherence and a narrow optical bandwidth.

In other cases, the term *laser sources* is justified by the fact that the source contains a laser, among other components. This is the case for combinations of lasers and amplifiers (*master oscillator power amplifier*), and also for sources based on nonlinear frequency conversion of laser radiation, e.g. with frequency doublers or optical parametric oscillators.

SAFETY ASPECTS

The work with lasers can raise serious safety issues. Some of those are directly related to the laser light, in particular to the high optical intensities achievable, but there are also various other hazards related to laser sources.

Laser Applications

There is an enormously wide range of applications for a great variety of different laser devices. They are largely based on various special properties of laser light, many of which cannot be achieved with any other kind of light sources. Particularly important application areas are laser material processing, optical data transmission and storage and optical metrology. See the article on laser applications for an overview.

Still, many potential laser applications cannot be practically realized so far because lasers are relatively expensive to make – or more precisely, because they are so far mostly made with relatively expensive methods. Most lasers are fabricated in relatively small volumes and with a limited degree of automation. Another aspect is that lasers are relatively sensitive in various respects, for example concerning the precise alignment of optical components, mechanical vibrations and dust particles. Therefore, there is ongoing research and development for finding more cost-effective and robust solutions.

For business success, it is often vital not just to develop lasers with high performance and low cost, but also to identify the best suited applications, or develop lasers which are best suited for particular applications. Also, the knowledge of the application details can be very important. For example, in laser material processing it is vital to know the exact requirements in terms of laser wavelength, beam quality, pulse energy, pulse duration etc. for optimum processing results.

OPTICAL EMISSION IN SEMICONDUCTOR MATERIALS

Fundamental radiation processes that occur within semiconductor materials are shown in Fig. 2. It also shows the response of a biased *p-n* junction diode for each process. Absorption produces electron-hole pairs and is the fundamental process exploited in semiconductor photodiodes where the junction is typically reverse biased. In contrast, electroluminescence (or electron-hole recombination radiation following electrical injection) occurs when a *p-n* junction is forward biased. This radiation is equivalent to spontaneous emission and leads to optical emission in an LED.

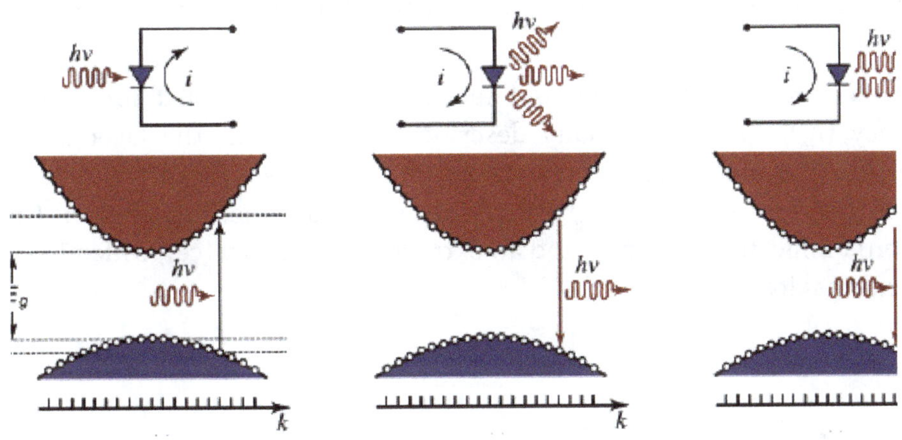

Fig. 2. Fundamental radiation processes that occur within a semiconductor p-n junction. Generation of an electron-hole pair by absorption (a) governs the operation of a photodiode. Electron-hole recombination processes give rise to spontaneous emission (b) in an LED and stimulated emission (c) in a laser diode.

This recombination radiation can also generate stimulated emission and, if the gain produced in the junction can overcome the optical losses (including absorption), lasing will occur in a laser diode. The evolution of this emission as a function of injection current, the mechanisms for efficient light extraction, and the typical device geometries employed for both LEDs and laser diodes are the subjects of this section.

Light-Current Curve

Electrical injection in the p-n junction of a laser diode provides the population inversion necessary to produce lasing. Like other laser systems, once this population is large enough so that the gain exceeds the losses, stimulated emission dominates and the system reaches a steady-state where the output light increases linearly with injection rate.

For a semiconductor laser diode with an operational wavelength of $ë_0$ (in µm), this can be expressed by a simple equation which gives the steady-state output power (P_o in W) as a function of the current applied to the diode (i, in A):

$$P_0 = R_d(i - i_t) \qquad \text{Eq. 1}$$

Where

$$R_d = \eta_d \frac{1.24}{\lambda_0} \equiv \eta_e \eta_i \frac{1.24}{\lambda_0}$$

This relationship is observed in the form of a light-current (L-I) curve in Figure 3. The current value at which the line intersects the x-axis and produces zero emission is the threshold current for lasing (i_t). The slope of this curve above threshold is known as the differential responsivity (R_d) and is typically expressed in units of W/A or mW/mA.

This differential responsivity is determined by the external differential quantum efficiency (η_d), which essentially describes how efficient the laser system is in converting injected electrons into output photons. The efficiency is made up of an internal quantum efficiency (η_i), which is the ratio of generated photons (inside the junction) to injected electrons, and an extraction efficiency (çe), which is associated with external losses.

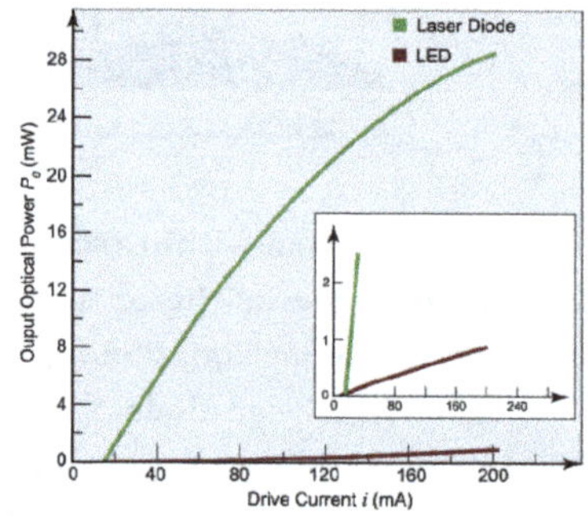

Fig. 3. Representative L-I curves for an LED and laser diode operating at a wavelength of 1.6 µm consisting of lnGaAsP / InP multiple QW structures. The inset provides an expanded view.

Equation (1) can also be used to describe the LED output power by setting i_t = 0 (since there is no threshold for spontaneous emission) R_d and η_d are replaced by the responsivity (R) and the external efficiency (η_{ex}). Figure 3 shows a L-I curve for an LED and laser diode. A laser diode exhibits the same response as an LED when it is operating below threshold; that is, it emits light spontaneously that propagates in all directions. However, when the laser reaches threshold, stimulated emission concentrates the light into a select number of spatial modes, giving rise to the sharp increase in the observed output power.

Extraction of Light

The overall efficiency of an LED (çex) is given by the product of η_i and η_e. While this device may be highly efficient at generating light within the semiconductor material, i.e., η_i approaching unity, the usefulness of the light source is predicated on efficiently extracting light from within the junction to outside the device. Spontaneous emission is isotropic, which implies that the photons generated in the junction of an LED are radiated uniformly in all directions.

However, as shown in Figure 4, only a fraction of this light will emerge from an LED with a planar surface. In addition to losses associated with absorption in the material, the large refractive index of the semiconductor (typically between 2.9 and 3.6) gives rise to losses from Fresnel reflections and from total internal reflection (TIR). The latter is particularly detrimental since any light outside of a cone defined by TIR (called the extraction cone) is not transmitted. Depending on the device design and junction geometry, virtually all the light can be either reflected or absorbed, yielding an extraction efficiency of less than 2%. Numerous mitigation approaches are used to enhance η_e.

One approach employs an epoxy material formed into a parabolic shape (illustrated in Figure 3), which exploits index-matching and geometry to reduce the impact of reflection losses. Transparent contacts and substrates are used to reduce absorption losses while imparting a texture to a surface can minimize impacts due to TIR. Another way to increase η_e is by incorporating a microresonator that confines light within a very small spatial region. This also creates angular confinement of the light such that it lies within the extraction cone.

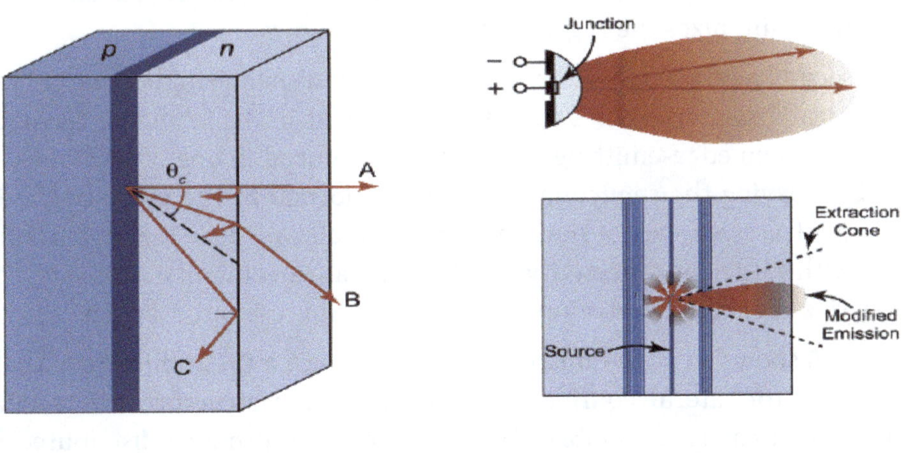

Figure 4. Optical emission from a surface-emitting LED (left) where light rays experience absorption and reflection (both A and B) while those outside of the critical angle undergo TIR (C) and are not extracted. Two methods for increased light extraction are shown on the right, including the use of a parabolic lens made of epoxy (upper right) and a microresonator that narrows the confinement

into the light extraction cone (lower right).

In Required Components for Lasing, the laser cavity end mirrors were a source of transmission loss. Since this loss represents useful laser light, the extraction efficiency in a laser diode is defined by the ratio of this loss to the total resonator loss. In this way, if the only losses are associated with laser light exiting the resonator mirror, the system would be considered highly efficient. Other loss sources include optical scattering from any element within the cavity, absorption due to impurities or charge carriers, as well as optical energy that lies outside the active region. Methods for addressing this optical confinement in different laser diode geometries are discussed below.

EDGE-EMITTING AND SURFACE-EMITTING

Semiconductor-based LEDs and laser diodes generally have two emission configurations: edge-emitting, where light exits from the edge of the active region of the device, or surface-emitting, where emission occurs through a face parallel to the plane of the active region. This section will focus on the configurations for laser diodes (see Fig. 5). The objective of laser diode design is to minimize the threshold injection current, which is accomplished by reducing the dimensions of the active region.

In the direction of the layers, this is achieved using DH or QW architectures. The means for lateral confinement (in the plane of the junction) as well as the method for creating the resonator cavity are specific to the emission configuration. Details regarding edge-emitting lasers will be discussed first. These lasers produce a beam in the plane of the junction region and two methods are used to determine the lateral confinement of the gain. The first method, referred to as gain-guiding, uses fabrication techniques to limit the lateral extent of the gain, e.g., a narrow electrical contact that minimizes the region where current flows.

The second method uses index-guiding, where a change in the index of refraction in the lateral direction creates optical confinement in the form of a waveguide. The laser cavity in an edge-emitting laser can be produced in one of two ways. The first way is by cleaving the semiconductor crystal normal to the plane of the junction. The large refractive index of the semiconductor allows these surfaces to function as mirrors with sufficient reflectivity (~ 30%) to create an etalon. In this way, the length of the active region of the *p-n* junction becomes the resonator length.

Figure 5 shows an edge-emitting laser diode with a DH architecture that utilizes gain-guiding for lateral confinement and an etalon structure for feedback. The second type of cavity uses fabrication techniques to produce distributed feedback (DFB) within the crystal or to create distributed Bragg reflector (DBR) structures at the ends of the crystal.

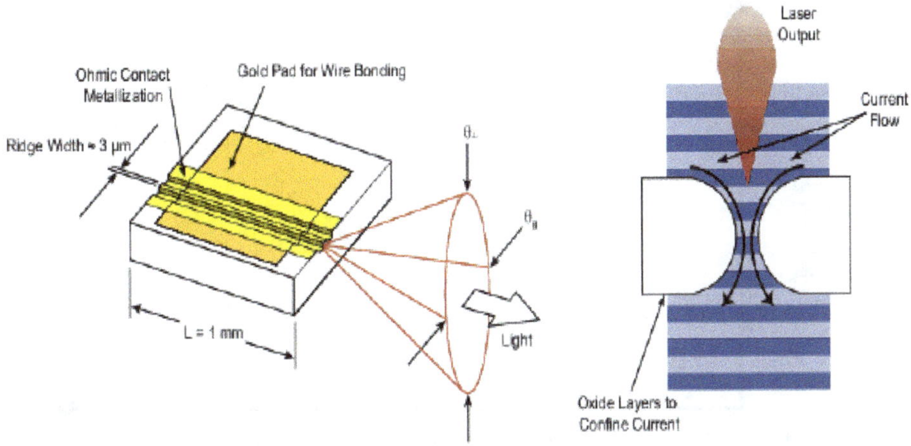

Fig. 5. Typical structure and optical emission from an edge-emitting laser diode (left) and a surface-emitting laser diode (right).

Surface-emitting lasers produce a beam in a direction perpendicular to the junction region. A commonly-employed structure is known as a vertical cavity surface emitting laser (VCSEL). Here, mirrors in the form of DBRs clad the active layer which results in a vertical cavity giving rise to a beam that propagates normal to the junction plane. Since the length of the gain medium is defined by the thickness of the active region, much smaller gains are involved compared to edge-emitting lasers. As a result, mirror reflectivities must be very high. Lateral confinement of the gain region is accomplished using dielectric materials which narrow the cross-sectional area of current flow (see Figure 4). The lower optical gain for VCSELs generally results in lower output powers compared to edge-emitting lasers. However, their monolithic construction allows for high packing density of two-dimensional arrays and provides important practical advantages over their edge-emitting counterparts.

SEMICONDUCTOR LASERS

Semiconductor lasers are solid-state lasers based on semiconductor gain media, where optical amplification is usually achieved by stimulated emission at an interband transition under conditions of a high carrier density in the conduction band.

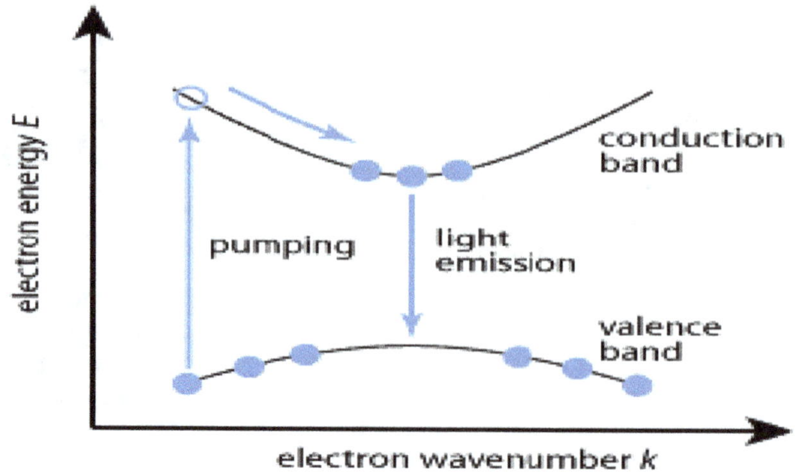

Fig. 6: Physical origin of gain in a semiconductor.

The physical origin of gain in an optically pumped semiconductor (for the usual case of an interband transition) is illustrated in Figure 1. Without pumping, most of the electrons are in the valence band. A pump beam with a photon energy slightly above the band gap energy can excite electrons into a higher state in the conduction band, from where they quickly decay to states near the bottom of the conduction band. At the same time, the holes generated in the valence band move to the top of the valence band.

Electrons in the conduction band can then recombine with these holes, emitting photons with an energy near the bandgap energy. This process can also be stimulated by incoming photons with suitable energy. A quantitative description can be based on the Fermi–Dirac distributions for electrons in both bands.

Note that this process works well only for a semiconductor with a direct band gap. In indirect band gap semiconductors (e.g. silicon), the conduction band electrons in the holes acquire substantially different wavenumbers, and that does not allow for optical transitions due to problems with momentum conservation.

Most semiconductor lasers are laser diodes, which are pumped with an electrical current in a region where an n-doped and a p-doped semiconductor material meet. However, there are also optically pumped semiconductor lasers, where carriers are generated by absorbed pump light, and quantum cascade lasers, where intraband transitions are utilized.

Common materials for semiconductor lasers (and for other optoelectronic devices), all having a direct band gap, are

- GaAs (gallium arsenide)
- AlGaAs (aluminum gallium arsenide)

- GaP (gallium phosphide)
- InGaP (indium gallium phosphide)
- GaN (gallium nitride)
- InGaAs (indium gallium arsenide)
- GaInNAs (indium gallium arsenide nitride)
- InP (indium phosphide)
- GaInP (gallium indium phosphide)

As the photon energy of a laser diode is close to the bandgap energy, compositions with different bandgap energies allow for different emission wavelengths. For the ternary and quaternary semiconductor compounds, the bandgap energy can be continuously varied in some substantial range. In AlGaAs = $Al_xGa_{1-x}As$, for example, an increased aluminum content (increased x) causes an increase in the bandgap energy.

While the most common semiconductor lasers are operating in the near-infrared spectral region, some others generate red light (e.g. in GaInP-based laser pointers) or blue or violet light (with gallium nitrides). For mid-infrared emission, there are e.g. lead selenide (PbSe) lasers (lead salt lasers) and quantum cascade lasers.

Apart from the above-mentioned inorganic semiconductors, organic semiconductor compounds might also be used for semiconductor lasers. The corresponding technology is by far not mature, but its development is pursued because of the attractive prospect of finding a way for cheap mass production of such lasers. So far, only optically pumped organic semiconductor lasers have been demonstrated, whereas for various reasons it is difficult to achieve a high efficiency with electrical pumping.

Types of Semiconductor Lasers

There is a great variety of different semiconductor lasers, spanning wide parameter regions and many different application areas:

- Small edge-emitting laser diodes generate a few milliwatts (or up to 0.5 W) of output power in a beam with high beam quality. They are used e.g. in laser pointers, in CD players, and for optical fiber communications.
- External cavity diode lasers contain a laser diode as the gain medium of a longer laser cavity. They are often wavelength-tunable and exhibit a small emission linewidth.
- Both monolithic and external-cavity low-power levels can also be mode-locked for ultrashort pulse generation.
- Broad area laser diodes generate up to a few watts of output power, but with significantly poorer beam quality.

- High-power diode bars contain an array of broad-area emitters, generating tens of watts with poor beam quality.
- High-power stacked diode bars can generate extremely high powers of hundreds or thousands of watts.
- Surface-emitting lasers (VCSELs) emit the laser radiation in a direction perpendicular to the wafer, delivering a few milliwatts with high beam quality.
- Optically pumped surface-emitting external-cavity semiconductor lasers (VECSELs) are capable of generating multi-watt output powers with excellent beam quality, even in mode-locked operation. Electrically pumped photonic crystal surface-emitting lasers promise to reach similar performance.
- Quantum cascade lasers operate on intraband transitions (rather than interband transitions) and usually emit in the mid-infrared region, sometimes in the terahertz region. They are used e.g. for trace gas analysis.

TYPICAL CHARACTERISTICS AND APPLICATIONS

Some typical aspects of semiconductor lasers are:

- Electrical pumping with moderate voltages and high efficiency is possible particularly for high-power diode lasers, and allows their use e.g. as pump sources for highly efficient solid-state lasers (*diode-pumped lasers*) and even as direct diode lasers.
- A wide range of wavelengths are accessible with different devices, covering much of the visible, near-infrared and mid-infrared spectral region. Some devices also allow for wavelength tuning.
- Small laser diodes allow fast switching and modulation of the optical power, allowing their use e.g. in transmitters of fiber-optic links.

Such characteristics have made semiconductor lasers the technologically most important type of lasers. Their applications are extremely widespread, including areas as diverse as optical data transmission, optical data storage, metrology, laser spectroscopy, laser material processing, pumping solid-state lasers (*diode-pumped lasers*), and various kinds of medical treatments.

Pulsed Output

Most semiconductor lasers generate a continuous output. Due to their very limited energy storage capability (low upper-state lifetime), they are not suitable for pulse generation with Q switching, but quasi-continuous-wave operation often allows for significantly enhanced powers.

Also, semiconductor lasers can be used for the generation of ultrashort pulses with mode locking or gain switching. The average output powers in short pulses are usually limited to at most a few milliwatts, except for optically pumped surface-emitting external-cavity semiconductor lasers (VECSELs), which can generate multi-watt average output powers in picosecond pulses with multi-gigahertz repetition rates.

Modulation and Stabilization

A particular advantage of the short upper-state lifetime is the capability of semiconductor lasers to be modulated with very high frequencies, which can be tens of gigahertz for VCSELs. This is exploited mainly in optical data transmission, but also in spectroscopy, for the frequency stabilization to reference cavities, etc.

Fiber-coupled Diode Lasers

For many applications, it is convenient to couple the output of a laser diode into an optical fiber in order to deliver the light to the place where it is needed. Fiber-coupled (also called fiber-integrated or fiber-pigtailed) diode lasers have several advantages:

- The light exiting the fiber has a circular and smooth (homogenized) intensity profile and a symmetric beam quality, which is in many cases very convenient. For example, less sophisticated optics are required for generating a circular pump spot for an end-pumped solid-state laser.
- It becomes possible to remove the laser diodes together with their cooling arrangements e.g. from a solid-state laser head, which can then be more compact, and more space is left for other parts at this place.
- Defective fiber-coupled diode lasers can easily be replaced without changing the alignment of the device where the light is used.
- Fiber-coupled devices can be easily combined with other fiber-optic components.

NARROW-LINEWIDTH AND WAVELENGTH-TUNABLE LASERS.

A number of laser applications (see below) require lasers with a very small optical linewidth, i.e., with a narrow optical spectrum. The term *narrow-linewidth lasers* usually applies to single-frequency lasers, i.e., lasers oscillating on a single resonator mode with low phase noise and thus with high spectral purity. Typically, such lasers also exhibit low intensity noise.

Some lasers do not only need to have a small linewidth, but also exhibit very high stability of the optical free. See the article on frequency-stabilized lasers.

Types of Narrow-linewidth Lasers

The most important types of narrow-linewidth lasers are the following:

- Among semiconductor lasers, distributed feedback laser diodes (DFB lasers) and distributed Bragg reflector lasers (DBR lasers), operating e.g. in the 1.5- or 1.0-μm wavelength region, are the most common. Typical operation characteristics are an output power of some tens of milliwatts (or possibly somewhat above 100 mW) and a linewidth of several megahertz.
- Significantly smaller linewidths from semiconductor lasers are possible e.g. by extending the resonator with a single-mode fiber containing a narrow-bandwidth fiber Bragg grating, or with other types of external-cavity diode lasers. In such a way, ultra-narrow linewidths of a few kilohertz or even below 1 kHz can be achieved.
- Small fiber lasers in the form of distributed feedback lasers (with the resonator formed essentially by a special fiber Bragg grating) can generate tens of milliwatts of output power with a linewidth in the region of a few kilohertz.
- Higher output powers can be generated with longer distributed Bragg reflector lasers (DBR fiber lasers) or unidirectional fiber ring lasers, and also by using a fiber amplifier.
- Diode-pumped solid-state bulk lasers, e.g. in the form of nonplanar ring oscillators, can also have linewidths of a few kilohertz, combined with relatively high output powers of the order of 1 W. Although a 1064-nm wavelength is typical, other wavelengths e.g. in the 1.3- or 1.5-μm wavelength regions are also possible.

Essential Factors for a Narrow Laser Linewidth

For achieving a narrow emission bandwidth (linewidth) from a laser, several issues of laser design have to be observed:

- First, single-frequency operation needs to be achieved. This is easiest when using a gain medium with small gain bandwidth and a laser resonator with short length (leading to a large free spectral range). The goal should be long-term stable single-frequency operation without mode hopping.
- Second, external noise influences must be minimized. This requires a stable resonator setup (preferably a monolithic one), possibly with special protection against mechanical vibrations. An electrically pumped laser should be operated with a low-noise voltage or current source, and an optically pumped laser should have a pump source with low intensity noise. Furthermore, any optical feedback must be avoided, e.g. by using

a Faraday isolator. Ideally, external noise influences will become lower than internal noise, e.g. from spontaneous emission in the gain medium. This is often easily achievable at high noise frequencies, but not at low noise frequencies which are most important for the linewidth.

- Third, the laser design should be optimized so that the laser noise and in particular the phase noise are minimized. A high intracavity optical power and long resonator can be beneficial, although stable single-frequency operation is more difficult to achieve with a longer resonator.

Of course, the design optimization requires that the relative importance of different noise sources is known, because different measures can be required depending on which noise source is dominant. For example, measures which minimize the linewidth according to the Schawlow–Townes equation will not necessarily minimize the actual linewidth, if this is determined e.g. by mechanical noise.

Physical Description of Narrow Linewidth Light

The light output of a narrow linewidth single-frequency laser may be described as follows:

- Since it is essentially in a single propagation mode, it can be fully described by a time-dependent complex amplitude of that mode. The spatial dependence is defined by the mode; the time- and location-dependent amplitude is the product of the mentioned time-dependent mode amplitude and the spatial mode profile.
- The evolution of that complex amplitude is random. There is some intensity noise, related to fluctuations of the modulus of the amplitude, and phase noise:
- The intensity noise can be described with an autocorrelation function of the intensity, the Fourier transform of which delivers the power spectral density of intensity noise; see also the article on relative intensity noise.
- For the phase noise, one has an autocorrelation function of the optical phase and again a power spectral density obtained by Fourier transform. The optical phase undergoes an unbounded random walk, since (in contrast to the intensity) there is no "restoring force" for the phase. The power spectral density is proportional to f^2 in simple random walk models.
- The laser linewidth is normally mostly determined by phase noise.
- The mathematical relation from phase noise spectrum to the field spectrum, from which one obtains the linewidth, is complicated. For the simple random walk as explained above, the field spectrum becomes Lorentzian.

If only the random walk of phase caused by spontaneous emission in the laser gain medium were present, one would obtain the Schawlow–Townes linewidth, which is usually very small. The real linewidth of a laser is usually far larger, and dominated by additional technical noise sources, e.g. related to mechanical vibrations and thermal fluctuations. Particularly if vibrations are important, the mathematical structure of the phase noise becomes more complicated.

For multimode lasers, the description is again more complicated, as one optical phase needs to be associated with each spectral component, and there are generally fluctuations of power and other interactions between those.

Noise Characterization and Specification

Both the characterization and the specification of the noise of narrow-linewidth lasers are far from trivial issues. Various measurement techniques are discussed in the article on linewidth, and particularly for linewidth values of a few kilohertz or less such measurements are demanding.

Furthermore, a linewidth value alone can not be considered a complete noise characterization; it is preferable to have a complete phase noise spectrum, apart from information on relative intensity noise.

At least, the linewidth value should be specified together with a measurement time, and possibly with some information concerning frequency drifts for longer time intervals.

Of course, different applications have different requirements, and it should be checked in detail how tight noise specifications should really be demanded in any particular case.

Applications of Narrow-linewidth Lasers

- A particularly important field of application is the area of sensors, e.g. fiber-optic sensors for strain and/or temperature, various types of interferometric sensing, trace gas detection with differential absorption LIDAR (DIAL), or wind speed measurements with Doppler LIDAR. Linewidths of only a few kilohertz are required for some fiber-optic sensors, whereas 100 kHz can be sufficient for, e.g., LIDAR measurements.
- Optical frequency metrology requires sources with very narrow linewidth, often achieved with stabilization techniques.
- Holography requires either continuous-wave or pulsed single-frequency lasers for generating highly coherent light.
- Normally less demanding in terms of linewidth are applications in optical fiber communications, e.g. in transmitters or for test and measurement purposes.

CHAPTER

OPTICAL SOURCES 2: THE LIGHT-EMITTING DIODE

7

INTRODUCTION

A light-emitting diode (LED) is an optoelectronic semiconductor device which generates light via electroluminescence. It contains a p–n junction, through which an electric current is sent. In the heterojunction, the current generates electrons and holes, which release their energy portions as photons when they recombine. Although the fundamental process of light generation is the same as in laser diodes, light-emitting diodes do not exhibit laser action, i.e., they usually do not exploit stimulated emission. Their optical spectrum is substantially wider.

NOBEL PRIZE IN PHYSICS 2014 FOR LIGHT-EMITTING DIODES

In 2014, the Nobel Prize in Physics 2014 was awarded to Isamu Akasaki, Hiroshi Amano and Shuji Nakamura "for the invention of efficient blue light-emitting diodes which has enabled bright and energy-saving white light sources". Before that work was done, efficient blue LEDs could not be made, and therefore also no white-emitting LEDs as usually needed for illumination purposes.

The researchers succeeded in developing efficient blue-emitting LEDs based on indium gallium nitride (InGaN). Methods had to be developed for producing such materials with a high quality. Surprisingly, a high device efficiency could be reached despite still substantial defect densities.

As a result of these developments, LED-based illumination has recently started to be used on a large scale, partly replacing traditional incandescent lamps. As LED-based illumination is far more energy-efficient, substantial reductions of emissions in power stations (CO_2, poisonous gases, heavy metals etc.) have become possible.

MATERIALS AND EMISSION WAVELENGTHS

The center wavelength and thus the emission color of an LED are largely determined by the band gap energy of the semiconductor material used. Essentially the whole visible wavelength region can be covered with LEDs, although the achievable power output and efficiency is not equally high for all wavelengths.

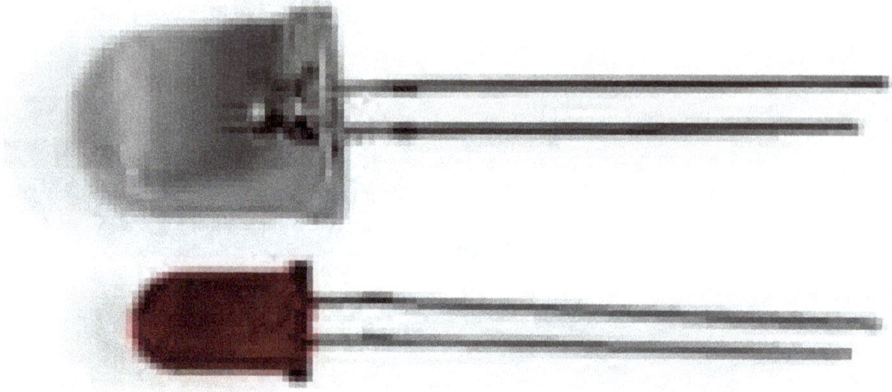

Fig. 1: Two light-emitting diodes, emitting white and red light.

Most LED chips are made of inorganic semiconductor materials. For deep red emission, aluminum gallium arsenide (AlGaAs) can be used, which is otherwise common for near-infrared laser diodes. Shorter wavelengths in the red spectral region are achieved with gallium arsenide phosphide (GaAsP) and aluminum indium gallium phosphide (AlInGaP). The internal quantum efficiency can be close to 100% for emission wavelengths around 650 nm, whereas high efficiencies are much more difficult to achieve at shorter wavelengths of e.g. 620 nm, where the human eye is more sensitive. Another challenge is that the temperature increase, which is hardly avoidable in high-power LEDs, decreases the quantum efficiency and increases the emission wavelength. Effective cooling methods are therefore essential for high-power LEDs. Note that although the thermal power involved is not very large, it is released within a very small volume and area. It is not uncommon to have several watts dissipated on one millimeter squared.

Indium gallium nitride (InGaN) is very suitable for blue and violet LEDs. Despite high defect densities in these materials, internal quantum efficiencies above 70% or even above 90% are achieved. Longer wavelengths (green and yellow) are obtained by increasing the indium (In) content, but the efficiency drops sharply as the wavelength is increased.

The technologically most difficult spectral region (in the visible range) is that of green–yellow–orange light. Intense research is being pursued to fill this gap, called the *green gap*. LEDs based on GaAsP have a lower efficiency than e.g. red LEDs. Zinc selenide (ZnSe) and zinc selenide telluride (ZnSeTe) have been developed for green emission, but the device lifetimes and efficiencies are not satisfactory for lighting purposes. Various others materials, including both modified II-VI alloys and oxychalcogenides such as LaCuOS, are therefore under consideration.

Table 1: Typical semiconductor materials and emission wavelengths of light-emitting diodes.White light can be generated either by mixing the outputs of e.g. red, green and blue LEDs, or by using a single blue LED and some phosphor, which converts part of the blue light into light with longer wavelengths. That conversion is normally done either with another semiconductor (e.g. ZnS), or with a scintillator crystal containing rare earth ions such as Eu^{2+} (europium) or Ce^{3+} (cerium). For example, Ce^{3+}:YAG can be used for converting blue light around 440–460 nm into yellow light around 520–640 nm. Modified hosts, e.g. with some of the yttrium in YAG being replaced with gadolinium (Gd^{3+}), lead to shifted emission ranges of the cerium ions. The color rendering index of such white LEDs is usually not very high, but it can be improved e.g. by combining yellow and red phosphors.There are also near-infrared LEDs, based e.g. on AlGaAs, and ultraviolet LEDs, using materials such as gallium nitride (GaN).

Table 1: Typical semiconductor materials and emission wavelengths of light-emitting diodes.

Material	Typical emission wavelengths
InGaN / GaN, ZnS	450–530 nm (blue – green)
GaP:N	565 nm (green)
AlInGaP	590–620 nm (orange)
GaAsP, GaAsP:N	610–650 nm (orange – red)
InGaP	660–680 nm (red)
AlGaAs, GaAs	680–860 nm (red)
InGaAsP	1000–1700 nm (infrared)

White light can be generated either by mixing the outputs of e.g. red, green and blue LEDs, or by using a single blue LED and some phosphor, which converts part of the blue light into light with longer wavelengths. That conversion is normally done either with another semiconductor (e.g. ZnS), or with a scintillator crystal containing rare earth ions such as Eu2+ (europium) or Ce^{3+} (cerium). For example, Ce^{3+}:YAG can be used for converting blue light around 440–460 nm into yellow light around 520–640 nm. Modified hosts, e.g. with some of the yttrium in YAG being replaced with gadolinium (Gd^{3+}), lead to shifted emission ranges of the cerium ions. The color rendering index of such white LEDs is usually not very high, but it can be improved e.g. by combining yellow and red phosphors.

There are also near-infrared LEDs, based e.g. on AlGaAs, and ultraviolet LEDs, using materials such as gallium nitride (GaN).

ORGANIC LEDS

A more recent development is that of LEDs based on organic semiconductors, called OLEDs (organic light-emitting diodes). Such devices are constructed with methods which are entirely different from those for inorganic LEDs; they are assembled from various thin elastic layers for functions like cathode and anode, transport layers, emissive layer etc.

It is even possible to use biological materials; for example, DNA molecules can exhibit highly efficient electroluminescence. Such materials have a potential for cheap mass fabrication of large and mechanically flexible devices with solution processing. They therefore appear to be very promising for future lighting applications.

However, further research and development are required for improving the efficiency and the device lifetime (the latter particularly for blue OLEDs), apart from reducing the cost. For example, OLED-based TVs have attractive looking displays with bright colors with little dependence on viewing direction, but they are expensive and can suffer from burn-in problems related to LED degradation.

DEVICE STRUCTURES

LEDs can be made as surface-emitting or edge-emitting devices. (That distinction is the same as for surface-emitting and edge-emitting semiconductor lasers.) Further sub-categories are discussed in the following sections.

Surface-emitting LEDs

Surface-emitting LEDs (SLEDs) have a thin active layer parallel to the surface from which the light extracted. In a simple flat-diode configuration (Figure 2), the active layer is just below the emitting surface, and the current is applied with a ring electrode.

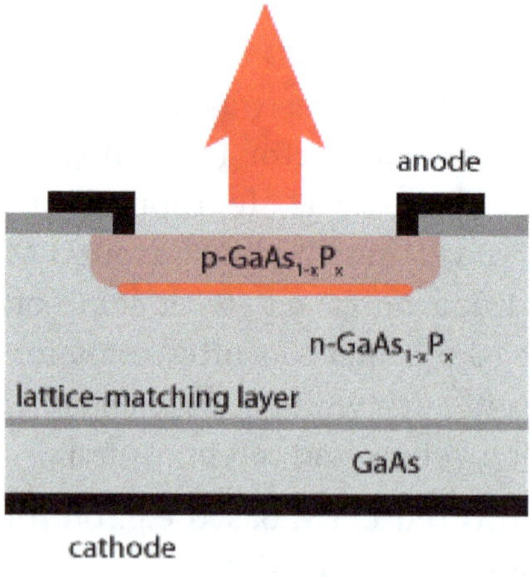

Fig 2: Structure of a flat-diode LED. The active area (shown in red) is slightly below the emitting surface.

The light emitted in the "wrong" direction is absorbed by the substrate. There are also devices where the substrate is made transparent, and the back electrode reflects back that light, so that at least some part of it can be used.

A variant is the Burrus-type LED, where the active layer can be deeper within the originally grown semiconductor structure, and a trench etched into the structure allows for more efficient light extraction. It is also possible to introduce a fiber into the trench in order to obtain a fiber-coupled LED.

A special type is the resonant-cavity LED (RC-LED), which has the light-emitting semiconductor junction embedded between two distributed Bragg reflectors (Bragg mirrors), i.e. in an optical resonator. Due to the moderate Q factor, lasing does not occur, but a larger directionality is obtained, compared with conventional LEDs.

This concept makes it easier to achieve efficient light extraction and increases the brightness of the output, from which e.g. applications in fiber-optic communications can profit. The external quantum efficiency can easily exceed 20%. The emission bandwidth is smaller than for other LEDs.

Edge-emitting LEDs

Edge-emitting LEDs have a structure similar to that of edge-emitting semiconductor lasers: they are emitting from the edge of a cleaved wafer, where the active region meets the cleaved surface. Such devices allow more efficient fiber coupling than surface-emitting LEDs. Applied in optical fiber communications, they allow higher data rates.

A variant of the edge-emitting LED is the superluminescent diode (SLD), where the spontaneous emission is substantially amplified within a waveguide. Here, the emission is much more directional, and as a consequence the brightness is much higher, even for SLDs with quite low output power.

Emission Properties

Light emitted by LEDs has a low spatial coherence. It is originally emitted in all directions. Even though many LED devices emit light preferentially in one direction (often via built-in reflecting structures), the focusability (beam quality) is very low, compared with, e.g., that of laser diodes.

The emission bandwidth is typically some tens of nanometers (e.g. 40 nm) or even >100 nm, i.e., much broader than for laser diodes, and comparable to that of superluminescent diodes. This means that the temporal coherence is much lower than that of a laser, although it is much higher than for, e.g., an incandescent bulb.

Efficiency

The internal process of generating light in an LED, as described above, can have a very high quantum efficiency and power efficiency, at least in the blue–violet and in the red spectral region. Nevertheless, the device efficiency of early light-emitting diodes was relatively poor. The reason is that it was not possible to extract the internally generated light efficiently; most of the generated light was absorbed inside the device. A key challenge is total internal reflection at the surface of the semiconductor material: due to the high refractive index, light can escape only for relatively small angles of incidence, and even then there is substantial Fresnel reflection. In some LEDs, there is also a problem with reabsorption of light in the substrate.

Fig. 3: A 4-W household LED lamp, producing 330 lm (efficacy: 82.5 lm/W).

In the 1990s, advanced LED designs were developed which allow fairly efficient light extraction and thus reach much higher device efficiencies. Examples for measures used are roughened surfaces and integrated photonic crystal structures. The luminous efficacy can now be well over 200 lm/W, i.e. substantially better than that of fluorescent lamps. Besides the efficacy of the light source itself, the directional emission makes it easier to achieve a high efficacy of the overall lighting system, comparing with omnidirectional emitters where a lot of light can be lost in a lamp housing.

A possible problem (called "efficiency droop") of GaInN-based blue LEDs, as also used in white LEDs, is that the efficacy tends to be decreased substantially when the LED is strongly pumped (for high-power operation). That problem appears to be caused by direct and indirect Auger effects.

DEVICE LIFETIME

Light-emitting diodes based on inorganic semiconductors can have very long lifetimes, which can exceed 100?000 hours. LEDs are therefore belonging to the most long-lived illumination devices.

On the other hand, LEDs are relatively sensitive to excessive reverse voltages, and can be destroyed by electrostatic discharge when improperly handled. Also, the lifetime is severely reduced for operation with a too high current and/or at too high ambient temperatures.

ELECTRICAL CHARACTERISTICS

As with any other semiconductor diode, a current can flow through an LED only from the p-doped part to the n-doped part (conventional current direction). A reverse voltage of a more than a few volts can destroy an LED.

In the forward direction, the current remains very small for low voltages and then rises very quickly (exponentially) with increasing voltage. Therefore, LEDs normally cannot be operated with a constant voltage; the current needs to be stabilized, e.g. by operation with a current source, or by using a simple series resistance for connecting to a constant voltage supply. The emitted optical power is proportional to the operation current, except if the induced increase of temperature decreases the quantum efficiency. The operation voltage is largely determined by the bandgap energy of the material, and thus by the emission wavelength; red LEDs may be operated with less than 2 V, while blue ones require of the order of 4 V.

Main Attractions

The main attractions of light-emitting diodes are:

The device efficiency (see above) can be very high, leading to a small electric power consumption and low heat generation. The effective efficiency of a lamp is often further improved by the more directed emission of LEDs, reducing the amount of light that is lost in the lamp housing.

- The device lifetime (see above) is very long. It normally ends with a gradual loss of brightness, rather than with abrupt failure.
- Larger LED lights contain multiple LEDs, and can remain largely functional when single LEDs fail. This is particularly important where safety is critical (e.g. for traffic lights).
- Light with specific colors (e.g. for traffic lights) can be directly generated. This is more efficient than using e.g. an incandescent lamp with a color filter, which has to absorb much of the generated optical power.
- LEDs can be used in very compact and lightweight packages.
- They are mechanically very robust and can tolerate even severe mechanical shocks.
- When white light is generated by using separate red, green and blue LEDs, the color tone can be adjusted by adjusting the relative operation currents.

- Dimming is possible by reducing the current or by quickly switching it on and off with a variable duty cycle.

In any case, the power efficiency is preserved when dimming an LED, and the color tone remains unchanged. Both would not be the case for a dimmed incandescent lamp.

The output power of a light-emitting diode can be very quickly modulated. Modulation frequencies of hundreds of megahertz are possible, since the carrier lifetime is only a few nanoseconds. This is useful e.g. for optical data transmission.

LEDs contain poisonous substances such as gallium arsenide, but only in small quantities.

LIMITATIONS

The cost per watt of output power of an LED for illumination is relatively high, but dropping rapidly with further advances of the technology – mainly by increasing the output power of the LED chips. Also, the higher cost may be offset by reduced electricity consumption and the long lifetime.

Although an LED produces less heat than an incandescent lamp with the same optical power, adequate heat sinking is necessary to prevent overheating, which would degrade the lifetime.

The color rendering index of some white LEDs is low, making them unsuitable for certain applications.

As the available electric supplies usually provide an (approximately) constant voltage, the operation current needs to be stabilized with some electronics. Fairly simple electronics may used, but often spoil the overall power efficiency.

APPLICATIONS OF LIGHT-EMITTING DIODES

Small LEDs are very widely used as small signal lights. Operated with a current of e.g. 5–20 mA, such devices produce enough light to be seen in normal ambient light conditions, and different colors can be used, e.g. to signal different states of a device.

As LEDs can be quickly modulated, they are suitable for optical fiber communications over short distances. While the poor directionality of their emission requires the use of multimode fibers and thus restricts the transmission distances, the cost is significantly lower than for a system with single-mode fibers and laser diode transmitters.

Moderately fast power modulation is also useful, e.g., for application in light barriers, as the modulated LED light is easily distinguished from the ambient light, and for remote controls.

LEDs are used as light transmitters in most opto-isolators.

The enormous progress in high-power LEDs has recently made it possible to use LEDs for larger signals and for lighting purposes. As the cost per watt is still relatively high, and the output power is limited, such applications started in areas such as traffic lights and stop lights for cars, where moderate optical powers are sufficient, the long lifetime is particularly important, and the efficiency advantage over a filtered incandescent lamp is very large.

LEDs are also widely used for the background illumination of liquid crystal displays, e.g. in screens of portable computers (laptops), media players and cell phones, where the reduced electricity consumption allows for a longer time with battery-powered operation.

Another advantage for computer and TV screens is that brighter colors can be achieved, comparing with fluorescent lamp backlights. The highest display quality is often achieved with OLEDs, which however increases the cost while decreasing the lifetime.

Further improved devices will soon make it possible to use high-power white LEDs for the main lights of cars, for residential lighting and street lights. In some areas such as airplanes, the compact package size and the low electricity consumption are particularly important.

LED LIGHT POWER & EFFICIENCY

LED Light Power

The power (P) of any electrical device including a LED light is measured in Watts (W), which is equal to the current or electricity drawn (I), measured in ampres, multiplied by the voltage (V).

$$P = V \times I$$

Therefore, the power of the LED light is proportional to voltage and/or current such that a device may have a low voltage but may still draw a very high current and have high power consumption. For example, the traditional 50W dichroic halogen downlight is only 12V AC but draws 4.167 Ampres.

LED lights by their nature are low voltage but also relatively low current making them lower in power and more efficient that traditional incandescent bulbs and halogen downlights. Generally we are talking about between 100 to 750 milli-amps depending on the forward voltage required to turn on the LED. In this regard, just because a LED light uses a higher current, it does not mean that it will be brighter. Rather it is dependent on the power which is proportional to the increase in voltage and/or current. There is some advantage to having higher voltage LEDs where long distances occur between the LED and power supply such as in Strip LED lighting. However, for most applications it does not really matter.

Typical power ranges for general purpose residential and commercial bulbs range from as little as 3W to 15W. Generally the higher the power the greater the current and therefore the greater the light output. However, this is not always the case and brings us to the concept of efficiency and power factor.

LED Light Efficiency

The efficiency of the LED light is measured in lumens per watt (Lm/W) which refers to the total quantity of light the LED lamp produces per 1 W of energy.

Efficiency = total lumen output / total power

Older LED chips found in older generation LED bulbs from as recent as 2008 – 2010 produce less light per watt than the 2011 – 2012 LED chips found in more modern LED bulbs. For example, a 2012 7W bulb with a CREE XT-E chip can produce more light or lumen output than a 12W bulb with an older CREE XP-E chip.

More modern LED bulbs also have better heat sink designs which enable higher light outputs.

The important message is that higher power does not always mean more light and "bigger is not always best". Ultimately for the consumer it is important to do your research or to "try before you buy". Consider referring to our LED buying guide checklist in the section on LED Life Time as a way to screen out potentially inefficient or unreliable products.

LED Efficiency versus Lamp Efficiency

As discussed in the section on Lumen levels in "Understanding LED Lighting" you must also be careful to ensure that the retailer information is specifying the Lamp Efficiency rather than the LED efficiency. Due to the inherent loses in the bulb the lamp efficiency will always be less than the LED efficiency depending on the design. This includes thermal effects, driver losses and optical inefficiencies which all combine to reduce the overall efficiency of the LED bulb or luminaire compared to the internal LED package or chip. Collectively, these losses can decrease the efficiency by more than 30%. In such cases a manufacturer may specify a LED MR16 bulb has having 720lm but in reality for the LED lamp it is only approximately 500lm.

LED LIGHTING AND POWER FACTOR

A further complication is Power Factor (PF) which is a value less than 1.0 which measures the efficiency of the LED driver or power supply. Essentially an electrical device may be rated at a Power of 100W but actually consume more than 100W due a phase delay between the instantaneous voltage and the instantaneous current. Remember that Mains power is AC or alternative current and it is made up of sunusiodal wave forms of oscillating voltage and oscillating current. Ideally these two wave forms are synchronous (PF=1) but due to the nature of electronics or inductive loads such as electric motors a delay emerges between the voltage waveform and the current waveform leading to a wasteage of electrical power or a Reactive Power which is incapable of doing any work. Therefore, a device may be rated at 1000W Real Power but consume 1500W Apparent or Active power due to a PF of 0.67 and end up wasting 500W or 1/3 of the total power consumed due to the current being out of phase. Noting that for the electrical device to make use of the current it must be in phase with the voltage given power is equal to voltage x current or P=VI.

PF is generally only a problem in commercial applications in inductive devices which use very high power such that the delays between current and voltage add up to produce significant power losses. Other components which will cause delays betweeen current and voltage include transformers and voltage regulators and ballasts in flourescent lighting. In residential settings such loses are relatively minimal and electricity companies will only charge for the real power anyway. However, there is still a loss so those energy conscious or green individuals may wish to examine the power factor of their LED lighting power supplies to ensure that have a PF of greater than 0.8 to ensure minimal energy loss. In fact US Department of Energy (DOE) Energy Star program mandates minimum acceptable power factors or 0.7 and 0.9 respectively for domestic and commercial LED lights.

Most power supply devicies these days will have some form of either passive or active power factor correction leading to PF of > 0.9 so minimal power losses can be achieved. One exception is ultra high dimming drivers which dim down to 1%. Due to the high capacitive loads required to stablise the current at very low dimming levels to avoid flicker PF is poor, generally approx 0.65 meaning that a 10W rated LED bulb will consume approximately 15.4W (or VA, apparent power) near full load. However, in practice this is not a big problem given these drivers are generally used in applications where the lamps will be dimmed down to low levels for most of their life such that the real power is 2 or 3W and the Apparent power still very low at up to 4.6W.

If dimming is only going to be occassional then we suggest comprimising on minimum dimming effect with a driver which has a PF > 0.9. Although as noted this is mainly an issue in commercial lighting where lamps are on for 8 to 24 hours per day. If you are a residential user you can be less concerned. Neverthless please note the PF ratings on the speicfications tab of our products to get an idea of efficiency.

If you are a commercial customer with a business that is particularly dominated by highly inductive loads such as electric motors or a large collection of capacitive loads with poor PF, then you should consider power factor correction (PFC) and visit our section on government rebates and schemes where such projects can be subsidised. If you are a very high energy user then PFC can leading to very large savings in electricity and power bills.

CHAPTER 8

OPTICAL DETECTORS

INTRODUCTION

A transducer is a device that converts input energy of one form into output energy of another. An optical detector is a transducer that converts an optical signal into an electrical signal. It does this by generating an electrical current proportional to the intensity of incident optical radiation. The relationship between the input optical radiation and the output electrical current is given by the detector responsivity. Responsivity is discussed later in this chapter.

Fiber optic communications systems require that optical detectors meet specific performance and compatibility requirements. Many of the requirements are similar to those of an optical source. Fiber optic systems require that optical detectors:

- Be compatible in size to low-loss optical fibers to allow for efficient coupling and easy packaging.
- Have a high sensitivity at the operating wavelength of the optical source.
- Have a sufficiently short response time (sufficiently wide bandwidth) to handle the system's data rate.
- Contribute low amounts of noise to the system.
- Maintain stable operation in changing environmental conditions, such as temperature.

Optical detectors that meet many of these requirements and are suitable for fiber optic systems are semiconductor photodiodes. The principal optical detectors used in fiber optic systems include semiconductor positive-intrinsic-negative (PIN) photodiodes and avalanche photodiodes (APDs).

SEMICONDUCTOR PHOTODIODES

Semiconductor photodiodes generate a current when they absorb photons (light). The amount of current generated depends on the following factors:

- The wavelengths of the incident light and the responsivity of the photodiode at those wavelengths
- The size of the photodiode active area relative to the fiber core size
- The alignment of the fiber and the photodiode

The optical fiber is coupled to semiconductor photodiodes similarly to the way optical sources are coupled to optical fibers. Fiber-to-photodiode coupling involves centering the flat fiber-end face over the photodiode active area. This is normally done directly by butt coupling the fiber up to the photodiode surface. As long as the photodiode active area is larger than that of the fiber core, fiber-to-detector coupling losses are very low. In some cases a lens may be used to couple the fiber end-face to the detector. However, this is not typically done.

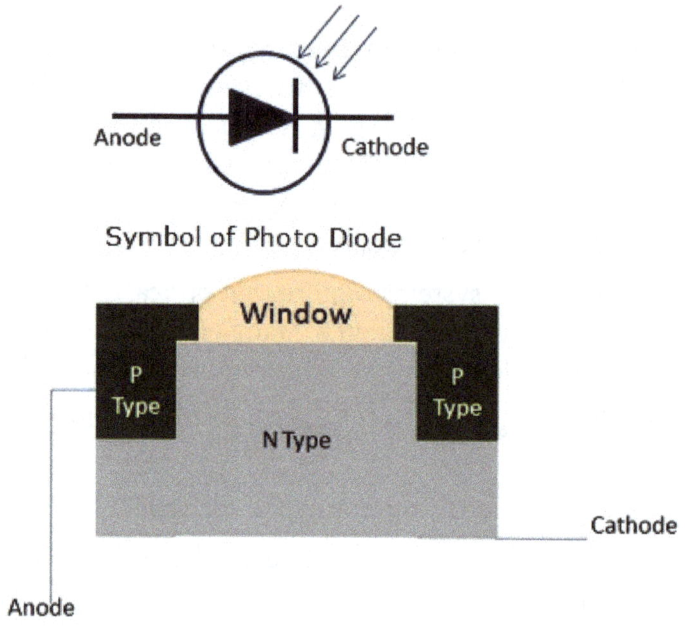

Fig. 1. Schematic symbol and constructional detail of a photo diode

Optical detectors or photodetectors are electronic devices which are employed to detect light. In some cases, these photo detectors can also be used to sense and measure other types of electromagnetic radiation that is incident on a specific device or circuitry. Thus, in simpler terms, optical detector or photo detectors are electromagnetic sensors that convert electromagnetic radiation into electric signals that can be measured through an appropriate device. The output electric signal thus obtained is proportional to the incident light or electromagnetic radiation.

Photo detectors have been a primary component of fiber optic communication setups along with other sectors such as healthcare, 'smart' electronic gadgets manufacturing sector, food production performance evaluation sector, etc. Along with this, advanced technologies like Artificial Intelligence and Internet of Things (IoT) have increased the scope of applicability of photo sensors in the last few years.

A photodiode is a P-N junction diode that will conduct current when exposed to light. This diode is actually designed to operate in the reverse bias mode. It means that larger the intensity of falling light, the greater will be the reverse bias current.

The following figure shows a schematic symbol and constructional detail of a photo diode.

WORKING OF A PHOTO DIODE

It is a reverse-biased diode. Reverse current increases as the intensity of incident light increases. This means that reverse current is directly proportional to the intensity of falling light.

It consists of a PN junction mounted on a P-type substrate and sealed in a metallic case. The junction point is made of transparent lens and it is the window where the light is supposed to fall.

As we know, when PN junction diode is reverse biased, a very small amount of reverse current flows. The reverse current is generated thermally by electron-hole pairs in the depletion region of the diode.

When light falls on PN junction, it is absorbed by the junction. This will generate more electron-hole pairs. Or we can say, characteristically, the amount of reverse current increases.

In other words, as the intensity of falling light increases, resistance of the PN junction diode decreases.

- This action makes the diode more conductive.
- These diodes have very fast response time
- These are used in high computing devices.
- It is also used in alarm circuits, counter circuits, etc.

TYPES OF PHOTODETECTORS

As the requirements for applications (see below) vary considerably, there are many types of photodetectors which may be appropriate in a particular case:

- Photodiodes are semiconductor devices with a p–n junction or p–i–n structure (i = intrinsic material) ('! p–i–n photodiodes), where light is absorbed in a depletion region and generates a photocurrent. Such devices can be very compact, fast, highly linear, and exhibit a high quantum efficiency (i.e., generate nearly one electron per incident photon) and a high dynamic range, provided that they are operated in combination with suitable electronics.

A particularly sensitive type is that of avalanche photodiodes, which are sometimes used even for photon counting.

- *Metal-semiconductor-metal (MSM) photodetectors* contain two Schottky contacts instead of a p–n junction. They are potentially faster than photodiodes, with bandwidths up to hundreds of gigahertz.
- *Phototransistors* are similar to photodiodes, but exploit internal amplification of the photocurrent. They are less frequently used than photodiodes.
- *Photoconductive detectors* are also based on certain semiconductors, e.g. cadmium sulfide (CdS). They are cheaper than photodiodes, but they are fairly slow, are not very sensitive, and exhibit a nonlinear response. On the other hand, they can respond to long-wavelength infrared light.
- *Phototubes* are vacuum tubes or gas-filled tubes where the photoelectric effect is exploited ('! *photoemissive detectors*).
- *Photomultipliers* are a special kind of phototubes, exploiting electron multiplication processes for obtaining a much increased responsivity. They can also have a high speed and large active area. Some of them are based on multichannel plates; they can be substantially more compact than traditional photomultipliers.
- Research is performed on novel photodetectors based on *carbon nanotubes (CNT) or graphene*, which can offer a very broad wavelength range and a very fast response. Ways for integrating such devices into optoelectronic chips are explored.

These devices are all based on the internal or external photoelectric effect; photoemissive detectors belong to the latter category.

Various kinds of photodetectors can be integrated into devices like power meters and optical power monitors. Others can be made in the form of large two-dimensional arrays, e.g. for imaging applications. They may be called *focal plane arrays*. For example, there are CCD and CMOS sensors which are used mainly in cameras.

IMPORTANT PROPERTIES OF PHOTODETECTORS

Depending on the application, a photodetector has to fulfill various requirements:

- It must be sensitive in a certain spectral region (range of optical wavelengths). In some cases, the responsivity should be constant or at least well defined within some wavelength range. It can also be important to have zero response in some other wavelength range; an example are *solar-blind photodetectors*, being sensitive only to short-wavelength ultraviolet light but not to visible sun light.

- The responsivity tells how much electrical signal is obtained per unit optical power. It depends on the optical wavelength.
- In some cases, not only a high responsivity, but also a high quantum efficiency is important, as otherwise additional quantum noise is introduced. This applies e.g. to the detection of squeezed states of light, and also affects the photon detection probability of photon counting detectors.
- The detector must be suitable for some range of optical powers. The maximum detected power can be limited e.g. by damage issues or by a nonlinear response, whereas the minimum power is normally determined by noise. The magnitude of the dynamic range (typically specified as the ratio of maximum and minimum detectable power, e.g. in decibels) is often most important. Some detectors (e.g. photodiodes) can exhibit high linearity over a dynamic range of more than 70 dB.
- The active area of a detector can be important e.g. when working with strongly divergent beams from laser diodes. A high enough uniformity of the responsivity may be important. For light sources with very high and/or non-constant beam divergence, it is hardly possible to get all the light onto the active area; an integrating sphere may then be used (with appropriate calibration) for measuring the total power.
- The detection bandwidth may begin at 0 Hz or some finite frequency (for AC-coupled detectors), and ends at some maximum frequency which may be limited by internal processes (e.g. the speed of electric carriers in a semiconductor material) or by the involved electronics (e.g. introducing some RC time constants). Some resonant detectors operate only in a narrow frequency range, and can be suitable e.g. for lock-in detection.
- Some detectors (such as pyroelectric detectors) are suitable only for detecting pulses, not for continuous-wave light.
- For detecting light pulses (possibly on a few-photon level), the timing precision may be of interest. Some detectors have a certain "dead time" after the detection of a pulse, where they are not sensitive.
- Different types of detectors require more or less complex electronics. Penalties in terms of size and cost may result e.g. from the requirement of applying a high voltage or detecting extremely small voltages.
- Particularly some mid-infrared detectors need to be cooled to fairly low temperatures. This makes their use under various circumstances impractical.
- For some applications, one-dimensional or two-dimensional *photodetector arrays* are needed – most often in the form of photodiode arrays. For

detector arrays, some different aspects come into play, such as cross-pixel interference, read-out techniques and detector uniformity.

- Finally, the size, robustness and cost are essential for many applications.

Different detector types, as listed above, differ very much in many of these properties. In typical application scenarios, some requirements totally rule out the use of certain detector types, and quickly lead to a fairly limited choice. Note also that there are some typical trade-offs.

For example, it is frequently difficult to combine a high detection bandwidth with a high sensitivity.

APPLICATIONS OF PHOTODETECTORS

Photodetectors have a very wide range of applications. Some examples:

- In radiometry and photometry, they can be used for measuring properties like optical power, luminous flux, optical intensity and irradiance, in conjunction with additional means also for properties like the radiance.
- They are used to measure optical powers e.g. in spectrometers, light barriers, optical data storage devices, autocorrelators, beam profilers, fluorescence microscopes, interferometers and various types of optical sensors.
- Particularly sensitive photodetectors are required for laser rangefinders, LIDAR, quantum optics experiments and night vision devices.
- Particularly fast photodetectors are used for optical fiber communications, optical frequency metrology and for the characterization of pulsed lasers or laser noise.
- Mostly two-dimensional arrays containing many identical photodetectors are used as focal plane arrays, mostly for imaging applications. For example, most cameras contain such devices as image sensors.

OPTICAL DETECTION PRINCIPLES

To detect optical radiation (photons) in the near-infrared region of the spectrum, both external and internal photoemission of electrons may be utilized.

- External photoemission devices typified by photomultiplier tubes and vacuum photodiodes meet some of the performance criteria but are too bulky, and require high voltages for operation
- Internal photoemission devices, especially semiconductor photodiodes with or without internal (avalanche) gain, provide good performance and compatibility with relatively low cost.
- The internal photoemission process may take place in both intrinsic and extrinsic semiconductors.

- With intrinsic absorption, the received photons excite electrons from the valence to the conduction bands in the semiconductor, whereas extrinsic absorption involves impurity centers created within the material
- For fast response coupled with efficient absorption of photons, the intrinsic absorption process is preferred and at presents all detectors for optical fiber communications use intrinsic photo detection
- Silicon photodiodes have high sensitivity over the 0.8–0.9 μm wavelength band with adequate speed negligible shunt conductance, low dark current and long-term stability
- They are therefore widely used in first-generation systems and are currently commercially available
- Their usefulness is limited to the first-generation wavelength region as silicon has indirect bandgap energy of 1.14 eV giving a loss in response above 1.09 μm
- Germanium and III–V alloys which give a good response at the longer wavelengths used in 2nd generation system.
- The development of advanced photodiode structures fabricated from III–V semiconductor alloys for operation at wavelengths of 1.3 and 1.55 μm

PHOTO DETECTION

- It is the process whereby optical power is detected and then converted to electrical power. Photo detector devices perform photo detection.
- The performance of an optical detector can be determined by its ability to detect smallest optical power possible and generate a maximum electric power at the output with an absolute minimum degree of distortion.
- Optical detection must also exhibit a comparatively wide Bandwidth and sharp response to accommodate high bit rate criteria for selecting particular photodiode is its ability to interface with optical cables, along operating life and cost highly sensitivity, fast response, low noise, high reliability
- The basic detection process in an intrinsic absorber is illustrated in Figure 8.1 which shows a p–n photodiode
- This device is reverse biased and the electric field developed across the p–n junction sweeps mobile carriers (holes and electrons) to their respective majority sides (p- and n-type material).
- A depletion region or layer is therefore created on either side of the junction.
- This barrier has the effect of stopping the majority carriers crossing the junction in the opposite direction to the field.

- However, the field accelerates minority carriers from both sides to the opposite side of the junction, forming the reverse leakage current of the diode.
- Thus intrinsic conditions are created in the depletion region.
- A photon incident in or near the depletion region of this device which has an energy greater than or equal to the bandgap energy Eg of the fabricating material (i.e. hf > Eg) will excite an electron from the valence band into the conduction band. This process leaves an empty hole in the valence band and is known as the photo generation of an electron–hole (carrier) pair.

QUANTUM EFFICIENCY

The quantum efficiency (or *quantum yield*) is often of interest for processes which convert light in some way. It is defined as the percentage of the input photons which contribute to the desired effect. Examples are:

- In a laser gain medium, the pump process may require the transfer of laser-active ions from one electronic level (into which the ions are pumped) to the upper level of the laser transition. This pump quantum efficiency is the fraction of the absorbed pump photons which contributes to the population of the upper laser level. This efficiency is close to unity (100%) for many laser gain media, but can be substantially smaller for others. It may depend on factors like the excitation density and parasitic absorption processes. It is not easy to measure, since the power conversion efficiency also depends on other factors, such as optical losses, which are not all easy to quantify.
- Similarly, the quantum efficiency of fluorescence can be defined. It can be reduced by non-radiative processes such as multi-phonon transitions and energy transfer processes. If such effects do not occur, it can be essentially 100%.
- In a photodiode (or some other photodetector), the quantum efficiency can be defined as the fraction of incident (or alternatively, of absorbed) photons which contribute to the external photocurrent. (See the article on photocurrent for equations.) In the visible and near-infrared region, photodiodes can have quantum efficiencies above 90%, although values between 40% and 80% are more common. Photomultipliers can have much lower quantum efficiencies, strongly depending on the wavelength region. In case of avalanche photodiodes in Geiger mode, dead time effects are not considered for the quantum efficiency.

Quantum Efficiencies Beyond 100%

In some special cases, the quantum efficiency of a laser or laser amplifier can be larger than unity. This is due to certain energy transfer processes between laser-active ions, which lead to a kind of *cross-relaxation*: starting with one ion in some excited state, a part of its energy is transferred to some other ion, which was originally in the electronic ground state, and both ions are finally in the upper laser level. This can, of course, only happen when the photon energy of the laser transition is lower than half that of the pump light, so that two generated photons together have less energy than one pump photon.

An example, illustrated in Figure 2, is that of thulium-doped 1.9-μm fiber lasers, where ions are pumped into the level $^3F_{2,4}$, and a cross-relaxation process (gray arrows) populates the upper laser level $^3H_4$. This could in principle lead to a quantum efficiency of up to 200%. Values well above 100% can be reached in practice.

The quantum efficiency should not be confused with the *quantum defect*.

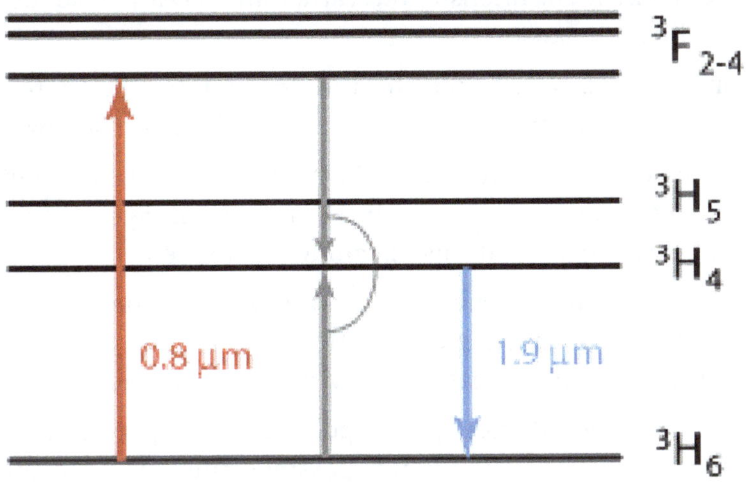

Fig 2: 1.9-μm emission in a thulium-doped fiber laser with > 100% quantum efficiency.

RESPONSIVITY

Responsivity is a measure of input-output gain of the detector in a fibre optic system. In the case of a photodetector, it is the measure of electrical output per optical input. Mostly photodetectors are linear functions of input incident power. The responsivity of a photodetector is typically expressed as units of amperes per watt or volts per watt of radiant power.

The responsivity (or *radiant sensitivity*) of a photodiode or some other kind of photodetector is the ratio of generated photocurrent and incident (or sometimes absorbed) optical power (neglecting noise influences), determined in the linear region of response. In the case of photodiodes, the responsivity is typically highest in a wavelength region where the photon energy is somewhat above the band gap energy, and declining sharply in the region of the bandgap, where the absorption decreases. It can be calculated according to

$$Dh^{(2)} = 2h_2 I^{(1)}$$

where hν is the photon energy, η is the quantum efficiency, and *e* the elementary charge.

For example, a silicon photodiode with 90% quantum efficiency at a wavelength of 800 nm, the responsivity would be ≈0.58 A/W. Values for other types of photodiode are basically always of that order of magnitude.

For avalanche photodiodes and photomultipliers, there is an additional factor for the internal current multiplication, so that values far above 1 A/W are possible. Note that the current multiplication is usually *not* subsumed in the quantum efficiency.

Note that the term *responsivity* cannot be directly applied to photoconductive detectors (photoresistors), where the response to incident light is not a photocurrent proportional to the incident intensity but rather an increase of conductivity – often with a quite nonlinear behavior.

The responsivity is usually defined for the steady state. The photodiode response typically falls off for signal frequencies above some detection bandwidth.

The term *sensitivity* is often used instead of *responsivity*, but that is not recommended, since the term can also have other meanings. It should be avoided particularly when a clear quantitative meaning is intended.

The responsivity is usually meant to be a wavelength-dependent quantity and thus considered for monochromatic light. However, one may consider an effective responsivity for non-monochromatic light with a certain spectral bandwidth.

A photodetector should ideally be operated in a spectral region where its responsivity is not far below the highest possible value, because this leads to the lowest possible detection noise and thus to a high signal-to-noise ratio and high sensitivity.

If some detector has a voltage rather than a current output, one can define its responsivity as the ratio of output voltage and optical power. This leads to units of V/W (volts per watt). If a photodiode is combined with some detector electronics generating a voltage output, the output voltage is the photocurrent times the so-called transimpedance of the electronics. In the simplest case, one uses a shunt resistor, and the transimpedance is then its resistance.

Thermal detectors usually have a responsivity with a weak wavelength dependence in a broad spectral range – in contrast to photodetectors like photodiodes, where the responsivity typically drops sharply for photon energies around the band gap energy.

LONG-WAVELENGTH CUTOFF

A cut of the wavelength of any photodetector can be defined as if any particular semiconductor can be absorbed photon over a limited wavelength range. The highest bandwidth is known as cut-off wavelength.

Cut off wavelength is determined by band gap energy Eg of a material.

Cut of wavelength = hc / E_g = 1.24 / E_g

Where,

E_g = Electron volts (eV)

A typical value of cut of wavelength for silicon is 1.06 micrometer and germanium is 1.6 micrometer.

For example :

In PIN photo diode E_g = 1.43 eV, hc = 1.24 , find cut of wavelength = ?

Cut of wavelength

= hc / E_g = 1.24 / E_g

= 1.24 / 1.43

= 0.867 micro meter

= 867 nm

A cut of the wavelength of any photodetector can be defined as if any particular semiconductor can be absorbed photon over a limited wavelength range. The highest bandwidth is known as cut-off wavelength.

Cut off wavelength is determined by band gap energy Eg of a material.

Cut of wavelength = hc / E_g = 1.24 / E_g

Where,

E_g = Electron volts (eV)

A typical value of cut of wavelength for silicon is 1.06 micrometer and germanium is 1.6 micrometer.

For example :

In PIN photo diode E_g = 1.43 eV, hc = 1.24 , find cut of wavelength = ?

Cut of wavelength = hc / E_g = 1.24 / E_g

= 1.24 / 1.43

= 0.867 micro meter

= 867 nm

SEMICONDUCTOR PHOTODIODES

Photodiodes are frequently used photodetectors, which have largely replaced the formerly used vacuum phototubes. They are semiconductor devices which contain a p–n junction, and often an intrinsic (undoped) layer between n and p layers. Devices with an intrinsic layer are called *p–i–n* or *PIN photodiodes*. Light absorbed in the depletion region or the intrinsic region generates electron–hole pairs, most of which contribute to a photocurrent. The photocurrent can be quite precisely proportional to the absorbed (or incident) light intensity over a wide range of optical powers.

Figure 3 schematically shows the typical design of the photodiode on p–i–n type. Here, one has an intrinsic region between an n-doped and a p-doped region, where most of the electric carriers are generated. Through the electrical contacts (anode and cathode), the generated photocurrent can be obtained. The anode may have a ring shape, so that the light can be injected through the hole. A large active area can be obtained with a correspondingly large ring, but that tends to increase the capacitance, thus reducing the detection bandwidth, and increases the dark current; also, the efficiency may drop if carriers are generated too far from the electrodes.

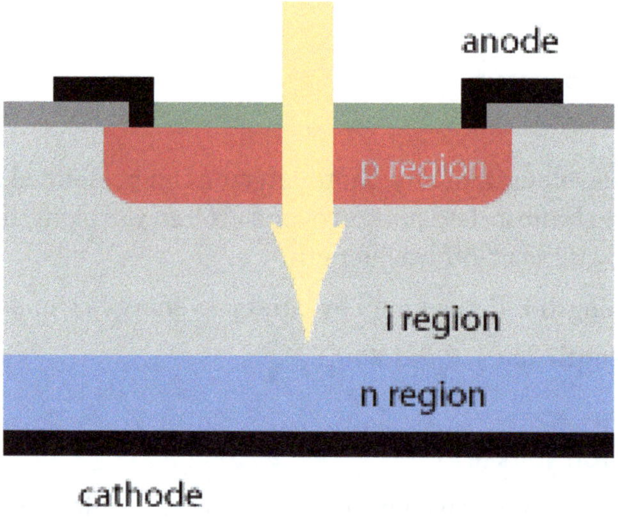

Fig. 3: Schematic drawing of a p-i-n photodiode. The green layer is an anti-reflection coating

For a high responsivity of a photodiode, one should have a material with a strong absorption for the optical wavelength of interest. When using a thicker layer for obtaining efficient absorption, one may lose a lot of the generated carriers and therefore still not substantially improve the responsivity.

Some photodiodes are available in the form of photodiode arrays of one-dimensional or two-dimensional kind. Two-dimensional detector arrays, e.g. for use as image sensors, may be realized with photodiodes or with other kinds of photodetectors.

For substantially increased responsivity, one may either use avalanche photodiodes or phototransistors; these are based on quite different operation principles.

OPERATION MODES OF PHOTODIODES

Photodiodes can be operated in two very different modes:

- *Photovoltaic mode*: like a solar cell, the illuminated photodiode generates a voltage which can be measured. However, the dependence of this voltage on the light power is nonlinear (see Figure 4), and the dynamic range is fairly small. Also, the maximum speed is not achieved.
- *Photoconductive mode*: here, a reverse voltage is applied to the diode (i.e., a voltage in the direction where the diode is not conducting without incident light) and measures the resulting photocurrent. The simplest solution for that *reversed-biased mode* is based on a voltage source and a load resistor, as shown in Figure 5. The dependence of the photocurrent on the light power can be very linear over six or more orders of magnitude of the light power, e.g. in a range from a few nanowatts to tens of milliwatts for a silicon p–i–n photodiode with an active area of a few mm². The magnitude of the reverse voltage has nearly no influence on the photocurrent and some influence on the (typically small) dark current (obtained without light). A higher reverse voltage tends to make the response faster but also increases the heating of the device, which can be a problem for high photocurrents.

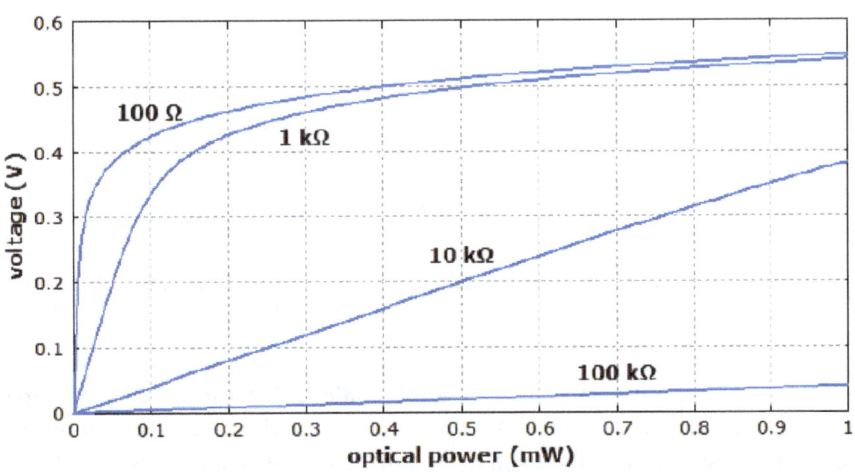

Fig. 4: Current-voltage characteristics of a photodiode in photovoltaic mode for different load resistors

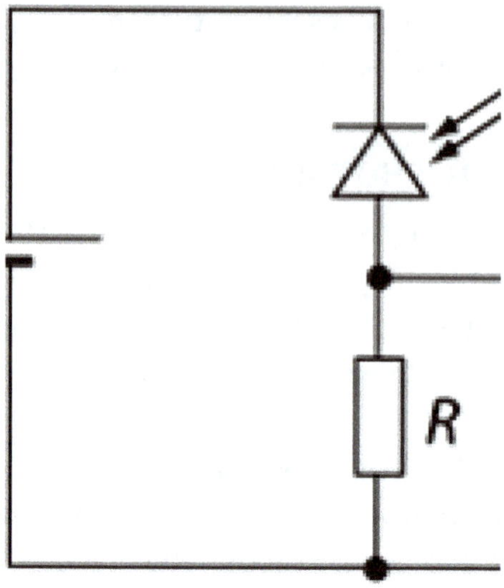

Fig. 5: A simple electronic circuit for a photodetector based on a photodiode.

Even when used in photoconductive mode, photodiodes are usually not understood to be photoconductive detectors, which have a significantly different operation principle.

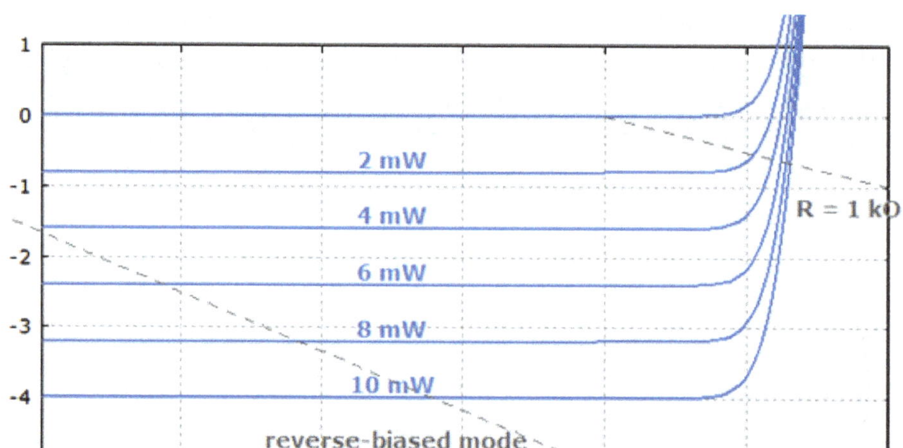

Figure 6: Current-voltage characteristics of a photodiode for different optical powers. In the photovoltaic mode (see the line for a 1-kÙ load resistor), the response is nonlinear. In the photoconductive mode, shown here for a simple circuit with a reverse bias applied through a load resistor, a very linear response is achieved. The same holds for a constant reverse bias (not shown).

In the simple circuit according to Figure 5, the magnitude of the bias voltage drops with increasing photocurrent due to the voltage drop at the load resistor. While that has little influence on the linearity, it leads to a charging or discharging of the photodiode's capacitance whenever the incident light intensity changes, so that the detection bandwidth is reduced; it may become RC-limited. That introduces a trade-off between detection bandwidth and responsivity: a high bandwidth requires a small load resistor, which leads to a low responsivity and also a higher noise-equivalent power, which is often limited by thermal noise (Johnson noise) of the load resistor.

In order to avoid that trade-off, one often uses a current amplifier (also called transimpedance amplifier). Such an amplifier, which is typically realized with an operational amplifier (op-amp), keeps the voltage at the diode nearly constant (e.g. near zero, or at some possibly adjustable reverse bias), so that the photodiode's capacitance loses much of its relevance. The residual voltage variations at the photodiode are inversely proportional to the gain of the used operational amplifier. Still, it is good to minimize the input capacitance when requiring a high detection bandwidth; for example, it is better to directly connect a photodiode to the current amplifier, instead of using a long cable connection.

Current amplifiers, which are also available as OEM devices, can also have very good noise properties. The relevant figure is the noise-equivalent input current, which can be well below 1 $pA/Hz^{1/2}$.

Commercially available laboratory current amplifiers help to make power measurements very flexible by providing many different sensitivity settings, and thus a huge dynamic range with low-noise performance, and also possibly a built-in display, adjustable bias voltage and signal offset, adjustable filters, etc.

Semiconductor Materials

Some photodiode materials and their typical characteristics are:

- silicon (Si): low dark current, high speed, good sensitivity between roughly 400 and 1000 nm (best around 800–900 nm)
- germanium (Ge): high dark current, slow speed due to large parasitic capacity, good sensitivity between roughly 900 and 1600 nm (best around 1400–1500 nm)
- indium gallium arsenide phosphide (InGaAsP): expensive, low dark current, high speed, good sensitivity roughly between 1000 and 1350 nm (best around 1100–1300 nm)
- indium gallium arsenide (InGaAs): expensive (particularly for large active areas), low dark current, high speed, good sensitivity roughly between 900 and 1700 nm (best around 1300–1600 nm)

The indicated wavelength ranges can sometimes be substantially exceeded by models with extended spectral response.

Key Properties of Photodiodes

The most important properties of photodiodes are:

- the responsivity, i.e., the photocurrent per unit optical power – related to the quantum efficiency, dependent on the wavelength
- the active area, i.e., the light-sensitive area
- the breakdown voltage, setting a limit to the usable bias voltage
- the maximum allowed photocurrent (usually limited by saturation, possibly lower for high bias voltages)
- the dark current (in photoconductive mode, dependent on the bias voltage, important for the detection of low light levels)
- the speed, i.e. the bandwidth (see below), related to the rise and fall time, often influenced by the electric capacitance

Additional quantities may be of interest

- The typically quite high shunt resistance contributes a small current when a bias voltage is applied. It also contributes some thermal noise current, which in some cases limits the sensitivity.
- The normally small series resistance causes an additional voltage drop proportional to the photocurrent, and can also somewhat contribute to the detection noise.

Bandwidth

The speed (bandwidth) of a photodiode is typically limited either by electrical parameters (capacitance and external resistor) or by internal effects such as carrier transit time in the depletion region. (In some cases, relatively slow diffusion of carriers generated outside the depletion region limits the bandwidth.) The highest bandwidths of tens of gigahertz are usually achieved with small active areas (diameters well below 1 mm) and small absorption volumes. Such small active areas are still practical particularly for fiber-coupled devices, but they limit the photocurrents achievable to the order of 1 mA or less, corresponding to optical powers of H"
2 mW or less. Higher photocurrents are actually desirable for suppression of shot noise and thermal noise. (Higher photocurrents increase shot noise in absolute terms, but decrease it relatively to the signal.) Larger active areas (with diameters up to the order of 1 cm) allow for handling of larger beams and for much higher photocurrents, but at the expense of lower speed.

The combination of high bandwidth (tens of gigahertz) and high photocurrents (tens of milliamperes) is achieved in velocity-matched photodetectors, containing several small-area photodetectors, which are weakly coupled to an optical waveguide and deliver their photocurrents into a common RF waveguide structure.

QUANTUM EFFICIENCY

The quantum efficiency of a photodiode is the fraction of the incident (or absorbed) photons which contribute to the photocurrent. For photodiodes without an avalanche effect, it is directly related to the responsivity *S*: the photocurrent is with the quantum efficiency η, the electron charge *e* and the photon energy *hí*. The quantum efficiency of a photodiode can be very high – in some cases more than 95% – but varies significantly with wavelength. Apart from a high internal efficiency, a high quantum efficiency requires the suppression of reflections e.g. with an anti-reflection coating.

A higher responsivity (although sometimes at the cost of lower quantum efficiency) can be achieved with avalanche photodiodes. These are operated with a relatively high reverse bias voltage so that secondary electrons can be generated (as in photomultipliers). The avalanche process increases the responsivity, so that noise influences of subsequent electronic amplifiers are minimized, whereas quantum noise becomes more important and multiplication noise is also introduced.

Additional Properties

In some cases, additional properties of photodiodes have to be observed, such as linearity of response over a wide dynamic range, the spatial uniformity of response, or the shape of the dynamic response (e.g. optimized for time domain or frequency domain), or the noise performance.

The noise performance of photodiodes can be very good. For high photocurrents, it can be limited by shot noise, although thermal noise in the electronics is often stronger than that. For the detection of very low light levels (e.g. for photon counting), the dark current can also play a role.

The electronics used in a photodiode-based photodetector can strongly influence the performance in terms of speed, linearity, and noise. As mentioned above, current amplifiers (transimpedance amplifiers) are often a good choice.

FAST PHOTODIODES

For a particularly high detection bandwidth in the gigahertz region, advanced photodiode designs are used. For example, some devices contain an optical resonator around the thin absorbing section. In that way, one can achieve efficient absorption and thus a high quantum efficiency despite a rather small thickness of the intrinsic region, as is chosen for reducing the drift time.

So-called waveguide photodiodes contain an optical waveguide which confines light along its path through the absorbing region. That region can then again be very thin, and nevertheless one can obtain efficient absorption in a short length. By minimizing the length of the active region, one can also minimize the electrical capacitance and reach a very high bandwidth.

In some cases, the electrode structure is made such that it forms an electrical waveguide, where the electric wave can propagate in parallel with the optical wave in the optical waveguide. Such *traveling-wave photodiodes* can reach a bandwidth well above 100 GHz.

Some semiconductor materials are intrinsically better suited than others for fast photodiodes. For example, indium gallium arsenide (InGaAs) is particularly suitable, because that direct band gap material (in contrast to silicon, for example) exhibits a rather short absorption length, allowing the realization of very thin absorbing layers, in which the photocarriers can be quickly collected. For fast avalanche photodiodes, it is also important to have a low ratio of the impact ionization coefficients for holes and electrons.

SANDWICH DETECTORS

There are so-called *sandwich detectors* or *two-color photodiodes* consisting of two (or more) photodiodes in a sequence. The top photodiode is made from the materials with the largest band gap energy and absorbs short-wavelength light while transmitting much of the light with longer wavelengths, which cannot be absorbed. That transmitted light then hits a further photodiode. The ratio of powers detected by the photodiodes depends on the wavelength.

The same principle may also be applied with photodiodes made of the same material, because at longer wavelengths (closer to the bandgap) the top photodiode will not absorb all light. One again gets a wavelength-dependent ratio of signals from two photodiodes.

Sandwich detectors may be used for remote temperature measurements, for example, where you uses the ratio of signals from two photodiodes: the higher the temperature, the higher is the relative amount of radiation at shorter wavelengths.

MULTI-SEGMENT PHOTODIODES AND PHOTODIODE ARRAYS

Photodiodes are available not only as single-segment detectors. There are dual and quadrant photodiodes, which can be used for precision sensing, and also one-dimensional and two-dimensional photodiode arrays. For more details, see the article on position-sensitive detectors.

A photodiode is sometimes integrated into the package of a laser diode. It may detect some light getting through the highly reflecting back facet, the power of which is proportional to the output power. The signal obtained can be used, e.g., to stabilize the output power, or to detect device degradation.

PHOTOTRANSISTORS

The phototransistor is a semiconductor device that is able to sense light levels and alter the current flowing between emitter and collector according to the level of light it receives.

Phototransistors and photodiodes can both be used for sensing light, but the phototransistor is more sensitive in view of the gain provided by the fact that it is a bipolar transistor. This makes phototransistors more suitable in a number of applications.

Fig. 7. A typical phototransistor

Although all bipolar transistors themselves are light sensitive, which is why most transistors are encapsulated in metal or plastic cans, phototransistors are a specialised form of bipolar transistor that have been optimised for its light sensitivity, making these electronic components ideal for many light sensing applications.

The idea of the phototransistor has been known for many years. William Shockley first proposed the idea in 1951, not long after the ordinary bipolar transistor had been discovered. It was then only two years before the photo-transistor was demonstrated.

Since their first introduction and use, phototransistors have been used in a variety of circuit designs and applications, and their development has continued ever since.

Phototransistors are widely available and can easily be obtained quite cheaply from electronic component distributors - in view of their use in many electronic circuits and applications, they are available as part of the standard semiconductor device inventory.

In view of the fact that these electronic components are cheap and easily available, it is necessary to ensure that the selected component meets the requirements of the circuit design in which it will be used. Understanding the various specifications is important for this.

Initial development of the phototransistor

The phototransistor invention followed on from the development of the first point contact transistor. A large number of semiconductor developments were being undertaken at Bell Labs around this time, and the phototransistor was development by one of the teams there.

Although the history of the phototransistor is not as well publicised as many of the other early semiconductor developments, it was certainly a very important development.

Phototransistor operation

The phototransistor uses the basic bipolar transistor concept as the basis of its operation. In fact a phototransistor can be made by exposing the semiconductor of an ordinary transistor to light. Very early photo transistors were made by not covering the plastic encapsulation of the bipolar transistor with black paint.

Fig. 8. A vintage OCP71 phototransistor - this was an OC71 PNP transistor with opaque encapsulation.

The photo-transistor operates because light striking the semiconductor frees electrons / holes and causes current to flow in the base region.

Photo-transistors are operated in their active regime, although the base connection is generally left open circuit or disconnected because it is often not required. The base of the photo transistor would only be used to bias the transistor so that additional collector current was flowing and this would mask any current flowing as a result of the photo-action. For operation the bias conditions are quite simple. The collector of an NPN transistor is made positive with respect to the emitter or negative for a PNP transistor.

The light enters the base region where it causes hole electron pairs to be generated. This generation mainly occurs in the reverse biased base-collector junction. The hole-electron pairs move under the influence of the electric field and provide the base current, causing electrons to be injected into the emitter. As a result the photodiode current is multiplied by the current gain â of the transistor.

The performance of the phototransistor can be superior to that of the photodiode for some applications in view of its gain. As a rough guide, where a photodiode may enable a current flow of around 1μA under typical room conditions, a phototransistor may allow a current of 100μA to flow. These are very rough approximations, but show the order of magnitude of the various values and comparisons.

One of the drawbacks of the phototransistor is that is particularly slow and its high frequency response is very poor. Photo-diodes are much faster electronic components and are used where speed is essential despite their inferior sensitivity.

Phototransistor circuit symbol

Standard circuit symbols are essential for each type of electronic component, enabling circuit diagrams to be drawn easily and recognisable by all. The phototransistor symbol consists of the basic bipolar transistor symbol with two arrows pointing towards the junction of the bipolar transistor. This diagrammatically represents the operation of the phototransistor.

Fig. 9. Phototransistor circuit symbol (for a device based around an NPN transistor)

Phototransistors can be based around both NPN transistors and PNP transistors and therefore it is perfectly possible to have a PNP phototransistor, and for this the direction of the arrow on the emitter is reversed in the normal way.

It can be seen that the phototransistor symbol shown does not give a base connection. Often the base is left disconnected as the light is used to enable the current flow through the phototransistor. In some instances the base may be biased to set the required operating point. In this case the base will be shown in the normal way on the phototransistor symbol.

Phototransistor structure

Although ordinary bipolar transistors exhibit the photosensitive effects if they are exposed to light, the structure of the phototransistor is specifically optimised for photo applications. The photo transistor has much larger base and collector areas than would be used for a normal transistor. These devices were generally made using diffusion or ion implantation.

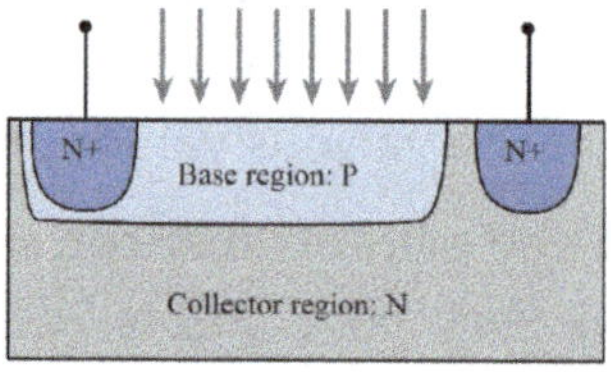

Fig. 10. Homojunction planar phototransistor structure.

Early photo transistors used germanium or silicon throughout the device giving a homo-junction structure. The more modern phototransistors use type III-V semiconductor materials such as gallium arsenide and the like. NPN transistor varieties are more popular in view of the fact that negative ground systems are used, and NPN transistors fit this mode of operation better.

Heterostructures that use different materials either side of the PN junction are also popular because they provide a high conversion efficiency. These are generally fabricated using epitaxial growth of materials that have matching lattice structures.

These photo transistors generally use a mesa structure. Sometimes a Schottky (metal semiconductor) junction can be used for the collector within a phototransistor, although this practice is less common these days because other structures offer better levels of performance.

In order to ensure the optimum conversion and hence sensitivity, the emitter contact is often offset within the phototransistor structure. This ensures that the maximum amount of light reaches the active region within the phototransistor.

PHOTOTRANSISTOR CHARACTERISTICS

As already mentioned the photo-transistor has a high level of gain resulting from the transistor action. For homo-structures, i.e. ones using the same material throughout the semiconductor device, this may be of the order of about 50 up to a few hundred.

However for the hetero-structure devices, the levels of gain may rise to ten thousand. Despite their high level of gain the hetero-structure devices are not widely used because these semiconductor devices are considerably more costly to manufacture. A further advantage of all phototransistors when compared to the avalanche photodiode, another device that offers gain, is that the phototransistor has a much lower level of noise. Avalanche diodes of all forms are known for the large levels of noise they generate as a result of the avalanche process.

One of the main disadvantages of the phototransistor is the fact that it does not have a particularly good high frequency response. This arises from the large capacitance associated with the base-collector junction. This junction is designed to be relatively large to enable it to pick up sufficient quantities of light. For a typical homo-structure device the bandwidth may be limited to about 250 kHz. Hetero-junction devices have a much higher limit and some can be operated at frequencies as high as 1 GHz.

The characteristics of the photo-transistor under different light intensities. They are very similar to the characteristics of a conventional bipolar transistor, but with the different levels of base current replaced by the different levels of light intensity.

There is a small amount of current that flows in the photo-transistor even when no light is present. This is called the dark current, and represents the small number of carriers that are injected into the emitter. Like the photo-generated carriers this is also subject to the amplification by the transistor action.

PHOTOTRANSISTOR APPLICATIONS

The fact that phototransistors are easy to use and perform well, within their limitations, means that these semiconductor devices are used in a wide variety of electronic circuit designs.

Often the circuits and applications are where a light beam is interrupted, but sometimes they can be used for light level detection.

- Encoders where a rotating disc with light and dark stripes rotates - this gives speed and direction or rotation.
- Card readers.
- Security systems
- Infra-red detectors.

- Lighting control.
- Opto-couplers
- Counting systems - a light or IR beam is interrupted for each item counted.
- Lighting control.

There are of course, many other applications where these electronic components are used.

Summary of advantages and disadvantages of phototransistors

Although these semiconductor devices are used in a huge number of electronic devices, circuits and applications, their advantages and disadvantages need to be weighed up to determine whether they are the right electronic component for the given application. Photoresistors or light dependent resistors LDRs; photodiodes; photodarlingtons, photo-FETs and even photo-thyristors and triacs are all available and may suit any given application.

Phototransistor advantages

- Have a relatively high gain and therefore they are relatively sensitive.
- These electronic components are relatively cheap as they are effectively a transistor that is open to light.
- They can be incorporated into an integrated circuit.
- Offer a reasonable speed.

Phototransistor disadvantages

- These devices cannot handle the high voltages of other semiconductor devices like photo-thyristors and triacs.
- In applications where they are exposed to transient voltage spikes and surges, they are open to damage
- Not as fast as other light sensitive electronic components like photo-diodes.
- These are some of the main advantages and disadvantages for these electronic components.
- Phototransistors are semiconductor devices based around the basic bipolar transistor, and they are available as NPN transistors or PNP transistors. Along with other electronic components and semiconductor devices they are available from virtually all electronic component distributors and often their cost is very low.

CHAPTER

9 OPTICAL AMPLIFICATION, WAVELENGTH CONVERSION AND REGENERATION

INTRODUCTION

Future optical networks may require both wavelength conversion and bit shape regeneration in an all-optical domain. The possibility of pulse reshaping while providing wavelength conversion may support new demands over medium and large distances links. Indeed, during propagation the optical data signal suffers deterioration due to the amplified spontaneous emission (ASE) from optical amplifiers, pulse distortion from intrinsic dispersion, crosstalk, and attenuation. All-optical regenerators may be important components for the restoration of these signals, providing complexity and cost reductions with the avoidance of optoelectronic conversions. The regeneration could be 2R (reamplification and reshaping) or 3R, which also provide retiming to solve jitter. Several 3R regenerators using the semiconductor optical amplifier (SOA) have been proposed, such as cascaded SOAs setups or SOA based MachZehnder interferometers (MZI).

In addition, the SOA is a helpful device for both 2R-regeneration and wavelength conversion. Several techniques for 2R-regeneration based on SOAs have been proposed and tested, for example by using four-wave mixing (FWM), cross-gain modulation (XGM), integration within MZI, multimode interferometric SOA, cross-phase modulation (XPM) with filtering, and feed forward technique. However, these techniques require complex designs and involve critical operation points, even the simplest ones based on XGM features. In addition, most of these techniques are not capable of wavelength conversion and regeneration simultaneously.

Although good results can be obtained, this technique demonstrates to be complex since it used optical delay line and two SOAs. In this chapter, we introduce a more simple technique based on XGM (using just one SOA, an optical isolator, an optical circulator, and a CW laser) with easy robust operation at high speed reconfiguration.

The regeneration is based on the abrupt profile of the SOA cross-gain modulation efficiency, which is compressed at high input optical powers. The two optical carriers are amplified in the counter-propagating mode allowing conversion to another or to the same wavelength. In addition, we present 2R-regeneration and conversion results for different kinds of deteriorated input signals. Experimental results such as eye diagrams and measured Qfactors are also shown, for various optical input powers, carriers detuning, bit rates and optical polarizations.

OPTICAL AMPLIFIER

With the rapid development of the optic communication networks, longer transmission lengths are required. Optical amplifier can satisfy the requirements of optical communication networks. An optical amplifier is a device that amplifies an optical signal directly, without the need to first convert it to an electrical signal.

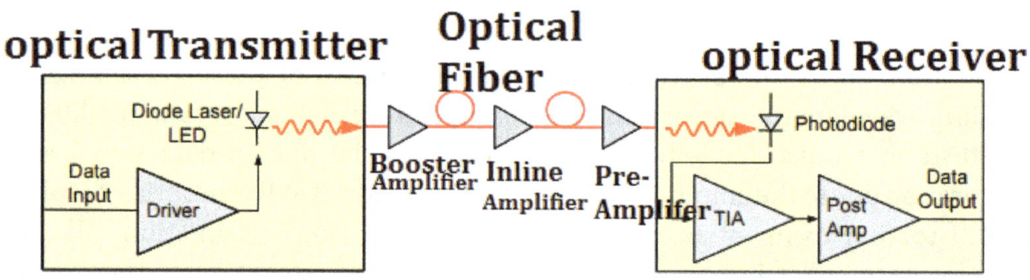

Figure 1. Optical amplifiers in fiber optic communication links

An optical amplifier may be considered as a laser without an optical cavity, or one in which feedback from the cavity is suppressed. This post is going to help you get a better understanding of optical amplifier.

Working Principles of Optical Amplifier

A basic optical communication link contains a transmitter and receiver, with an optical fiber cable connecting them. Although signals transmitting in optical fiber suffer far less attenuation than in other mediums, such as copper, there is still a limitation about 100 km on the distance the signals can travel before becoming too noisy to be detected.

Optical amplifiers are widely used in fiber optic data links. Figure 1 shows three ways in which optical amplifiers can be used to strengthen the performance of optical data links. A booster amplifier is used to increase the optical output of an optical transmitter just before the signal enters an optical fiber. The optical signal is attenuated as it travels in the optical fiber.

An inline amplifier is utilized to restore (regenerate) the optical signal to its original power level. An optical pre-amplifier is operated at the end of the optical fiber link in order to increase the sensitivity of an optical receiver.

Features of Optical Amplifier

- Ratio of output power to input power
- Gain as a fuction of inpout power
- Range of wavelengths over which the amplifier is effective
- Maxmum output power, beyond which no amplification is reached
- undesired signal due to physical processing in amplifier

Types of Optical Amplifier

There are three most commonly used types of optical amplifiers, as shown from left to right: erbium-doped fiber amplifier, the semiconductor optical amplifier, and the fiber Raman amplifier.

The amplifying medium is a glass optical fiber doped with erbium ions. The erbium is pumped to a state of population inversion with a separate optical input. The erbium-doped glass optical gain medium amplifies light at wavelengths that are in the neighborhood of 1550nm – the optical wavelengths that suffer minimum attenuation in optical fibers. The erbium-doped fiber amplifier (EDFA) is the most deployed fiber amplifier. Erbium-doped optical fiber amplifiers (EDFAs) have low noise and can amplify many wavelengths simultaneously, making the EDFA the amplifier of choice for most applications in optical communications.

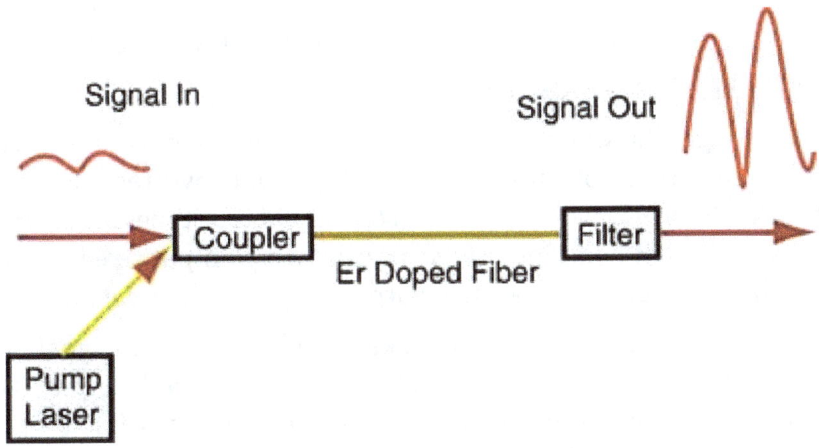

Fig. 2. Erbium-doped fiber amplifier working principle

The gain medium is undoped InGaAsP. This material can be tailored to provide optical amplification at wavelengths near 1.3 μm or near 1.5 μm – important wavelengths for optical communications. Other semiconductors can be used to amplify optical signals at other wavelengths. The input and output faces of the amplifier are antireflection coated in order to prevent optical feedback to the gain medium and lasing. Semiconductor Optical Amplifier with its features of small package, low-cost applications and potential use for optical switching, it can be a great choice to suit most customers.

In a Raman amplifier, the signal is intensified by Raman amplification. Unlike the EDFA and SOA the amplification effect is obtained by a nonlinear interaction between the signal and a pump laser within an optical fiber. There are two types of Raman amplifier: distributed and lumped. A distributed Raman amplifier is one in which the transmission fiber is utilized as the gain medium by multiplexing a pump wavelength with signal wavelength, while a lumped Raman amplifier utilizes a dedicated, shorter length of fiber to provide amplification.

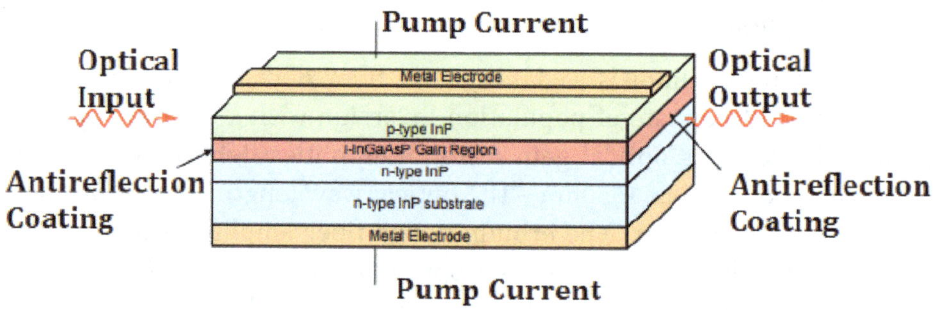

Fig. 3. Semiconductor optical amplifier working principle

WAVELENGTH CONVERSION

As the name suggests wavelength converter converts optical signal with one wavelength to the other optical signal. These types of converters are widely used in WDM networks. Optical wavelength converters are of four types based on their input and output wavelength handling limits. They are fixed input fixed output, variable input fixed output, fixed input variable output and variable input variable output.

Fixed input fixed output converter takes fixed wavelength signal as input and converts it to the fixed wavelength output signal. Variable input fixed output converter takes arbitrary wavelength as input and provides fixed wavelength as output. Fixed input, variable output takes fixed wavelength signal as input and converts it to the variety of wavelengths as output signal. Variable input, variable output type converts variable wavelength signal input to the variable output signal.

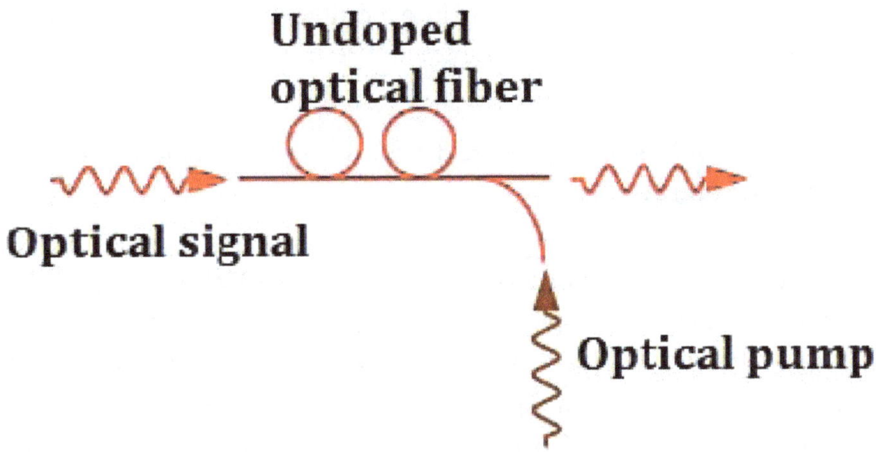

Fig. 4. Raman amplifier working principle

Wavelength conversion methods

Wavelength conversion methods are optoelectronic, optical gating, interferometric, and wave mixing.

Optoelectronic: This type of optical wavelength converter is variable input, fixed output type as mentioned above. It converts the input signal first to electrical form, regenerates it and transmits it using the laser device. Optical Receiver takes either 1310 or 1550 nm wavelength as input. Tunable laser device is used to obtain the desired wavelength.

Optical gating: This method utilizes cross gain modulation and exploits nonlinear effect of the SOA (Semiconductor Optical Amplifier). It also takes variable input and will provide either fixed or variable optical signal as output. This optical gating method utilizes optical device whose characteristics varies based on intensity of the input optical signal to obtain the output.

Interferometric: In this method also phase distortion of cross gain modulation is utilized to obtain wavelength signal conversion. This method is also referred as cross phase modulation. Carrier density varies based on input signal and in turn change refractive index and consecutively modulates the phase of the probe signal.

Wave mixing: Here nonlinear phenomenon of the transmission medium is utilized to obtain wavelength conversion. Four wave mixing produces f1,f2,f3 and f1+f2-f3. This is exploited to obtain the conversion in this type of optical wavelength converter with the help of optical filter. If we take signal frequency as fs and probe frequency as fp then we will have 2fs-fp and 2fp-fs.

CROSS-PHASE MODULATION

Cross-phase modulation is the change of the optical phase of a light beam caused by the interaction with another beam in a nonlinear medium, specifically a Kerr medium. This can be described as a change in the refractive index:

$$\Delta\eta^{(2)} = 2\eta_2 I^{(1)}$$

where n_2 is the nonlinear index. Here, the intensity $I^{(1)}$ of beam 1 causes a refractive index change for beam 2.

Compared with the corresponding equation for self-phase modulation, there is an additional factor of 2. This factor 2 is valid for beams with the same polarization; for cross-polarized beams in isotropic media (e.g. glasses), it must be replaced with 2/3. A more complicated correction is required in case of a birefringent medium.

A fundamental description of cross-phase modulation effects refers to the nonlinear polarization caused in the medium, based on the $\div^{(3)}$ nonlinearity. For example, the above-mentioned factor 2 can be understood on that basis.

Effects of Cross-phase Modulation

Cross-phase modulation can be relevant under different circumstances:

- It leads to an interaction of laser pulses in a medium, which allows e.g. the measurement of the optical intensity of one pulse by monitoring a phase change of the other one (without absorbing any photons of the first beam). This is basis of a scheme for *quantum nondemolition* (QND) *measurements.*
- The effect can also be used for synchronizing two mode-locked lasers using the same laser gain medium, in which the pulses overlap and experience cross-phase modulation.
- In optical fiber communications, cross-phase modulation in fibers can lead to problems with channel cross-talk.
- Cross-phase modulation is also sometimes mentioned as a mechanism for channel translation (wavelength conversion), but in this context the term typically refers to a kind of cross-phase modulation which is not based on the Kerr effect, but rather on changes in the refractive index via the carrier density in a semiconductor optical amplifier.

CROSS-PHASE MODULATION WAVELENGTH CONVERTER

During the process of XGM in an SOA the carrier density variation gives rise to a change in the refractive index of the nonlinear medium which is proportional to the amplitude of the input signal and inversely proportional to carrier density variations. These refractive index variations produce phase modulation when a CW probe signal is coupled to the SOA. When these phase modulations are converted into amplitude modulation using an interferometric arrangement, the resultant probe signal contains an exact copy of the intensity-modulated pattern and the converter is then referred to as a cross-phase modulation (XPM) wavelength converter.

The basic structure of an XPM wavelength converter is illustrated in Fig 5. It is similar to the XGM wavelength converter but it incorporates an additional phase modulation–amplitude modulation (PM–AM) element with another path to it from the probe signal creating an interferometric structure. Hence the CW probe signal splits into two parts and each is applied to the SOA and the PM–AM element. The probe signal at wavelength λ_{cw} propagates through the optical filter acquiring the refractive-index-induced phase modulation in the active cavity of the SOA while the input signal at ëin is blocked by this optical filter. Hence the CW probe signal combines with the phase-modulated probe signal in the PM–AM element when the interference is constructive. The output at the PM–AM element therefore incorporates a probe signal that contains a copy of the data of the input signal but it is now at a different wavelength, being that of the probe signal, thus giving XPM wavelength conversion.

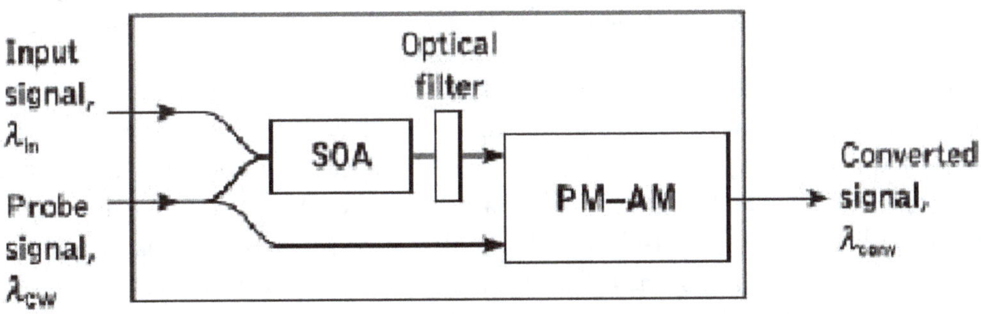

Fig. 5 Structure of a cross-phase modulation wavelength converter

There are several possiblities for realizing an XPM wavelength converter utilizing SOAs in an interferometeric configuration. Such a wavelength converter based on a Mach-Zehnder interferometer (MZI) configuration is depicted in Figure 6 which comprises two SOAs located in the two paths or arms. The operation is similar to the basic XPM wavelength converter while the use of another SOA (i.e. SOA 2) in the lower path increases the gain of the converted probe signal. Phase to amplitude modulation can be obtained when a relative phase difference is introduced between the interferometer paths which can be provided by setting different path lengths. Alternatively, a phase-shifting element may be placed in the lower path of the interferometer before the optical filter at which the probe signal is recombined producing constructive interference. In the copropagating scheme shown in Figure 6(a) the CW probe and the input signals are launched into SOA 1 and both the CW probe and input signals travel in the same direction, whereas in the counterpropagation scheme displayed in Figure 6(b) the input signal at wavelength ëin and CW probe signal at λ_{cw} are applied to SOA 1 from opposite directions. Although the operation of the wavelength converter is similar for both schemes, the counterpropagating scheme has the advantage of reduced noise as the CW probe signal does not interact and hence an optical filter may not be required.

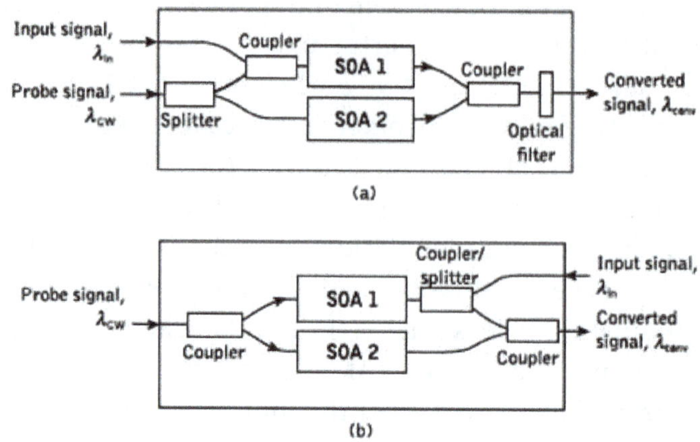

Fig. 6 Cross-phase modulation (XPM) wavelength converter using semiconductor optical amplifiers (SOAs) in a Mach-Zehnder interferometer (MZI) arrangement: (a) copropagating scheme; (b) counterpropagating scheme

During the process of XGM and XPM another phenomenon, namely that of instantaneous frequency variation, occurs which is commonly referred to as frequency (or wavelength) chirp. It is defined as the deviation in the emission frequency with respect to time when a laser is driven by a time-varying current source (i.e. intensity-modulated digital signal). Since refractive index is related to frequency then a variation in the refractive index produces a variation in frequency at each instant. This chirp is commonly observed in the intensity-modulated converted probe signal affecting the leading and trailing edges of the pulses when the refractive index varies suddenly due to an instantaneous change in pulse amplitude.

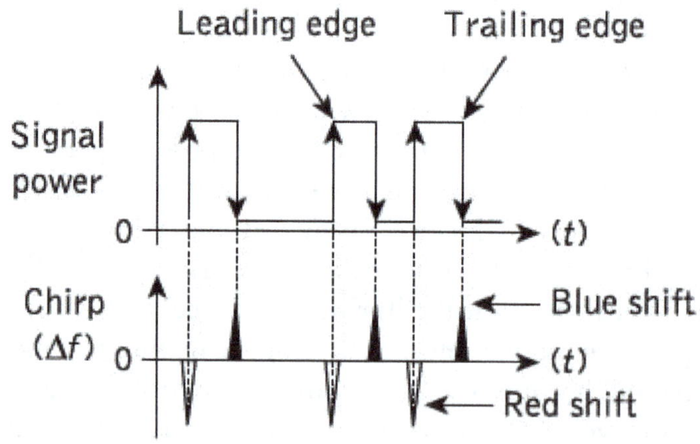

Fig. 7 Frequency chirp: intensity-modulated signal (top) and the corresponding positive chirp (bottom)

Figure 7 illustrates the concept of frequency chirp associated with the intensity-modulated signal shown in the top diagram, while the diagram underneath depicts the corresponding frequency chirp at the leading and trailing edges. When frequency chirp shifts the optical frequency towards the shorter wavelength it is known as blue shift, while a shift of optical frequency towards the longer wavelength is called red shift. The sign of the chirp is said to be positive when the leading edge of the pulse is red shifted in relation to the central wavelength and the trailing edge is blue shifted, and when the shifts are opposite to the foregoing then the sign of chirp will be negative. Hence, Fig. 7 displays a positive chirp condition.

2R-REGENERATOR AND WAVELENGTH CONVERTER WORKING PRINCIPLE

The quality of a 2R regenerator depends on its ability to suppress optical noise and to improve the extinction ratio. The ideal regenerative properties are provided by a system with a transfer function as close as possible to an ideal "S" like behavior. This refers to a characteristic function with the following properties: wide and flat dynamic range gain for bit levels "1" and "0" in order to suppress the noise, and a linear increasing curve which determines the discrimination between bit levels "0" and "1".

The 2R-regenerator presented in this chapter has a characteristic function similar to the "S" like behavior. The response of the device here is similar, not equal to the "S", since it presents gain compression only for the bit level "1". This is due to the fact that just the power level of this bit (at the high level "1") can saturate the SOA gain. The compression of the level "1" is noticed in cases where an overshoot is presented. After the regenerator, the overshoot is removed because SOA gain is deeply saturated. In this way, the 2R-converter works as a "low-pass filter", removing "higher frequencies" present in the overshoot. Therefore, saturated gain acts as a power equalizer for the fluctuation in the bit level "1" reducing the noise. On the other way, the bit level "0" has much less improvement than bit level "1", since its low signal power level cannot saturate the SOA.

The 2R-converter of this section is based on XGM effect in SOAs, and this non-linear effect always degrades the extinction ratio due to the ASE noise added by the SOA. Consequently, there is no extinction ratio improvement. However, it will be showed in this chapter that the improvements caused by the 2R-converter and quantified by Q-factor improvement can be higher than the ER degradation, at least for some cases.

The pattern dependence reduction caused by operating the 2R-converter in its optimum optical input powers could improve both the format of the bit levels "1" and "0". As it will be shown, there is an optimum relationship between the modulated input signal power (at λ_1) and the CW signal power (at λ_2), which should be maintained for different cases in order to kept the pattern dependence effects at the same level.

As mentioned before, the regeneration in the 2R-converter occurs simultaneously with the wavelength conversion via XGM. The regenerative effects are associated with the wavelength conversion efficiency, and XGM is the simplest technique using SOA to implement wavelength conversion. In this process, a strong modulated input signal at λ_1 saturates the amplifier. A continuous wave pump signal at λ_2, injected simultaneously with the modulated signal, is modulated by the gain saturation while being concurrently amplified. The output signal, properly filtered, is a replica of the modulated input signal at a different wavelength with a phase inversion of 180°. In this way, methods to obtain the data without phase inversion are needed. One way is to use a proper software control. Commands given to the receiver force it to associate a bit "1" (in its input) to a bit "0" (in the original data). Another way is to use others devices like an additional SOA, but it will be enlarge the ER degradation. The exploiting of the phase changes provoked by the XGM effect using an optical filter could invert the output signal without ER degradation. Others forms are for example, the use of: delay interferometer and quantum dots SOAs with narrow optical filters.

In this manner, the commentaries made before about the deterioration of the bit levels "1" and "0" refer to bit levels of modulated input signal. Otherwise, the improvement observed in the output signal is commented about the inverted bit levels, that is: for deterioration in the bit level "1" of the modulated input signal, the improvement is noted in the level "0" after the 2R-converter; and in a similar way for the deterioration in the level "0" of the modulated input signal.

The gain saturation of the SOA is related to the XGM wavelength conversion and to the gain compression. The last one is responsible for the noise reduction in the bit level "1" and for the overshoot elimination. The SOA gain must be deeply saturated to obtain the gain compression effects.

This much clear output signal was obtained for the regeneration case where it was observed a Q-factor improvement for the converted signal at λ_2 equal to 1.5. Due to the higher level power of the modulated and CW input signals, and the high SOA bias current (300 mA), the SOA in the 2R-converter presents a saturated gain.

In this manner, the eye diagram for the output signal after the regenerator presents compressed eye. The overshoots and undershoots observed in the output signal might be associated to a self-phase modulation (SPM) and/or changes in the signal phase around the narrow band optical filter.

These results of compressed eye diagrams were observed for other values of the modulated input signal power. Therefore it confirms that the gain of SOA used in the 2R-converter is saturated, inducing the compression needed to the regenerative effects occurrence.

SEMICONDUCTOR OPTICAL AMPLIFIER (SOA)

Optical amplifier, with the introduction in the 1990s, conquered the regenerator technology and opened doors to the WDM technology. It is mainly used to amplify an optical signal directly, without the need to first convert it to an electrical signal. There are many types of optical amplifiers, namely Raman amplifiers, erbium doped fiber amplifiers (EDFAs), and semiconductor optical amplifier (SOA). This section will make a clearer introduction to SOA amplifier, and analyze its advantages and disadvantages.

Fig. 8. Semiconductor Optical Amplifier

THE BASICS OF SEMICONDUCTOR OPTICAL AMPLIFIER (SOA)

SOA optical amplifiers use the semiconductor as the gain medium, which is designed to be used in general applications to increase optical launch power to compensate for the loss of other optical devices. Semiconductor optical amplifiers are often adopted in telecommunication systems in the form of fiber-pigtailed components, operating at signal wavelengths between 0.85 μm and 1.6 μm and generating gains of up to 30 dB. Semiconductor optical amplifier, available in 1310nm, 1400nm, 1500nm, 1600nm wavelength, can be used with single mode or polarization maintaining fiber input/output.

Key Points of SOA Amplifier

- 1310 nm, 1400 nm, 1550 nm and 1610 nm wavelength selectable
- A high fiber-to-fiber gain of 20 dB
- Up to 16 dBm output
- 1 MHz with 10 ns pulse width (optional)
- PM Panda fiber input/output (optional)

- Similar to lasers, but with non-reflecting ends and broad wavelength emission
- Incoming optical signal stimulates emission of light at its own wavelength
- The process continues through the cavity to amplify the signal

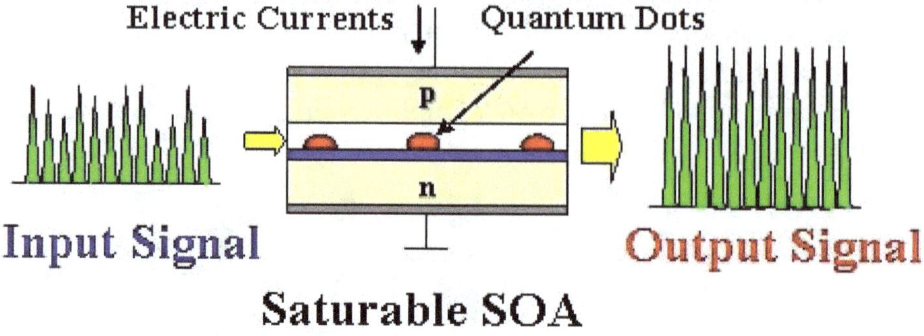

Fig. 9. Working principle of SOA amplifier

Working principle of SOA amplifier

The basic working principle of an SOA is the same as a semiconductor laser but without feedback. SOAs amplify incident light through stimulated emission. When the light traveling through the active region, it causes these electrons to lose energy in the form of photons and get back to the ground state. Those stimulated photons have the same wavelength as the optical signal, thus amplifying the optical signal.

Comparison with Erbium-doped Fiber Amplifiers

The technology of semiconductor amplifiers competes with that of erbium-doped fiber amplifiers (EDFAs). The main differences compared with EDFAs are:

- The setup is much more compact, containing only a small semiconductor chip with electrical and fiber connections. The fabrication cost is accordingly much lower. A compact housing may contain additional polarization-insensitive Faraday isolators at the input, output, or both ports.
- The output powers are substantially smaller, particularly when comparing with high-power fiber amplifiers.
- The required operation power is also smaller; typical drive currents are a few hundred milliamperes.
- The gain bandwidth can be similar. Devices operating in different wavelength regions can be made.
- The upper-state lifetime and thus the stored energy are much smaller, so that the gain reacts to changes in pump power or signal power within nanoseconds (instead of milliseconds). Changes in gain also cause phase changes.
- SOAs exhibit much stronger nonlinear distortions in the form of self-phase modulation and four-wave mixing already at moderate power levels. These

distortions are often unwanted, but nonlinear effects can also be used e.g. for optical signal processing.

- The noise figure is typically higher.
- The amplification is intrinsically polarization-dependent, as far as that is not avoided with special designs.

Tapered Amplifiers

Although most SOAs are fairly limited in output power, there are tapered amplifiers which are optimized for high output powers – hundreds of milliwatts or even several watts. The amplification begins in a single-mode waveguide of limited transverse dimensions, and the waveguide area is expanded along the propagation direction.

Vertical-cavity SOAs

A special type of SOAs is that of the vertical-cavity SOA (VCSOA). This is very similar to a vertical cavity surface-emitting laser, but the top mirror reflectivity is reduced so that the laser threshold is not reached. The still significant top reflectivity is required for obtaining a reasonably high gain, since the single-pass gain through a few quantum wells is fairly low, but it also reduces substantially the gain bandwidth. Comparing with standard edge-emitting SOAs, VCSOAs can be made significantly smaller and cheaper, and can be operated with lower drive currents. (Typically, the order of 10 mA should be sufficient for 20 dB of gain.) Also, they can be fabricated in arrays.

Application in Channel Translation

Amplification is actually not the only application of SOAs: there are also applications in optical fiber communications based on nonlinearities such as gain saturation, or a kind of cross-phase modulation, which is associated with changes in the refractive index via the carrier density in a SOA. Such effects can be used for channel translation (wavelength conversion) in wavelength division multiplexing systems, for modulation format conversion, clock recovery, signal regeneration, and pattern recognition.

FIBER AMPLIFIERS

Fiber amplifiers are optical amplifiers based on optical fibers as laser gain media. In most cases, the gain medium is a glass fiber doped with rare earth ions such as erbium (EDFA = *erbium-doped fiber amplifier*), neodymium, ytterbium (*YDFA*), praseodymium, or thulium. This active dopant is pumped (provided with energy) with light from a laser, such as a fiber-coupled diode laser; in almost all cases, the pump light propagates through the fiber core together with the signal to be amplified. A special type of fiber amplifiers, not, are Raman amplifiers.

The originally dominating application of fiber amplifiers was in optical fiber communications over large distances, where signals need to be periodically amplified. Still, one typically uses erbium-doped fiber amplifiers with signals of moderate optical power (output power e.g. a few hundred milliwatts) in the 1.5-μm spectral regions. Often, many different wavelength channels are simultaneously amplified in a single fiber amplifier.

Other important application areas of fiber amplifiers have been developed later. In particular, high-power fiber amplifiers have been developed which can produce output powers of hundreds of watts or even many kilowatts. Such amplifier are now widely used in laser material processing, replacing solid-state bulk lasers and CO_2 lasers, for example. They are mostly based on ytterbium-doped double-clad fibers for signals in the spectral region of 1.03–1.1 μm.

It is often very convenient (e.g. in optical fiber communications) that light from a fiber can easily be sent into a fiber amplifier, and the amplified light can be sent into further transmission fiber. Besides the active (amplifying) fiber, some additional components may be used, e.g. dichroic fiber couplers and Faraday isolators (see Fig. 10).

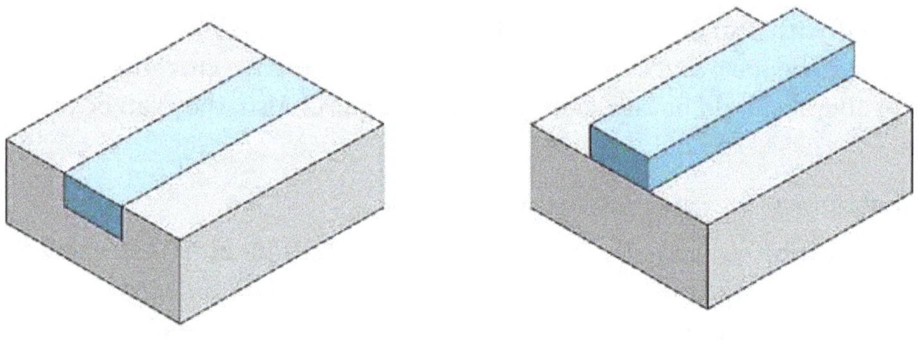

Fig. 10: Schematic setup of a simple erbium-doped fiber amplifier. Two laser diodes (LDs) provide the pump power for the erbium-doped fiber, allowing it to amplify light with wavelength around 1550 nm. Two pig-tailed Faraday isolators strongly reduce the sensitivity of the device to back-reflections.

Gain and Output Power

Due to the possible small mode area and long length of an optical fiber, a high gain of e.g. 30 or 40 dB can be achieved with a moderate pump power, i.e., the gain efficiency can be very high (particularly for low-power amplifiers). The gain achievable is often limited by ASE.

The high surface-to-volume ratio and the robust single-mode guidance also allow for very high output powers with diffraction-limited beam quality, particularly when double-clad fibers are used. However, high-power fiber amplifiers usually have a moderate gain in the final stage, partly due to power efficiency issues; one then uses amplifier chains where the preamplifier provides most of the gain and a final stage the high output power.

Amplification of Light Pulses

Fiber amplifiers cannot only amplify continuous signals, but also short or even ultrashort light pulses. Here, one can utilize that such an amplifier stores a substantial amount of energy, which can then be released within a very short while through an intense pulse. Further, rather short pulse durations (e.g. 100 fs or less) are possible due to the substantial gain bandwidth. However, there are limitations for pulse amplification in fiber amplifiers:

- The saturation energy and thus the possible pulse energy is substantially lower than for a bulk amplifiers based on similar materials, essentially because the mode area is much smaller (even when using large mode area fibers).
- The high peak powers of short pulses (even for moderate pulse energies) in conjunction with the smaller mode area can lead to strong nonlinear effects, which can be detrimental in various ways.

Saturation Characteristics

In terms of gain saturation, fiber amplifiers are very different from semiconductor optical amplifiers (SOAs). Due to the small transition cross sections, the saturation energy is fairly high, e.g. some tens of microjoules for a typical erbium-doped telecom amplifier, or hundreds of microjoules for a large mode area ytterbium-doped amplifier. At the same time, the upper-state lifetime is rather long – e.g. around 8 ms for erbium-doped amplifiers. As a result, significant energy (sometimes several millijoules) can be stored in a fiber amplifier, and can later be extracted e.g. by a single short pulse. Only for output pulse energies of the order of the saturation energy or higher, pulse distortions through saturation become significant. For amplifying the output of a mode-locked laser, where the pulse energy is quite low, the gain saturation is normally the same as for a continuous-wave laser with the same average power.

These saturation characteristics are also important for optical fiber communications, because they prevent any intersymbol interference as can occur with semiconductor optical amplifiers.

In terms of average power, fiber amplifiers are often operated in the strongly saturated regime. This allows for the highest average output power and power conversion efficiency, and the effect of slight variations of pump power on the signal output power is substantially reduced.

ASE and Noise

The gain achievable is often limited not by the available pump power, but by amplified spontaneous emission (ASE). This can become relevant for gains roughly exceeding 40 dB. High-gain amplifiers also need to be protected from any parasitic reflections, because these could lead to parasitic laser oscillation or even to fiber damage, and are therefore often equipped with optical isolators at the output and possibly also at the input.

ASE also determines the fundamental limitation of the amplifier noise properties. While the excess noise of a loss-less four-level amplifier can approach the theoretical limit (the *standard quantum limit*), corresponding e.g. to a noise figure of 3 dB in the case of high gain, the excess noise can be substantially stronger for the usual quasi-three-level laser gain media and in the presence of extra losses. Note that ASE and excess noise are often stronger in backward-pumped amplifiers.

Noise introduced by the pump source may also be an issue. It can directly affect the gain and thus the signal output power, but not for noise frequencies substantially above the inverse upper-state lifetime. (The laser-active ions act as an energy reservoir which can effectively reduce the effect of fast power fluctuations.) A variable pump power can also lead to temperature-dependent heating which translates into phase noise.

ASE itself may be utilized for superluminescent sources with very low temporal coherence, as required e.g. for optical coherence tomography. A superluminescent source has to contain little more than a high-gain fiber amplifier.

Amplifier Modeling

It is possible to model the essential performance aspects of fiber amplifiers in various ways, normally using suitable fiber simulation software. Part of such a model is typically a set of rate equations, with which the population densities for given signal and pump intensities can be calculated. Such a rate equation model may be incorporated in a more comprehensive model which then calculates the optical powers along the fiber.

Applications of amplifier models are manifold. For example, it is possible to quantify various detrimental effects on the amplifier performance, and use such results for optimizing the fiber parameters or other aspects of the amplifier design.

Although basic properties of a fiber amplifier can be calculated analytically, a full quantitative understanding is normally possible only with numerical simulations. These can take into account various details such as the quasi-three-level behavior, strong gain saturation (with optical powers often being far higher than the saturation powers), amplified spontaneous emission (ASE) due to the high optical gain, and possibly energy transfer process for sophisticated situations e.g. with erbium-ytterbium-doped fibers.

Even in simple cases, relatively complex behavior can result. Figure 2 shows an example, where a simple ytterbium-doped fiber amplifier is pumped at 940 nm. The decay of pump power in the device is first quite fast, then slower, and finally faster again. This results from gain saturation by ASE. Forward ASE is partially reabsorbed before it reaches the right end.

Neodymium and Ytterbium Fiber Amplifiers

Fiber amplifiers based on ytterbium- or neodymium-doped double-clad fibers can be used to boost the output power of 1-μm laser sources to very high levels of up to several kilowatts (*high-power fiber lasers and amplifiers*). The broad gain bandwidth is also suitable for the amplification of ultrashort pulses (*ultrafast amplifiers*); limitations arise from fiber nonlinearities such as the Kerr effect and Raman effect. Single-frequency signals can also be amplified to high powers; in this case, stimulated Brillouin scattering usually sets the limits.

Erbium Fiber Amplifiers

Fiber amplifiers based on erbium-doped single-mode fibers (EDFAs) are widely used in long-range optical fiber communication systems for compensating the loss of long fiber spans. High-power versions based on double-clad fibers also exist, although they are significantly more limited in performance than ytterbium amplifiers. See the article on erbium-doped fiber amplifiers for more details.

Thulium Fiber Amplifiers

Thulium-doped fluoride fibers (*TDFA = thulium-doped amplifier*) pumped around 1047 or 1400 nm can be used for amplification in the telecom S band around 1460–1530 nm, or even around 1.65 μm. Combined thulium–erbium amplifiers can thus provide optical amplification in a very wide wavelength range.

There are also thulium-doped amplifiers for the first telecom window, operating at ≈800–850 nm.

Praseodymium Fiber Amplifiers

Fiber amplifiers for the second telecom window around 1.3 μm are also available, but have a lower performance compared with that of erbium-doped amplifiers. They can be based on praseodymium-doped fluoride fibers (*PDFA = praseodymium-doped amplifier*), which are pumped around 1020 nm (a relatively inconvenient pump wavelength) or at 1047 nm (with a YLF laser).

Some Design Issues

Forward and Backward Pumping

Fiber amplifiers can be pumped in forward direction (i.e. with a pump wave copropagating with the signal wave), in the backward direction, or bidirectionally. The direction of the pump wave does not influence the small-signal gain, but the power efficiency of the saturated amplifier and also the noise characteristics. Bidirectional pumping can be a way not only to apply a high pump power, but also to achieve a low noise figure and a high power efficiency at the same time.

Quasi-three-level Gain Media

Most fiber amplifiers (e.g. those based on erbium and ytterbium) operate on quasi-three-level transitions. (Neodymium-doped amplifiers are a notable exception.) This means that in the unpumped state such amplifiers exhibit some signal attenuation caused by the active ions; only when a certain excitation level is exceeded, actual amplification takes place. The quasi-three-level nature also has implications for amplifier noise, in particular an increased noise figure, which however can be minimized by certain design optimizations.

Optical Nonlinearities

Optical nonlinearities such as the Kerr effect can be significant in fiber amplifiers, particularly for those amplifying ultrashort pulses ('! *ultrafast amplifiers*). This can lead to strong self-phase modulation, but also to excessive Raman gain and thus to the generation of a strong first-order Stokes wave at a wavelength some tens of nanometers longer than that of the amplified signal. For single-frequency operation, stimulated Brillouin scattering is the most important nonlinearity.

The effect of the nonlinearity can be reduced e.g. by increasing the fiber mode area (but at the expense of a lower gain efficiency and possibly worse beam quality) or by decreasing the fiber length. The latter measure becomes possible when using a fiber with higher doping concentration, but that can lead to concentration quenching.

Chirped-pulse Amplification

A technique for radically reducing nonlinear effects is chirped-pulse amplification, where pulses are strongly dispersively stretched before entering the amplifier and subsequently compressed again. All-fiber setups using that technique (with a fiber Bragg grating as the compressor) can generate femtosecond pulses only with energies well below 1 μJ, leading to peak powers in the kilowatt region. Substantially higher peak powers of the order of 100 MW are possible when using bulk-optical elements in addition to a fiber amplifier based on large mode area fiber, but some characteristic advantages of fiber-optic systems are lost with that approach. Compared with fully bulk-optical systems, the use of amplifying fibers has the advantage that sufficient single-pass gain is achievable so that the principle of a regenerative amplifier does not need to be used.

Multi-stage Amplifiers

In some cases, a multi-stage amplifier, i.e., an amplifier chain, needs to be realized. This allows e.g. ASE suppression with filters or modulators between the stages, an optimized power efficiency and noise figure, and possibly a modular approach which increases the flexibility for further amplifier developments.

Polarization Issues

Most fiber amplifiers are not made of polarization-maintaining fibers, so they do not preserve the polarization state of the input. However, a wide range of polarization-maintaining active fibers is also available.

The amplification process itself is normally not polarization-dependent; this is an advantage over semiconductor optical amplifiers for use in telecommunications. In some cases, however, polarization hole burning can cause problems.

Fiber Amplifier Modules

Some companies offer OEM fiber amplifier modules which can be convenient for system integrators. Input and output are then often done through fiber connectors. A compact module contains not only the actual fiber amplifier(s), but also the control electronics for the pump diodes, and possibly extras such as an input and/or output optical power monitor, power stabilization, alarms, gain-flattening filters, etc. Such amplifier modules are available based on erbium-doped fibers, ytterbium-doped fibers, and others, and for various power levels.

Raman Fiber Amplifiers

Raman amplifiers are based not on a laser amplification process, but on Raman scattering in a fiber. They differ in various respects from rare-earth-doped amplifiers.

CHAPTER 10
INTEGRATED OPTICS AND PHOTONICS

The goal of integrated optics (IO) is to develop miniaturized optical devices of high functionality on a common substrate. The state-of-the-art of integrated optics is still far behind its electronic counterpart. Today, only a few basic functions are commercially feasible. However, there exists a growing interest in the development of more and more complex integrated optical devices.

In IO we distinguish between optical integrated circuits, which perform functions similar to electronic circuits in communications systems, and planar optical devices, which are integrated optical systems other than communication systems.

Integrated optics is a technology which aims at constructing so-called integrated optical devices or photonic integrated circuits or planar lightwave circuits, containing several or many optical components which are combined to fulfill some more or less complex functions. Such components can e.g. be optical filters, modulators, amplifiers, lasers and photodetectors. They can, e.g., be fabricated on the surface of some crystalline material (such as silicon, silica, or $LiNbO_3$) and connected with waveguides.

The original inspiration of integrated optics came from the technology of electronic integrated circuits, which has shown rapid development over several decades and has led to amazing achievements, such as complex and powerful microprocessors containing many millions of transistors, specialized signal processors and computer memory chips with huge data storage capacity. Unfortunately, integrated optics has not been able to match the progress of microelectronics in terms of the complexity of possible devices. This results from a number of technical limitations:

- While electronic circuits can contain extremely small wires, optical components need to be connected via waveguides, the dimensions of which usually cannot be much smaller than the wavelength, and which often cannot tolerate very sharp bends. (This limitation might be eliminated by using waveguides with very high index contrast, e.g. nanofibers or photonic bandgap waveguides.)

- Optical connections, e.g. between waveguides, and couplers are significantly more critical than electrical connections.
- Waveguides, device connections and passive optical components exhibit optical losses, which often need to be compensated with optical amplifiers. These are larger and more complex than electronic amplifiers based on transistors.
- Some types of optical components can hardly be miniaturized.

There are different material platforms, which differ substantially in terms of capabilities and limitations, and it is not always easy to find a platform which can meet all requirements of an application. In contrast to that, the CMOS platform for microelectronics can cover most requirements of that area.

For these reasons, integrated optical circuits have not reached by far the complexity of electronic integrated circuits. However, devices of moderate complexity can still be useful for example for optical fiber communications, where they can host multiple data transmitters and/or receivers, consisting of distributed feedback lasers, optical modulators, photodiodes, and optical filters (e.g. in the form of arrayed waveguide gratings). Recently, new hope for a powerful and cost-effective integrated optical technology has arisen from developments in silicon photonics.

OPTICAL INTEGRATED CIRCUITS

A Photonic Integrated Circuit (PIC) is analogous to an electronic integrated circuit, where photons are manipulated at sub-micron scale. Ability to confine and control flow of light in a sub-micron feature such as an integrated optical waveguide allows us to study and exploit the properties of light. Some of the fundamental devices developed are wavelength selective filters, light-chip couplers, optical modulators, photodetectors and light source, all these components are either monolithically or heterogeneously integrated on a single chip.

PHOTONIC INTEGRATED CIRCUITS

Photonic integrated circuits (also called *planar lightwave circuits* = PLC or integrated optoelectronic devices) are devices on which several or even many optical (and often also electronic) components are integrated. The technology of such devices is called integrated optics. Photonic integrated circuits are usually fabricated with a wafer-scale technology (involving lithography) on substrates (often called chips) of silicon, silica, or a nonlinear crystal material such as lithium niobate ($LiNbO_3$). The substrate material already determines a number of features and limitations of the technology:

- *Silica-on-silicon integrated optics* builds on silicon wafers, for which many aspects of the powerful microelectronics technology can be used. Silica waveguides allow the realization of couplers, filters (e.g. for multiplexers and demultiplexers in wavelength division multiplexing technology), power splitters and combiners, and even active elements with optical gain. They can also be connected to optical fibers. A natural solution for coupling multiple waveguides is to use fiber arrays.
- An area of strong current interest is *silicon photonics*, where photonic functions are implemented directly on silicon chips.
- An already commercialized photonic integrated circuits technology is based on indium phosphide (InP); it is used mainly in optical fiber communications.
- The relatively new silicon nitride (Si_3N_4) platform is suitable for photonic devices operating in the 1-μm spectral region or even at shorter wavelengths. Low-loss waveguides with small bend radius can be made, also various types of photonic components such as couplers, filters, arrayed waveguide gratings and others. The high nonlinear coefficient gives the potential for nonlinear signal processing.
- Waveguides can be fabricated on *silica glass* (fused silica) e.g. with lithographic techniques involving chemical processing or indiffusion of dopants, or with laser micromachining. The latter techniques can be used for fabricating waveguides far below the surface (embedded waveguides), so that three-dimensional circuit designs become possible. Amplifiers and lasers can be made by using rare-earth-doped glasses.
- *Lithium niobate* ($LiNbO_3$) as a nonlinear crystal material is suitable for devices performing nonlinear functions, for example electro-optic modulators or acousto-optic transducers. Waveguides can be fabricated on lithium niobate substrates e.g. via proton exchange or by indiffusion of titanium, in any case controlled by a lithographic method. Doping with rare earth ions makes possible amplifiers and lasers. The birefringence of this material creates opportunities for polarization control, which may then be used e.g. for filtering purposes. On the other hand, the birefringence makes it more difficult to obtain polarization-independent devices, as are often required for optical fiber communications.

Photonic integrated circuits can either host large arrays of identical components, or contain complex circuit configurations. However, for various reasons the complexity achievable is not nearly as high as for electronic integrated circuits.

Their main application is in the area of optical fiber communications, particularly in fiber-optic networks, but they can also be used for, e.g., optical sensors and in metrology.

An important distinction is that between devices with smaller or larger mode areas:

- Some waveguides (e.g. made in silicon-on-insulator technology) exhibit strong confinement, leading to small effective mode areas and allowing for tight bends without excessive bend losses. They are therefore potentially suitable for chips with a very high level of integration. However, such devices are essentially always polarization dependent, having a strong built-it birefringence. Polarization-insensitive designs would be possible in principle, but would introduce unrealistic fabrication tolerances.
- Other waveguides exhibit much weaker guidance and can be made in polarization-insensitive form. However, such waveguides do not allow tight bends and thus prevent a high level of integration.

Application Areas

Photonic integrated circuits can find applications in different area; some examples:

- Optical fiber communications and free-space optical communications can utilize circuits for signal generation, detection, regeneration and other processing.
- Optical metrology e.g. in the form of LIDAR and fiber-optic sensors can profit from integrated circuits where even delicate devices such as interferometers can be realized in a compact and stable manner. Optical frequency metrology can utilize highly compact frequency comb sources as optical frequency synthesizers.
- Terahertz imaging usually involves photonic elements for generating and detecting terahertz waves and processing the signals.
- Quantum cryptography and quantum computing are another field of application.

Very often, photonic integrated circuits are specially designed and fabricated for a specific application. They may then be called ASPIC = application-specific photonic integrated circuits.

Fabrication in Foundries

In microelectronics, the model of foundries has been widely accepted. This means a separation between a company designing and later selling an integrated circuit for a specific purpose and another company (the foundry) for fabricating it. Only the foundry needs to have the complex machinery and detailed know-how for fabrication.

The same model is also suitable for photonic integrated circuits. Again, a complex technology is required, which is mastered by some foundries, and their customers can focus on designing circuits for specific purposes and bringing those into the markets. A suitable interface needs to be developed, where the foundry exactly describes its capabilities and receives the designs to be fabricated. For the designs, certain elements (e.g. for realizing certain device functions like waveguides, couplers, resonators, modulators, photodetectors etc.) may already be predefined and can be appropriately connected by the circuits designer. The foundry may fabricate bare chips or possibly also offer packaging solutions.

A foundry may support different technology platforms, such as indium phosphide (InP), gallium arsenide (GaAs), silicon on insulator (SOI), silica on silicon, etc. Note that these can substantially differ in technical details of the required machinery. It can also be important to over the combination with other technologies such as micro-electronics and micro-electromechanical systems (MEMS).

A real world example of Photonic IC design

The following steps describe the workflow used by a customer to design a PIC for an optical transceiver.

- Install the foundry's process design kit (PDK) for Synopsys tools.
- Make a schematic design in OptoCompiler.
- Simulate the circuit and optimize the optical performance in OptSim.
- Use Photonic Device Compiler to design custom devices when the standard PDK from the foundry doesn't meet the design request. These custom devices can be used together with the foundry PDK components.
- Translate the schematic design into a layout by using the Schematic Driven Layout (SDL) capabilities of OptoCompiler.
- Adjust the layout in OptoDesigner based on the footprint requirement and finish the waveguide and electrical routing.
- Back-annotate the layout changes and confirm changes made to the schematic design.
- Re-simulate the circuit in OptSim Circuit to evaluate the impact of the changes.
- Adjust the schematic and/or layout design if necessary.
- Run a design rule check (DRC) on the layout implementation with IC Validator from the OptoCompiler cockpit.
- Run a layout versus schematic (LVS) check on the final layout and implement and compare the netlist of the schematic design with IC Validator.

- Stream out the mask layout and submit to the foundry, together with the DRC and LVS reports.

INTEGRATED OPTICS AND PHOTONICS TECHNOLOGIES

Integrated technology for optical devices has developed within optical fiber communications so that it is now possible to fabricate a complete system onto a single chip. Integration for such devices has become a confluence of several optical or photonic disciplines. Both IO and IP technologies are referred to in the above where the control of the optical devices distinguishes one technology from the other. Electronic control of the optical devices determines the terminology of IO whereas photons control the operation of IP devices.

In addition, IP does not involve any optoelectronic conversion of optical signals and hence this technology is also termed as 'all-optical'. Both IO and IP use planar waveguide technology to provide interconnections between optical components including the basic components for guiding and control of optical signals. IP technology, however, enables the fabrication of subsystems and systems to be realized onto a single substrate.

Hence when both active and passive devices are monolithically integrated onto a single substrate in a multilayered integration then these devices are normally referred to as IP devices, while when both active and passive elements fabricated as individual devices are interconnected together they form larger IO devices or circuits. Thus IP can also be seen as a process for the miniaturization and integration of optical systems on a single substrate, and therefore it may be considered as a further enhancement of IO technology, not necessarily as an alternative technology.

Both IO and IP seek to provide an alternative to the conversion of an optical signal back into the electrical regime prior to signal processing by allowing such processing to be performed on the optical signal. Hence thin transparent dielectric layers on planar substrates which act as optical waveguides are used in IO and IP to produce small-scale and miniature optical components and circuits.

The birth of IO may be traced back to basic ideas outlined by Anderson in 1966. He suggested that a microfabrication technology could be developed for single-mode optical devices with semiconductor and dielectric materials in a similar manner to that which had taken place with electronic circuits. It was in 1969, however, after Miller had introduced the term integrated optics when discussing the long-term outlook in the area, that research began to gain momentum.

Developments in IO have passed the stage where both signal processing and logic functions can be physically realized. Furthermore, such devices may form the building blocks for future digital optical computers. Nevertheless, a number of these advances combine to be closely linked with developments in lightwave communication employing optical fibers.

A major factor in the development of IO is that it is essentially based on single-mode optical waveguides and therefore tends to be incompatible with multimode fiber systems. Hence IO did not make a significant contribution to early deployed optical fiber systems. The advent, however, of single-mode transmission technology further stimulated work in IO in order to provide devices and circuits for these more advanced third-generation systems.

Furthermore, the continued expansion of single-mode optical fiber communications has created a growing market for such IO components. It is also likely that new generations of optical fiber communication systems employing coherent and possibly soliton transmission will lean heavily on IO and IP techniques for their implementation.

The proposals for IO and IP devices and circuits which in many cases involve reinventions of electronic devices and circuits exhibit major advantages other than solely a compatibility with optical fiber communications. Electronic circuits have a practical limitation on speed of operation at a frequency of around 1010 Hz resulting from their use of metallic conductors to transport electronic charges and build up signals. The large transmission bandwidths (over 100 GHz) currently under investigation for optical fiber communications are already causing difficulties for electronic signal processing within the terminal equipment. The use of light with its property as an electromagnetic wave of extremely high frequency (10^{14} to 10^{15} Hz) offers the possibility of high-speed operation around 10^4 times faster than that conceivable employing electronic circuits. Interaction of light with materials

INTEGRATED OPTICS

- Integrated technology for optical devices has developed within optical fiber communications so that it is now possible to fabricate a complete system onto a single chip.
- Integration for such devices has become a confluence of several optical or photonic disciplines.
- Both IO and IP technologies are referred to in the above where the control of the optical devices distinguishes one technology from the other.
- Electronic control of the optical devices determines the terminology of IO whereas photons control the operation of IP devices.
- IP does not involve any optoelectronic conversion of optical signals and hence this technology is also termed as 'all-optical'.

- Both IO and IP use planar waveguide technology to provide inter-connections between optical components including the basic components for guiding and control of optical signals.
- IP technology, however, enables the fabrication of subsystems and systems to be realized onto a single substrate.
- Hence when both active and passive devices are monolithically integrated onto a single substrate in a multilayered integration then these devices are normally referred to as IP devices, while when both active and passive elements fabricated as individual devices are interconnected together they form larger IO devices or circuits.
- Thus IP can also be seen as a process for the miniaturization and integration of optical systems on a single substrate
- Both IO and IP seek to provide an alternative to the conversion of an optical signal back into the electrical regime prior to signal processing by allowing such processing to be performed on the optical signal.
- Hence thin transparent dielectric layers on planar substrates which act as optical waveguides are used in IO and IP to produce small-scale and miniature optical components and circuits.
- Developments in IO have passed the stage where both signal processing and logic functions can be physically realized.
- uch devices may form the building blocks for future digital optical computers.
- A number of these advances combine to be closely linked with developments in light wave communication employing optical fibers.
- A major factor in the development of IO is that it is essentially based on single-mode optical waveguides and therefore tends to be incompatible with multimode fiber systems.
- Hence IO did not make a significant contribution to early deployed optical fiber systems.
- Furthermore, the continued expansion of single-mode optical fiber communications has created a growing market for such IO components.
- The proposals for IO and IP devices and circuits which in many cases involve reinventions of electronic devices and circuits exhibit major advantages other than solely a compatibility with optical fiber communications.
- The other major attribute provided by optical signals interacting within a responsive medium is the ability to utilize light waves of different frequencies (or wavelengths) within the same guided wave channel or device.

- The devices of interest in IO and IP are often the counterparts of microwave or bulk optical devices.
- These include junctions and directional couplers, switches and modulators, filters and wavelength multiplexers, lasers and amplifiers, detectors and bistable elements.

Design and fabrication of IP circuits:

- The technology associated with the design and fabrication of IP circuits and devices depends upon different factors that mainly result from the characteristics of the substrate material on which the various devices are to be fabricated.
- The IP process may require serial, parallel or hybrid integration of independent devices.
- In serial integration of devices, different elements of the optical chip can be interconnected in a consecutive manner and therefore side-, or edge-emitting, or conducting optical devices can be readily integrated on the same substrate.
- In the parallel case the chip is constructed by developing columns of devices in which surface- or bottom-emitting devices can effectively be used whereas in hybrid integration IP technology the devices are fabricated using both serial and parallel integration on the same substrate.
- To gain control of the optical signals, however, additional elements can be developed separately or be directly attached to the IP circuit, or be interconnected to it by optical fibers.
- In addition, both active and passive devices may be required to be located on the same substrate and therefore hybrid IP integration demands multilayered IP circuits and components to be produced on a single substrate such that they must be compatible with three-dimensional structures of other IO or IP devices
- The enabling technologies for IP mainly rely on silica-on-silicon (SOS) where the waveguide structures comprise three layers, namely the buffer, the core and the cladding.
- Due to its refractive index match to silica-based optical fiber, vertical light confinement in SOS can be achieved by increasing the refractive index of the core layer relative to the surrounding glass, whereas lateral confinement is obtained through structuring of the core layer.
- The real benefit of SOS is the ability to apply wafer-scale, planar lithography and processing techniques to integrate substantial numbers of functions

either as arrays of identical devices or in the form of customized circuit configurations on single or multiple chips.

- This integration capability offers an efficient platform for the implementation of typical fiber-based functionalities such as optical power splitters or combiners, couplers, wavelength-selective couplers, multiplexers/de-multiplexers and optical gain elements.

BASIC PRINCIPLES OF IO FOR INTERFEROMETRIC SENSING

There is a growing interest in finding detection technologies enabling real-time and on-line monitoring of many elements having remarkable, (in some cases dramatic), impacts on our everyday lives. Consequently, a huge effort has been devoted to designing and building instruments with capabilities that would not even be thought of a few years ago. Nowadays are required instruments to analyse many fields of high interest ranging from environmental pollutants or pathogens in air and water, up to the Security and Safety in the urban ambient or in the logistic chain of food production and transport, as well as many other applications.

In most cases, the realisation of this kind of device involves analytical techniques that should avoid chemical reactants, difficult to be used outside specialised laboratories. Moreover, the device has to be routinely applied by almost untrained people that should also be able to read and understand the measurement result. At the same time, the analytical equipment must be self-contained and available at a very low cost.

Nowadays, the only available solution suitable to satisfy the whole of these requirements is to develop highly miniature integrated equipment to allow mass production using technologies likewise the planar technologies driven by the development of microelectronics. At the present state of the art, most of these requirements can be satisfied when it is possible to develop sensing techniques belonging to the family of the optical integrated microdevices.

Optical detection can concern different properties of light such as Intensity, (in the case of photometric analyses), Wavelength, (in the case of spectroscopic analyses), Refractive Index, (in the case of Index change due to the presence of the analyte molecules at the Sensor Surface), etc.

The development at industrial quality level of active optical materials, (particularly Lithium Niobate, $LiNbO_3$), joined to the use of planar technology, has allowed the fabrication of monolithic integrated optical microdevices having sensing capabilities comparable to the correspondent standard laboratory equipment.

As highlighted in the next paragraphs, IO and Photonics microfabrication techniques, offer the possibility to generate in monolithic and miniaturised systems, nearly all the equivalents of classical optical components (mirrors, splitters, combiners, phase shifters, etc.). Furthermore, such IO components can be integrated into and interfaced with guided wave circuits, such as integrated optical waveguides and optical fibres. These characteristics often strongly simplify all issues related to optical alignment and maintenance, the flow of the luminous signal, carrying the information, being firmly confined within fixed and well defined optical channels. In the following an overview of the characteristics and potentialities of IO devices is presented, in particular for a wide range of interferometric sensing applications.

Basic principles of IO for interferometric sensing

The basic element of any Integrated Optic device is the optical waveguide that can be generated by tailoring the refractive index (n) in the near-surface region of the base material. It must be remembered that the light transmission is confined in the regions where the refractive index is higher than in the surroundings.

Several techniques have been developed to obtain local variation of the refractive index creating the possibility to obtain an integrated optical waveguide. In this chapter, we will only mention some of the most common techniques used for $LiNbO_3$ substrate, which is one of the most used materials in optical device fabrication. One possibility is to use local doping processes obtained by photolithographic definition of the desired waveguide geometry associated with the dopant thermal diffusion (typically, Ti diffusion or Proton exchange processes), to increase the refractive index in the doped region.

Alternatively, it is possible to create a waveguide by lowering the refractive index of the base material in the regions outside the waveguide introducing lattice damage through ion bombardment. A third possibility is to pattern the surface with the desired geometry, and etching the surrounding region to obtain a ridge waveguide protruding from the surface. Figure 1 reports a sketch of the two geometries.

In general, the geometries of the integrated microsystems are designed to reproduce the same physical effects obtained by optical instrumental architectures created in laboratory, assembling several optical elements like mirrors, beam splitters, etc.

Probably, the most frequently used integrated micro-spectrometric devices are based on the Mach-Zehnder Interferometric, (MZI), geometry or on Young Interferometer, (YI) or, more recently, on the Staircase Micro Diffractive Gratings, (MDG) first developed by Michelson.

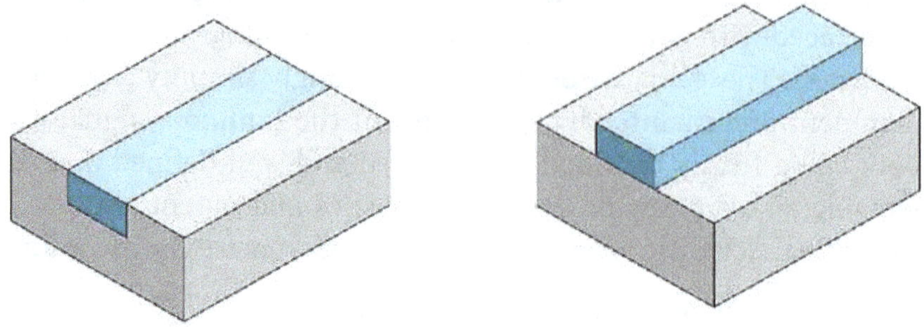

Fig. 1. Two basic geometries of integrated optics waveguides: (a) buried waveguide, (b) ridge waveguide.

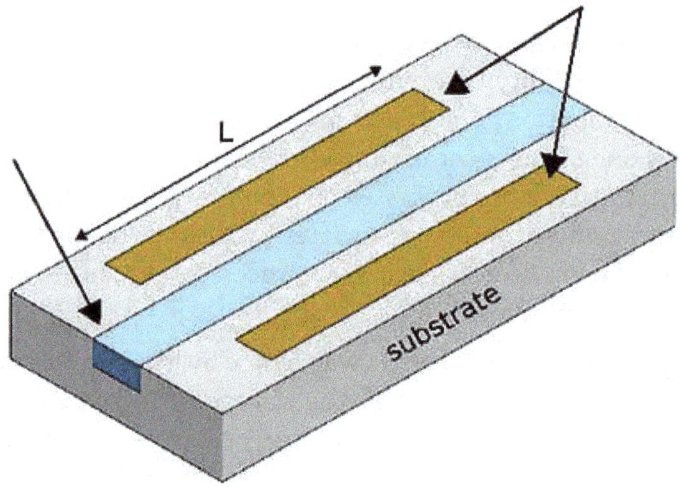

Fig 2. Integrated version of a Pockels cell to be used as a phase modulator.

All these devices take advantage of the Electro-Optic properties of the $LiNbO_3$ substrate. In fact, the optoelectronic properties of the substrate allow to locally controlling the refractive index of a waveguide by applying a suitable electric field, so creating a Pockels cell that induces a phase modulation in the light transmitted in the specific waveguide (See Fig. 2).

In particular, it can be useful to recall that, in the case of spectroscopic analysis techniques, the wavelength-dispersive systems, such as prisms or gratings, spatially spread the light wavelengths at different angles allowing the direct measurements of the relative intensities, (wavelength spectrum), by using suitable photodetectors at the corresponding angles.

On the contrary, in the Fourier Transform Spectroscopy the intensity of the total light beam that contains the whole ensemble of wavelength, at the same time, is measured. In this case, the measurement with a traditional Mach-Zehnder instrument is performed by splitting the light beam into two branches that are then recombined giving rise to an interference pattern.

The light intensity of the recombined beams is measured as a function of the phase shift, generated by the respective different optical path-lengths $l = nL$, where n is the refractive index and L is the geometrical path length.

Figure 3(a), reports a typical example of Mach-Zehnder equipment, whereas Figure 3(b) is shown the equivalent device fabricated with planar technology.

Following the optical interference laws, the intensity measured by the photodetector (PD) depends on the phase shift between the two optical paths. Inducing a variation in the optical paths, l, of one of the two arms the phase shift changes and, in turn, the correspondent variation of the light intensity is measured by the PD.

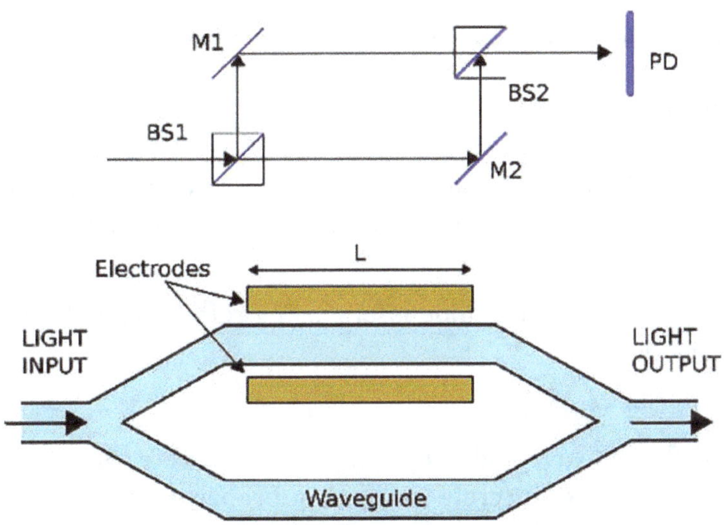

Fig. 3. Upper: Traditional Mach-Zehnder interferometer geometry. The light beam first crosses the BS1 beam splitter then, through the M1 and M2 mirrors, the two light beams are recombined in BS2 and then arrive in the photodetector PD. lower: Equivalent integrated optic device: The voltage applied to the electrodes controls the phase shift in the light beams propagating in the two arms.

PLANAR WAVEGUIDES

Planar waveguides, also called *slab waveguides*, are waveguides with a planar geometry, which guide light only in one dimension. They are often fabricated in the form of a thin transparent film with increased refractive index on some substrate, or possibly embedded between two substrate layers. For example, a thin neodymium-doped YAG layer can be embedded in undoped YAG with slightly lower refractive index (see Fig. 4). Advantages of that more symmetric design are a symmetric shape of the waveguide mode and the easier realization of single-mode waveguiding.

A waveguide function can also be achieved by diffusion of an index-raising agent into a surface, often achieved by treatment in an oven. The resulting refractive index profile is then usually not a step-index profile, but one with a smooth shape.

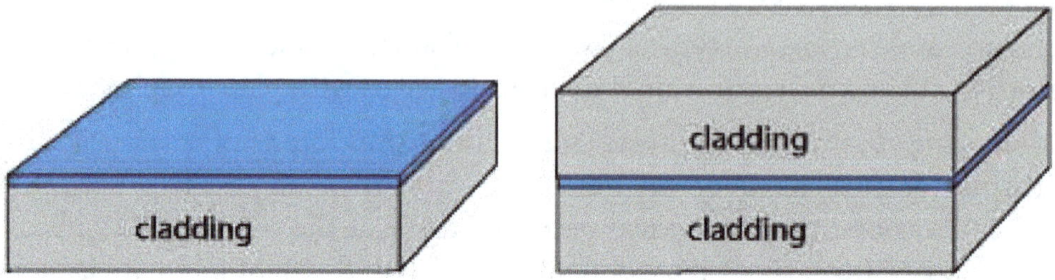

Fig 4: A planar waveguide made on a crystal or glass piece (left side), or embedded between two layers (right side).

Active planar waveguides are sometimes used e.g. for optical amplifiers with high gain (compared with that of bulk amplifiers), relatively high output power (at least multiple watts) and high beam quality at least in one dimension. Some of these devices can be side pumped with a proximity-coupled laser diode, not requiring any pump optics.

There are also planar waveguide lasers, using reflecting end surfaces, e.g. using the Fresnel reflections.

Planar Waveguide Modes

The modes of planar waveguides represent the discrete waves characterized by discrete amplitudes and discrete velocities propagating through a waveguide structure. The number of supported modes depends on the thickness of a planar waveguide. There are two independent electromagnetic modes: the transverse electric (TE) mode and the transverse magnetic (TM) mode.

In guided TE mode, the electric field component along the z-axis is equal to zero (Ez=0). The only non-vanishing components in the guided TE mode are H_x, E_y, and H_z. Similarly, in guided TM mode, the magnetic field component along the z-axis is equal to zero (H_z=0). The only non-vanishing components in the guided TM mode are E_x, H_y, and E_z.

Due to their high sensitivity, imperviousness to electromagnetic interference, short detection time, compactness, low cost, and ease of interface, planar waveguide optical sensors are often used for sensing applications. Cadence's suite of design and analysis tools can help in developing integrated electronic/photonic design automation (EPDA) environments suitable for designing integrated optical sensors based on planar waveguides as well as optical fibers.

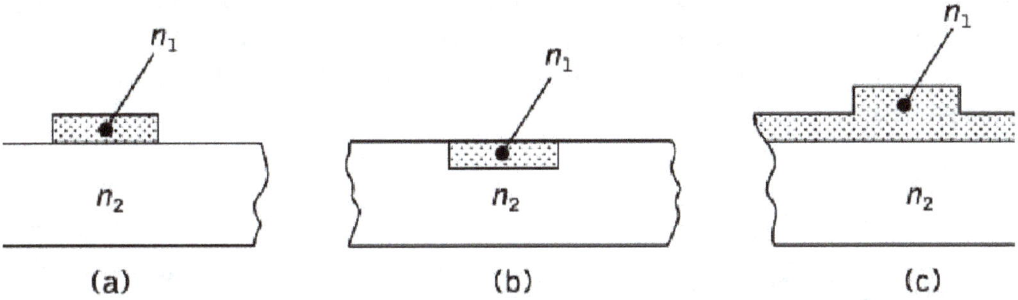

Fig. 5. Cross-section of some strip waveguide structures: (a) ridge guide; (b) diffused channel (embedded strip) guide; (c) rib guide

The planar slab waveguide confines light in only one direction, allowing it to spread across the guiding layer. In many instances it is useful to confine the light in two dimensions to a particular path on the surface of the substrate. This is achieved by defining the high-index guiding region as a thin strip (strip guide) where total internal reflection will prevent the spread of the light beam across the substrate. In addition the strips can be curved or branched as required. Examples of such strip waveguide structures are shown in Fig. 5.

They may be formed as either a ridge on the surface of the substrate or by diffusion to provide a region of higher refractive index below the substrate, or a rib of increased thickness within a thin planar slab. Techniques employed to obtain the strip pattern include electron and laser beam lithography as well as photolithography.

The rectangular waveguide configurations illustrated in Fig. 5 prove very suitable for use with electro-optic deflectors and modulators giving a reduction in the voltage required to achieve a particular field strength. In addition they allow a number of optical paths to be provided on a given substrate.

A trade-off also exists between the minimum radius of curvature which is required for high-density integration and the ease of fabrication. When the change is large, the dimensions of the waveguide may be reduced, even though the scattering losses become larger. As the maximum confinement of the single-mode guide occurs when it is operated near to the cutoff of the second-order mode, then when the refractive index change is large, the radius of curvature of the waveguide can also be made very small. It is therefore necessary to find a compromise for the waveguide material used.

The symmetrical waveguide structure which is depicted in Fig. 6(a) comprises two waveguides (i.e. waveguide 1 and 2) in which the layers are stacked by epitaxial growth in the y direction and the guided waves travel along the x axis and therefore the refractive indices of both waveguides must be closely matched. This requirement does not permit the production of dissimilar elements (i.e. active and passive devices) on the same substrate, hence necessitating a separation layer to isolate them. Figure 6(b) displays an asymmetrical waveguide structure where both the active and/or passive devices can be formed in separate, vertically displaced waveguides on the same substrate.

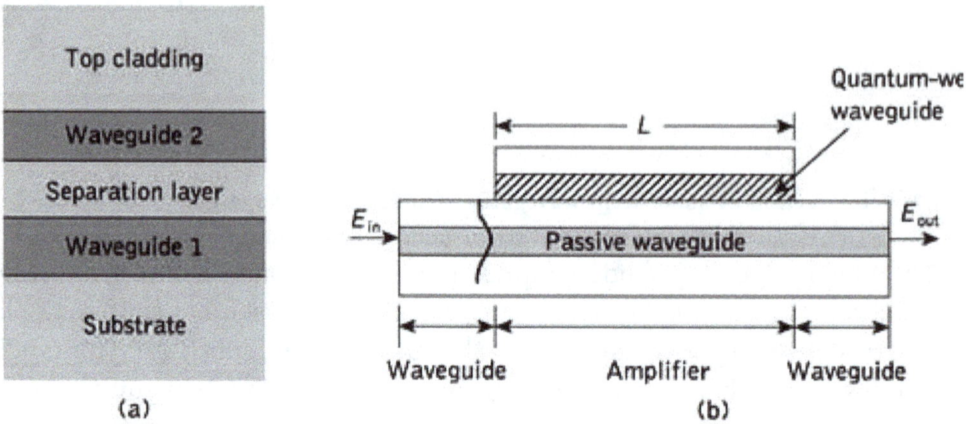

Fig. 6 Twin-waveguide structures: (a) symmetric waveguides; (b) asymmetric waveguides incorporating an optical amplifier on the same substrate

In this twin-waveguide structure the quantumwell active waveguide provides a gain section while the central section of the device of length La functions as an amplifier. This approach therefore introduces an asymmetry in waveguide refractive index profiles and hence the technique can be used to overcome the problems associated with monolithic integration of dissimilar active and/or passive devices on the same substrate. Based on this method various active/passive devices have been successfully demonstrated.

INTEGRATED OPTICAL DEVICES

Integrated optical devices are promising solutions for many applications, such as medical diagnosis, health monitoring and light therapies. Despite the many existing reviews focusing on the materials that these devices are made from, a systematic review that relates these devices to the various materials, fabrication processes, sensing methods and medical applications is still seldom seen.

This work is intended to link these multidisciplinary fields, and to provide a comprehensive review of the recent advances of these devices. Firstly, the optical and mechanical properties of optical waveguides based on glass, polymers and heterogeneous materials and fabricated via various processes are thoroughly discussed, together with their applications for medical purposes.

Then, the fabrication processes and medical implementations of integrated passive and active optical devices with sensing modules are introduced, which can be used in many medical fields such as drug delivery and cardiovascular healthcare. Thirdly, wearable optical sensing devices based on light sensing methods such as colorimetry, fluorescence and luminescence are discussed. Additionally, the wearable optical devices for light therapies are introduced. The review concludes with a comprehensive summary of these optical devices, in terms of their forms, materials, light sources and applications.

Beam Splitters

A beam splitter (or *beamsplitter, power splitter*) is an optical device which can split an incident light beam (e.g. a laser beam) into two (or sometimes more) beams, which may or may not have the same optical power (radiant flux).

Different types of beam splitters exist, as described in the following; the most important ones are plate and cube beam splitters. They are used for very different purposes. For example, beam splitters are required for various interferometers, autocorrelators, photo cameras, projectors and laser systems. The wide range of applications implies widely varying requirements, which can be fulfilled with different types of splitters.

Important Properties

Apart from the characteristics concerning the basic function of a beam splitter – the splitting ratio – other properties of beam splitters can be important in applications:

- Some beam splitters are polarizing, others are non-polarizing. There are also devices designed for use with only one polarization direction – for example, with a laser beam as the input, which is in most cases linearly polarized.
- While some devices work only in a narrow wavelength region (e.g. around a common laser line), others are designed for broadband operation, e.g. working throughout the whole visible wavelength region. Similarly, beam splitters may operate properly only with a finite range of incidence angles.
- The optical losses vary significantly between different types of devices. For example, beam splitters with metallic coatings exhibit relatively high losses, whereas devices with dichroic coatings may have negligible losses: the total output power nearly equals the input power.
- The losses may also be related to the damage threshold, which can be important particularly for use with Q-switched lasers.
- The spatial configuration can be important for applications. Some require the output ports to be at 0° and 90° relative to the input beam (possibly without any beam offset of the transmitted beam), while others require two parallel outputs or some other configuration.

- For bulk-optical devices, a large open aperture is sometimes needed.

Plate Beam Splitters Based on Dielectric Mirrors

Any partially reflecting mirror can be used for splitting light beams. In laser technology, dielectric mirrors are often used for such purposes, and they are called *plate beam splitters* to distinguish them from *cube beam splitters* (see below). The angle of incidence may be 45° (as in Fig. 7), leading to a 90° deflection of one of the output beams, as is often convenient. However, one can design such beam splitters for other deflection angles; they will usually work only for a limited range of angles. A wide range of power splitting ratios can be achieved via different designs of the dielectric coating.

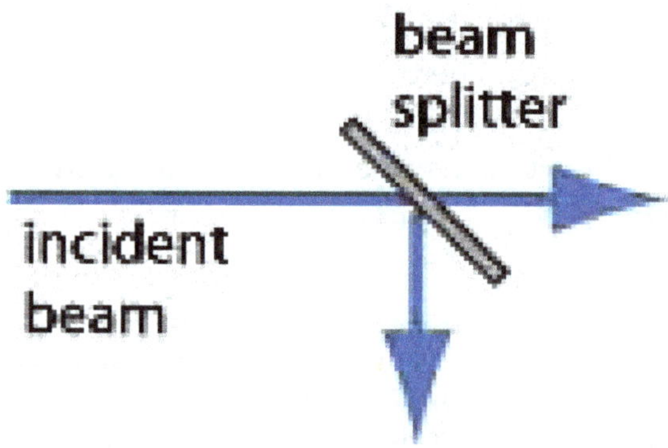

Fig. 7: A partially reflecting mirror, used as a beam splitter.

The transmitted beam always experiences an offset (spatial shift), the magnitude of which depends of the thickness and the refractive index of the substrate. This is a problem for some applications.

For infrared applications (e.g. infrared spectroscopy), the absorption of the substrate is often a limiting factor. One often uses beam splitters with calcium fluoride (CaF_2) substrates for wavelengths up to 8 μm. KBr-based beam splitters with a germanium-based coating can be used up to 25 μm wavelength, but that material is hygroscopic and must therefore be carefully protected against moisture. For the far infrared, polymer films are available.

In general, the reflectance of a dichroic mirror depends substantially on the polarization state of the beam. Such a device can even be optimized to function as a *thin-film polarizer*, where in some wavelength range a beam with a certain polarization can be nearly totally reflected, while a beam with different polarization is largely transmitted. On the other hand, it is also possible to optimize for a minimized polarization dependence to obtain a *non-polarizing beam splitter* within a limited wavelength range. This is most easily achieved for near normal incidence.

Dielectric beam splitters usually have a strongly wavelength-dependent reflectance. This can be used for *dichroic beam splitters* ('! *dichroic mirrors*), which can separate spectral components of a beam. For example, such a device may be used after a frequency doubler for separating the harmonic beam from residual pump light. The separation may occur based on the difference in wavelength or polarization.

A beam splitter as shown in Figure 1 will always lead to a transverse offset of the transmitted beam, which is proportional to the thickness of the used substrate. There are so-called *pellicle beam splitters* with a very thin substrate, minimizing that beam offset. Note, however, that parasitic reflections from the back side (which occur even if that side is anti-reflection coated) may lead to disturbing interferences, and therefore it is often better to use some larger thickness, so that the two reflections are spatially well separated.

Beam Splitter Cubes

Many beam splitters have the form of a cube, where the beam separation occurs at an interface within the cube. Such a cube is often made of two triangular glass prisms which are glued together with some transparent resin or cement. The thickness of that layer can be used to adjust the power splitting ratio for a given wavelength. One may also use some dielectric multilayer coating or a thin metal coating on one or both of the prisms to modify the optical properties, e.g. in terms of operation bandwidth or polarizing properties.

Beam splitter cubes can be used not only for simple light beams, but also for beams carrying images, e.g. in various types of cameras and projectors.

Generally, cube beam splitters cannot tolerate a high optical powers as plate beam splitters, although optically contacted cubes can also exhibits substantial power handling capabilities. Concerning durability and handling, cube beam splitters are often preferred over plates.

NON-POLARIZING BEAM SPLITTER CUBES

Non-polarizing beam splitter cubes can be made by refining the design, normally via a multilayer coating between the prisms. The substantial angle of incidence will naturally introduce a substantial polarization dependence, but there are certain design principles which can be used to minimize such effects at least within some limited optical bandwidth.

Note that "non-polarizing" usually does not imply that such a cube is polarization-preserving. For example, if an input beam is polarized at 45° against the axis, it can generally *not* be expected that the output beam is still linearly polarized, since the two polarization components will in general have different phase delays, apart from somewhat different amplitudes.

Polarizing Beam Splitter Cubes

Instead of glass, crystalline media can be used, which can be birefringent. This allows the construction of various types of polarizing beam splitter cubes (polarizers) such as *Wollaston prisms* and *Nomarski prisms*, where the two output beams emerge from the same face, and the angle between these beams is typically between 15° and 45°. Other types are the *Glan–Thompson prism*, and the *Nicol prism*, the latter having a rhombohedral form (i.e., not that of a cube).

Beam Splitters with Geometric Splitting

It is also possible to split beams geometrically (*aperture splitting*), e.g. by inserting a highly reflecting mirror only partially into a light beam, so that some part of the light can pass. One may also use other means, such as pattern of reflecting stripes or dots on a glass surface. A common design with dots is the *Polka dot plate beamsplitter*.

Ad advantage over dichroic beam splitters is the small wavelength dependence of the splitting ratio. The resulting modification of the intensity profile can be tolerated in some applications (but generally not for imaging).

Beam Splitters with Multiple Outputs

While most beam splitters have only two output ports, there are also beam splitters with multiple outputs. They may be realized, for example, based on diffractive optics. Another option is to use multiple cascaded beam splitters.

There are devices which produce some number of output beams of quite similar optical powers with a certain spatial pattern (e.g. all in one row, four at the edges of a square, etc.).

Fiber-optic Beam Splitters

Various types of fiber couplers can be used as fiber-optic beam splitters. Such a device can be made by fusion-combining fibers, and may have two or more output ports. As for bulk devices, the splitting ratio may or may not strongly depend on the wavelength and polarization of the input.

Fiber-optic splitters are required for fiber-optic interferometers, as used e.g. for optical coherence tomography. Splitters with many outputs are required for the distribution of data from a single source to many subscribers in a fiber-optic network, e.g. for cable-TV.

Other Types

Other types of beam splitters are:

- Metal-coated mirrors (e.g. half-silvered mirrors), where the metallic coating is made thin enough to obtain partial reflectance

- Pellicles, which are thin membranes, sometimes used in cameras
- Micro-optic beam splitters, often used for generating multiple output beams
- Waveguide beam splitters, used in photonic integrated circuits

Beam Splitters in Quantum Optics

In quantum optics, a beam splitter cannot be regarded as a device where the optical amplitudes at the outputs are simply given by constant factors times the input amplitude. This is essentially because there is always a second input port; even if that port remains unused, it must be considered as an input for vacuum fluctuations of the optical field. In a semi-classical picture, one can consider those vacuum fluctuations to influence the output beams, adding intensity noise and phase noise to the outputs. In a photon picture, one can see the amplitude noise in the form of *partition noise* – noise which results from the random "decisions" of the device to send an input photon to one output or the other one. This is also related to the fact that the shot noise level of the outputs, measured relative to the average powers ('! relative intensity noise), is increased. Similar effects occur for other types of linear attenuation of optical beams, e.g. by partial absorption.

Combining Beams

Any beam splitter may in principle also be used for combining beams to a single beam. This can be considered as operation with the reversed direction of time. However, the output power is then not necessarily the sum of input powers, and may strongly depend on details like tiny path length differences, since interference occurs. Such effects can of course not occur e.g. when the different beams have different wavelengths or polarization.

DIRECTIONAL COUPLER

A directional coupler is an electronic component having four-port circuits with one port being isolated from the input port and another being considered as a through port. The device is normally used to split the input signal and distributed power. The device couples part of the transmission power by a specific factor through one port. Directional couplers are used in a wide range of applications which involve measurement, power monitoring and other utilities.

Output Couplers

An *output coupler* is a semi-transparent dielectric mirror used in a laser resonator. Its function is to transmit part of the circulating intracavity optical power in order to generate a useful output of the laser.

As for basically all laser mirrors, the reflecting coating of an output coupler mirror is on the inner side, i.e., on the side of the mirror substrate which is oriented towards the laser resonator. This is because otherwise one would have a serious impact of losses by parasitic reflections on that inner side, which is exposed to the full circulating optical power.

Output Coupler Transmissivity

An important property of an output coupler is its transmissivity (the degree of its transmission of optical power). A low output coupler transmissivity leads to a low threshold pump power, but also possibly to a poor laser efficiency if the losses due to output coupling do not dominate over other (parasitic) losses in the laser resonator (*low output coupling efficiency*). The output coupler transmissivity is often chosen to maximize the output power, although its optimum value may be lower or higher if there are other design goals, e.g. minimizing the intracavity intensities or suppressing Q-switching instabilities in a passively mode-locked laser.

Flat and Curved Output Couplers

Output coupler mirrors are often mirrors with a flat surface, since it is often convenient to obtain a collimated beam at the output. (For a linear resonator, the wavefronts of the resonator modes must match the mirror surface, so that a flat output coupler results in a focus at that place.) Also, misalignment of any optical element other than the output coupler itself will then lead only to a parallel shift of the output beam, but not to a change in direction ('! *beam pointing fluctuations*). In the case of an output coupler with a curved (e.g. concave) surface, one also has to take into account that the mirror substrate will usually have a focusing or defocusing effect on the output beam.

Reflections from the Back Side

Note also that the back side of the glass substrate of an output coupler mirror has some reflectivity, which may have an effect on the laser performance. Even the residual reflectivity of an anti-reflection-coated back side may create an interference (etalon) effect (if the surfaces are flat and parallel), which modulates the effective transmission of the device as a function of wavelength and can thus influence the laser bandwidth. In a mode-locked laser, such effects (even at a very low level) can cause instabilities. A way to avoid such problems is to use a slightly wedged output coupler, so that the reflected light from the back side does not interfere with the laser modes.

Variable Reflectivity Mirrors

In the case of *stable* optical resonators, the output coupler mirror usually has a spatially constant output coupler transmission (see below). Output couplers of *unstable* resonators can be very different: they are often variable reflectivity mirrors, having a transversely variable reflectance e.g. with a Gaussian or super-Gaussian profile, possibly even with total reflection within some radius and total transmission outside that radius.

SWITCHES

Switches may also be fabricated by placing two parallel strip waveguides in close proximity to each other as illustrated in Figure 8. The evanescent fields generated outside the guiding region allow transverse coupling between the guides. When the two waveguide modes have equal propagation constants â with amplitudes *A* and *B* (Figure 8), then the coupled mode equations may be written as

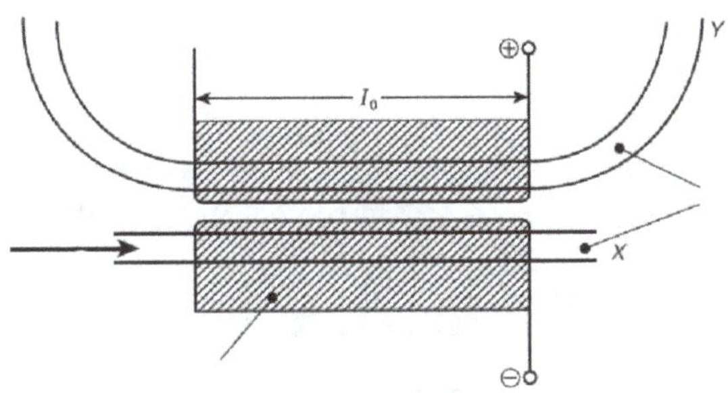

Figure 8 Electro-optically switched directional coupler. The COBRA configuration uses two electrodes.

where *C* is the coupling coefficient per unit length. In this case, assuming no losses, all the energy from waveguide *X* will be transferred to waveguide *Y* over a coupling length l_0.

Furthermore it can be shown that for this complete energy transfer *l*0 is given by ð/2*C*. If the waveguide modes have different propagation constants, however, only part of the energy from guide X will be coupled into guide Y, and this energy will be subsequently recoupled back into X.

It is also noted that when the propagation constants differ, the coupling length *l* is reduced from the matched value l_0 and although less energy is transferred, the exchange occurs more rapidly. This property may be utilized to good effect in the formation of an optical switch. The mismatch in propagation constants can be adjusted such that the coupling length *l* is reduced to $l_0/2$. In this case, energy coupled from one guide into the other over a distance $l_0/2$ will be recoupled into the original guide over a similar distance. Hence two distinct cases exist for a switch of length *l*0, namely the matched case whereby all the energy is transferred from one guide to the other and the mismatched case when $l = l_0 /2$ where over a distance *l*0 the energy is recoupled into the original guide.

Optical switches of the above type use electrodes placed on the top of each matched waveguide (Fig. 8) so that the refractive indices of the guides are differentially altered to produce the differing propagation constants for the mismatched case. A widely used switch utilizing this technique is called the COBRA (*Commutateur Optique*

Binaire Rapide) and is normally formed from titanium-diffused lithium niobate. Fabrication of the device, however, is critical in order to provide a coupling length which is exactly l_0 or an odd multiple of l_0. An electrode structure which avoids this problem by dividing the electrodes into halves with opposite polarities on each half is shown in Figure 9. With this device, which is called the stepped Δβ reversal coupler, it is always possible to obtain both the matched and mismatched cases described previously by applying suitable values of the reversed voltage. Hence the fabricated coupling length is no longer critical as the effective coupling length of the device may be adjusted electrically to achieve l_0.

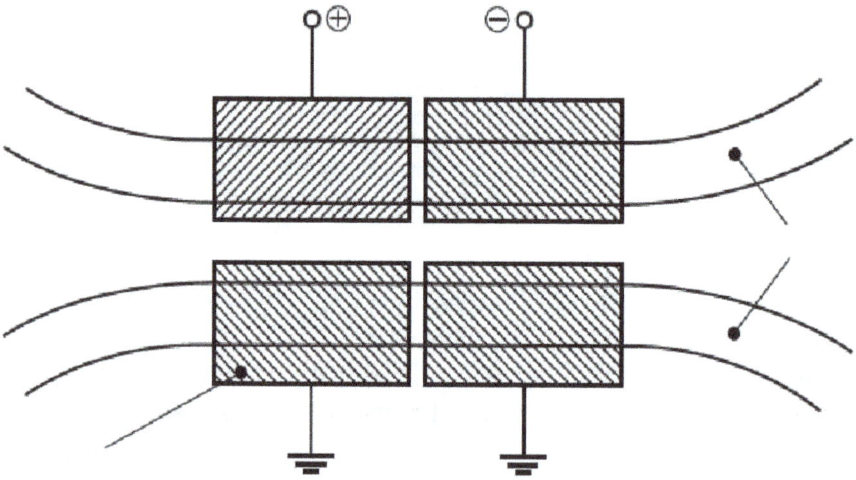

Fig. 9 The stepped Δβ reversal coupler switch.

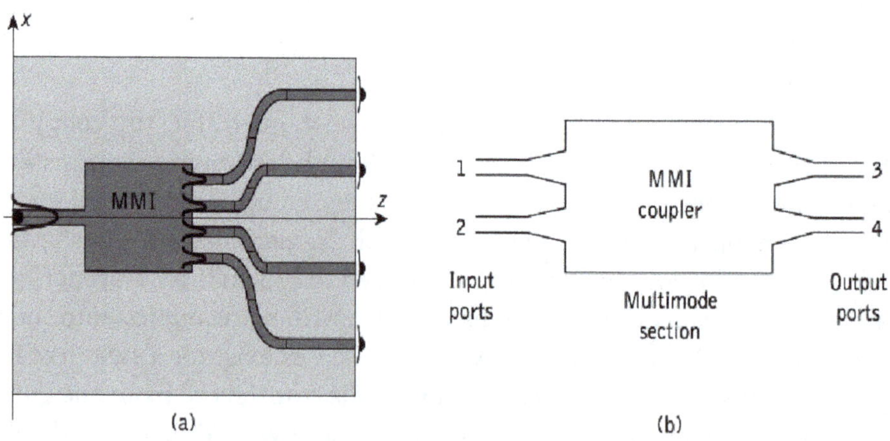

Fig. 10 Multimode interference (MMI) coupler: (a) concept of reproducing self-images to produce a coupler; (b) schematic diagram of a 2 × 2 planar device

A multimode interference (MMI) coupler similar to a fused fiber coupler can be produced using planar waveguides. Multimode interference filter devices are based on the principle of self-imaging operation as shown in Fig. 10(a) where an input field profile in a waveguide is reproduced at the output terminal and the device operates as a directional coupler. Single or multiple images (depending on the type of waveguide and number of input/output ports) at periodic intervals along the propagation direction of the waveguide, as a result of beating of different modes in the waveguide, can be obtained. In this case the signal is shown split into four equal signal levels appearing at each of the output ports. A simple 2 × 2 MMI coupler is illustrated in Fig. 10(b).

OPTICAL MODULATORS

An optical modulator is a device which can be used for manipulating a property of light – often of an optical beam, e.g. a laser beam. Depending on which property of light is controlled, modulators are called intensity modulators, phase modulators, polarization modulators, spatial light modulators, etc.

A wide range of optical modulators are used in very different application areas, such as in optical fiber communications, displays, for active Q switching or mode locking of lasers, and in optical metrology.

Types of Optical Modulators

There are very different kinds of optical modulators:

- *Acousto-optic modulators* are based on the acousto-optic effect. They are used for switching or continuously adjusting the amplitude of a laser beam, for shifting its optical frequency, or its spatial direction.
- *Electro-optic modulators* exploit the electro-optic effect in a Pockels cell. They can be used for modifying the polarization, phase or power of a beam, or for pulse picking in the context of ultrashort pulse amplifiers.
- *Plasmonic modulators* are a special type of electro-optic modulators which exploit the formation of plasmons (a special type of electromagnetic excitation) at metal surfaces, which lead to surface plasmon polaritons (SPPs). They can be extremely fast while having a low energy consumption.
- *Electroabsorption modulators* are semiconductor-based intensity modulators, used e.g. for data transmitters in optical fiber communications.
- *Interferometric modulators*, e.g. Mach–Zehnder modulators, are mostly exploiting the electro-optic effect in conjunction with interference. They are often realized in photonic integrated circuits for optical data transmission.
- *Liquid crystal modulators* are suitable for, e.g., optical displays and ultrafast pulse shapers. They are also available as spatial light modulators, i.e. with a spatially varying transmission, e.g. for displays.

- *Micromechanical modulators* (which are microelectromechanical systems = MEMS), e.g. silicon-based light valves and two-dimensional mirror arrays, are particularly useful for projection displays.
- *Chopper wheels* can periodically switch the optical power of a light beam, as required for certain optical measurements (e.g. those using a lock-in amplifier), and may thus be considered as optical modulators in a wider sense. Of course, that kind of device cannot provide arbitrary modulation controlled with an electrical input signal.

Bulk-optical modulators, e.g. of the electro-optic type, can be used with large beam areas, and handle correspondingly large optical powers. On the other hand, there are fiber-coupled modulators, often realized as a waveguide modulator with fiber pigtails, which can easily be integrated into fiber-optic systems.

A simple phase modulator may therefore be realized on a strip waveguide in which the ratio L/d is large as shown in Figure 10. These devices, when, for example, fabricated by diffusion of Nb into $LiTaO_3$, provide a phase change of π radians with an applied voltage in the range 5 to 10 V.

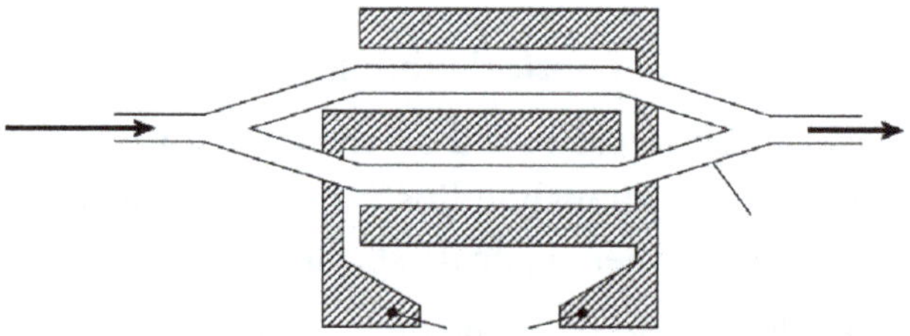

Fig. 11. A Y-junction interferometric modulator based on the Mach–Zehnder

INTERFEROMETER

However, because the electro-optic refractive index change is small this is rarely the case and its effect on the optical phase velocity is dependent on the overlap integral of the electrical and optical fields. The consequence of these nonuniform fields can be incorporated into an overlap integral α, having a value between 0 and 1 which gives a measure of the overlap between the electrical and optical fields.

Interference Filters

Interference filters are optical filters which are based on exploiting optical interference effects – in contrast to absorbing filters, which utilize wavelength-dependent absorption.

Most interference filters are dielectric multilayer mirrors, where the total reflected light arises from the interference of reflections at several or even many optical surfaces between different thin layers; the same holds for the transmitted light. See the article on dielectric mirrors for more details on fabrication and the operation principle.

There are also multilayer filters containing both dielectric and metal films. With that technology, one can work with fewer layers (e.g. just two metal films with a dielectric film in between), still obtaining substantial reflections, but the absorption losses are substantially higher than for purely dielectric mirrors.

Some interference filters exploit the principle of Fabry–Pérot interferometers. This is possible with monolithic dielectric or metal/dielectric mirror structures, but also with composite structures, involving two separate mirror devices, which are in most cases rigidly fixed to each other such that the spacing, which is essential for interference effect, is accurately maintained.

The function of an interference filter can be that of a bandpass filter, notch filter, high-pass or low-pass filter, for example, depending on the used design type. For example, Fabry–Pérot designs naturally lead to bandpass characteristics in transmission and notch filter characteristics in reflection. High- and low-pass filters can be designed as somewhat modified Bragg mirrors. The optimization of interference filter designs often involves numerical multilayer thin-film software.

General Properties

Some general properties of interference filters are briefly summarized in the following:

- Through their design, one may realize a wide range of properties. As a substantial number of layer pairs are usually used, there are many degrees of freedom.
- In contrast to absorbing filters, interference filters can usually be operated with relatively high optical intensities without being damaged.
- There can be a significant temperature dependence on the detailed optical properties, since the temperature affects the refractive indices.
- Because longer propagation lengths provide larger phase differences even for small differences in optical wavelength, tentatively thicker devices are required for improved wavelength resolution. However, resonance effects can substantially improve resolution without using a large thickness.
- The filter characteristics substantially depend on the angle of incidence of the input light, because that influences the optical phase differences between reflections from different interfaces. Generally, larger angles of incidence lead to smaller phase differences, although the opposite may be expected. As a result, spectral features tend to move to shorter wavelengths as the angle of incidence is increased.

Periodic optical filters used for WDM devices based on thin-film or arrayed-waveguide grating (AWG) result in the loss of channels because of the need for a guard band in between adjacent channels. For example, an experimental demonstration of an eight-band filter with each band comprising four channels and leaving a single channel between each band (i.e. 4-skip-1) allowed a total of 32 channels, each spaced by 100 GHz, spanning an overall optical bandwidth of 4 THz. Clearly it is important to use the full bandwidth potential which can be achieved by employing periodic waveband filters, where instead of demultiplexing single-wavelength channels specified numbers of wavelength channels can be grouped together into bands thereby minimizing bandwidth usage. Such devices, known as interleaver waveband filters, can then be used to separate the required wavelength signals.

Fig. 12 illustrates the different functions of an interleaver waveband filter performing as a wavelength division demultiplexer. A demultiplexing scheme where even and odd channels are separated onto the two different output ports identified is shown in Figure 12(a). Further separation of specific wavelength channels can be obtained as indicated in Fig. 12(b) where a multiplexed input signal is demultiplexed into four wavelength channels. It is also possible to achieve a bandpass filter operation while isolating the required number of channels in each band. This process is illustrated in Fig. 12(c) where the banded interleaver splits into an even and odd band of channels with each band containing four discrete channels. In addition, a demultiplexing filter can be used to isolate a single required channel from a WDM comb. This situation is represented in Fig. 12(d) where an asymmetric interleaver is shown to separate one channel from a WDM comb signal containing eight wavelength multiplexed signals. For example, a silica-based planar circuit utilizing a bandpass interleaver with a five-band 8-skip-0 waveband filter has demonstrated flexible waveband optical networking performance for 40 channels without guard bands at a transmission rate of 40 Gbit s^{-1}

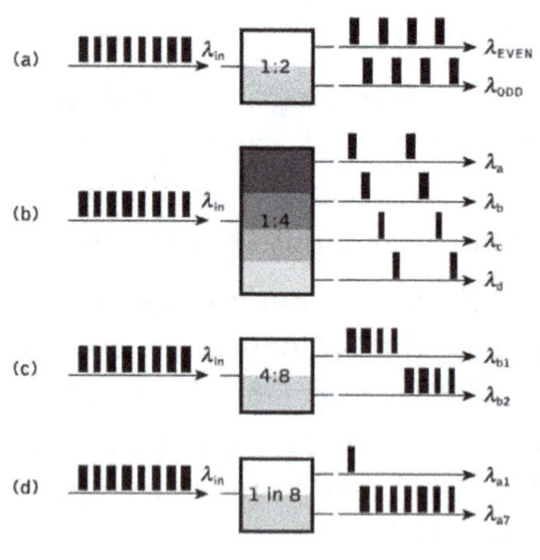

Fig. 12 Interleaver waveband filter functions: (a) with even and odd channels being separated onto two different ports; (b) separation of channels out to 1:4 or higher; (c) banded Interleaver which

separates even and odd bands of channels with a 4:8 combination; (d) asymmetric interleaver separating one channel in eight.

INTEGRATED INJECTION LASER SYSTEM

The integrated ILS must maintain its reliability under a variety of pulse shapes, energies, durations, and modulation bandwidth conditions. In order to assure high reliability without damage to the laser system, a set of checks and balances has been employed. The MOR is constantly monitoring the pulse shape and energy. The monitoring occurs with approximately 100 pulse averages occurring at the 960 Hz MOR pulse repetition frequency. Thus the response to an incorrect pulse shape is not instantaneous, but rather on the order of 100 ms. Slow degradation of any MOR pulse shape would not damage the other optics in the system due to additional checks described below.

The only realistic way that damage could occur in the Regen would be during pulse shaping. The controlling software is used to verify that all Regen shutters are closed during pulse shaping operations. Once pulse shaping is complete, the active control of the FET's is stopped.

The amplitude is adjusted with a separate control to the DC bias on the amplitude modulator. Once the Regen shutter is open, the output energy is actively monitored after the variable attenuator is used to set the NIF shot energy. If that energy is more than 2% outside of the required energy for the shot, the shot is aborted. This, combined with the MOR pulse shape monitoring, makes the system very reliable for rod shots (PAM MPA active, Main Amplifier dormant). After each rod shot, the energy, the pulse shape, and the near field profile are monitored at the input sensor package (ISP) and compared to the requested configuration by the Laser Performance Optical Model (LPOM).

All critical devices are reserved (cannot be moved) before each rod shot. If the rod shot pulse shape at the ISP meets all laser performance criteria , the software will proceed on to the system shot (PAM MPA and Main Amplifier both active). With all critical devices reserved, the MOR pulse shape checks, and the Regen energy monitoring we guarantee with high confidence that the appropriate spatial pulse shape, temporal profile, and energy are obtained on the system shot.

The electroluminescent properties of the forward-biased p–n junction diode have been considered in the preceding sections. Stimulated emission by the recombination of the injected carriers is encouraged in the semiconductor injection laser (also called the injection laser diode (ILD) or simply the injection laser) by the provision of an optical cavity in the crystal structure in order to provide the feedback of photons. This gives the injection laser several major advantages over other semiconductor sources (e.g. LEDs) that may be used for optical communications. These are as follows:

- High radiance due to the amplifying effect of stimulated emission. Injection lasers will generally supply milliwatts of optical output power.
- Narrow linewidth on the order of 1 nm (10 Å) or less which is useful in minimizing the effects of material dispersion.
- Modulation capabilities which at present extend up into the gigahertz range and will undoubtedly be improved upon.
- Relative temporal coherence which is considered essential to allow heterodyne (coherent) detection in high-capacity systems, but at present is primarily of use in single-mode systems
- Good spatial coherence which allows the output to be focused by a lens into a spot which has a greater intensity than the dispersed unfocused emission.

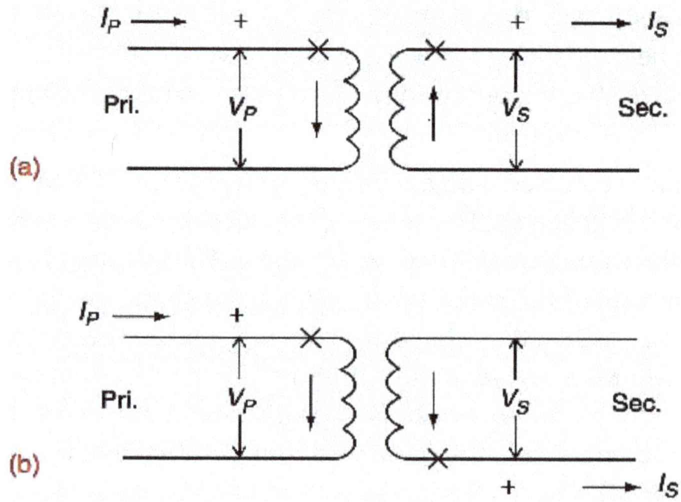

Fig. 13. Polarity definitions for transformers: (a) Subtractive polarity (b) Additive polarity

This permits efficient coupling of the optical output power into the fiber even for fibers with low numerical aperture. The spatial fold matching to the optical fiber which may be obtained with the laser source is not possible with an incoherent emitter and, consequently, coupling efficiencies are much reduced These advantages, together with the compatibility of the injection laser with optical fibers (e.g. size), led to the early developments of the device in the 1960s. Early injection lasers had the form of a Fabry–Pérot cavity often fabricated in gallium arsenide which was the major III–V compound semiconductor with electroluminescent properties at the appropriate wavelength for first-generation systems. However, as mentioned previously these devices had a high threshold current density (greater than 10^4 A cm^{-2}) due to their lack of carrier containment and proved inefficient light sources.

High current densities required dictated that these devices when operated at 300 K were largely utilized in a pulsed mode in order to minimize the junction temperature and thus avert damage. Improved carrier containment and thus lower threshold current densities (around 10^3 A cm^{-2}) were achieved using heterojunction structures.

The DH injection laser fabricated from lattice-matched III–V alloys provided both carrier and optical confinement on both sides of the p–n junction, giving the injection laser a greatly enhanced performance. This enabled these devices with the appropriate heat sinking to be operated in a CW mode at 300 K with obvious advantages for optical communications (e.g. analog transmission). However, in order to provide reliable CW operation of the DH injection laser it was necessary to provide further carrier and optical confinement which led to the introduction of stripe geometry DH laser configurations. Prior to discussion of this structure, however, it is useful to consider the efficiency of the semiconductor injection laser as an optical source.

POLARIZATION TRANSFORMERS AND WAVELENGTH CONVERTERS

Polarity is very important for the operation of transformers and protection equipment. A clear understanding of polarity is useful in understanding and analyzing transformer connections and operations as well as testing protection relays and systems.

UNDERSTANDING POLARITY

Polarity is very important for the operation of transformers and protection equipment. A clear understanding of polarity is useful in understanding and analyzing transformer connections and operations as well as testing protection relays and systems.

It's also essential in understanding power system performance during both normal and abnormal operations.

Transformer polarity

The polarity indications for transformers are well established by standards that apply to all types of transformers. There are two varieties of polarity: subtractive and additive. Both follow the same rules.

Power and instrument transformers are subtractive, whereas some distribution transformers are additive. The polarity marking can be a dot, a square, or an X, or it can be indicated by the standardized transformer terminal markings, the practices varying over the years.

The two fundamental rules of transformer polarity, illustrated in Fig. 13 applying to both varieties are the following:

- Current flowing in at the polarity mark of one winding flows out of the polarity mark of the other winding. Both currents are substantially in-phase.
- The voltage drop from polarity to non-polarity across one winding is essentially in phase with the voltage drop from polarity to non-polarity across the other winding(s).

The currents through and the voltages across the transformers are substantially in-phase because the magnetizing current and the impedance drop through the transformers are very small and can be considered negligible. This is normal and practical for these definitions.

The current transformer (CT) polarity markings are shown in Figure 14.

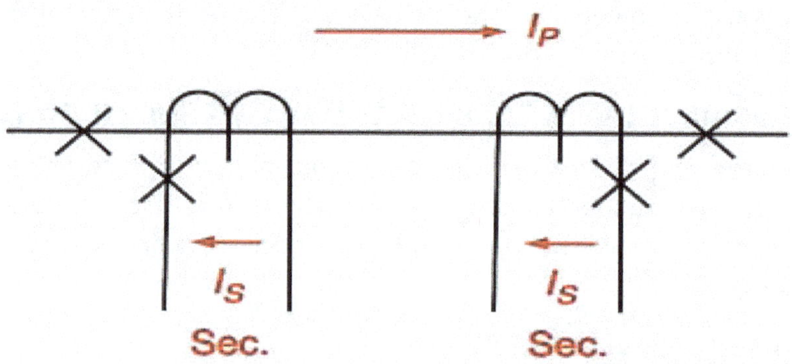

Fig. 14. Polarity markings for CTs

Note that the direction of the secondary current is the same, independent of whether the polarity marks are together on one side or on the other.

For CTs associated with circuit breakers and transformer banks, it is a common practice for the polarity marks to be located on the side away from the associated equipment.

The voltage-drop rule is often omitted in the definition of transformer polarity, but it is an extremely useful tool to check the phase relations through wye-delta transformer banks, or in connecting up a transformer bank for a specific phase shift required by the power system.

The ANSI/IEEE standard for transformers states that the high voltage should lead the low voltage by 30° with wye-delta or delta-wye banks. Thus, different connections are required if the high side is wye than if the high side is delta.

The connections for these two cases are shown in Figures 15 and 16. The diagrams below the three-phase transformer connection illustrate the use of the voltage-drop rule to provide or check the connections.

Arrows on these voltage drops have been omitted (preferably not used), for they are not necessary and can cause confusion.

In Fig. 15, the check is made by noting that a to n from polarity to non-polarity on the left-side winding is in phase with A to B from polarity to non-polarity on the right-side winding.

Accordingly, the wye side would be the high-voltage side if this is an ANSI/IEEE standard transformer.

This same technique of applying voltage drops to Fig. 16 shows that for this three-phase bank connection the voltage-drop polarity to non-polarity or phase a to n is in phase with the voltage-drop polarity to non-polarity or phase A to phase C.

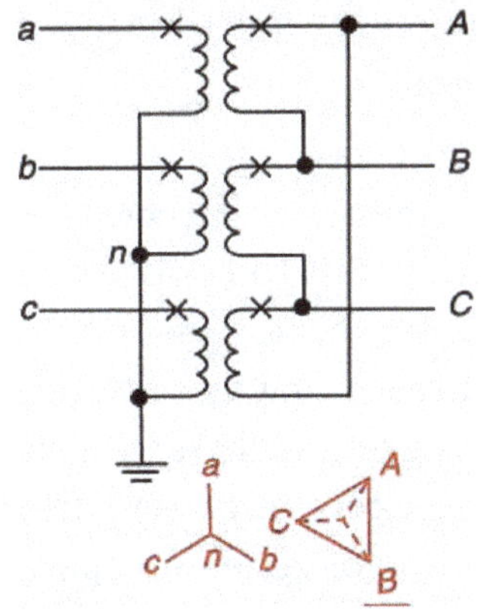

Fig. 15. Voltage-drop polarity rule useful in checking or connecting wye-delta transformer banks: Wye-connected side leads, delta-connected side 30°.

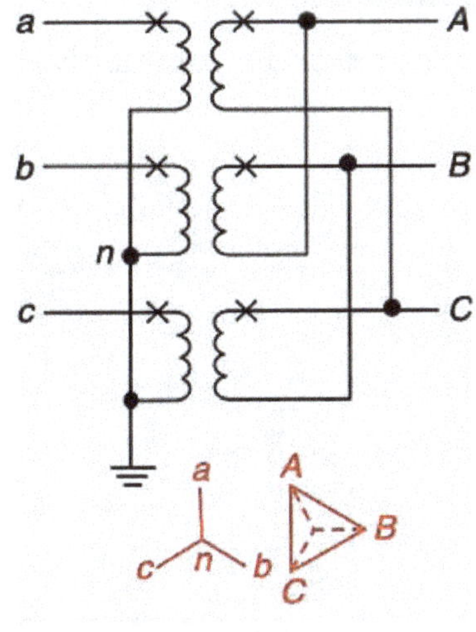

Fig. 16. Voltage-drop polarity rule useful in checking or connecting wye-delta transformer banks: Delta-connected side leads, wye-connected side 30°.

Similarly, voltage-drop across phase b to n is in phase with voltage-drop phase B to phase A, and voltage-drop phase c to n is in phase with voltage-drop across phase C to phase B.

By comparing similar voltages on the two sides of the transformer, phase-A-to-neutral voltage drop leads the phase-a-to-n voltage drop by 30°, so the delta winding would be the high-voltage side if this is an ANSI/IEEE standard transformer bank.

Relay polarity

Relays involving interaction between two input quantities from the power system may have the polarity marking that is necessary for their correct operation.

There are no standards in this area, so if the polarity of the relay connections is important, the relay manufacturer must both specify the polarity markings and clearly document their meaning.

Directional units are usually not applied alone, but rather, in combinations with other units, such as fault sensors or detectors. A common practice is to use the output of the directional-sensing unit to control the operation of the fault sensors, which often is an instantaneous or an inverse-time–overcurrent unit, or both units together. Thus, if the current flow is in the desired operating direction (trip direction) and its magnitude is greater than the fault sensor's minimum-operating current (pickup), the relay can operate. If the current is in the opposite direction (non-trip or non-operate direction or zone), no operation can occur even though the magnitude of the current is higher than the pickup threshold current.

A directional-sensing unit requires a reference quantity that is reasonably constant, against which the current in the protected circuit can be compared.

Typical polarity indications for three commonly used directional-sensing units are shown in Fig. 17.

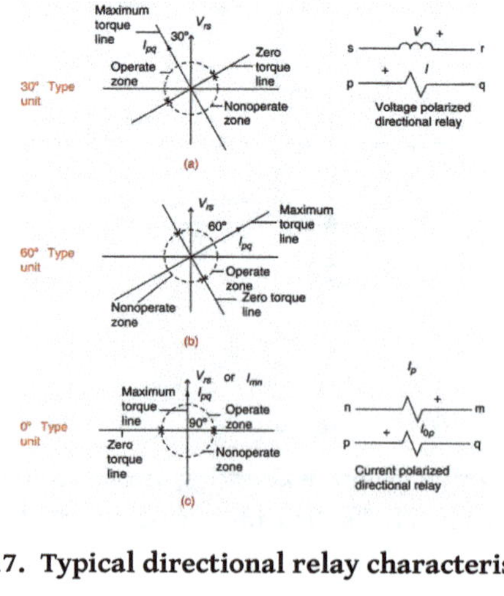

Fig. 17. Typical directional relay characteristics

This uses the custom of showing several loops for voltage coils and a single loop for current coils, of placing the reference circuit or voltage circuit above the current circuit, and of placing the polarity markings diagonally, all as shown on the relay schematics in Fig 17.

The reference quantity is commonly called the "polarizing" quantity, especially for ground-fault relaying, where either current polarizing or voltage polarizing is used, or both.

The polarity marks (Fig 17) are small plus symbols (+) placed, as illustrated, above one end of each coil, diagonally as shown, or on the opposite diagonal.

As shown in Fig. 14 above, relay operation is not affected whether the polarity marks are on one diagonal or the other. The meaning of the polarity for a specific relay must be stated clearly in words or by a diagram, such as the one shown in Fig. 17. These show the basic design characteristics of an individual relay, independent of any connection or association with the power system.

The interpretation of relay polarity is illustrated in Fig. 17 for three typical electromechanical units.

Solid-state units can have adjustments for (1) the maximum-torque angle and (2) the angle limits of the operate zone, but the application and operation is the same for both types.

In Fig. 17a, the maximum-operating torque or energy occurs when the current flows from polarity to non-polarity (I_{pq}) and leads by 30° the voltage drop from polarity to non-polarity (V_{rs}). The minimum pickup of the directional unit is specified as the maximum-torque or operating condition

As seen, the unit will operate for currents from almost 60° lagging the reference voltage V_{rs} to almost 120° leading. The operate (trip, contact close) zone or area is represented by the half plane, bordered on one side by the zero-torque (non-operating) line and extending in the direction that contains both the reference (polarizing) and operating quantities.

The operating torque at any angle is a function of the cosine of the angle between the current (I_{pq}) and the maximum-torque line, as well as the magnitudes of the operating quantities. For ground-fault protection, the 60° unit in Fig. 17b is used with a 3 V_0 reference and the zero (watt) unit of Fig. 17c with a 3 I_0 current reference. The unit in Fig. 17c is also used for power or var applications.

A similar type of electromechanical directional unit, as in Fig. 17a, has its maximum-torque angle at 45° leading, instead of 30° leading. Both units are in wide use for phase-fault protection.

Solid-state units with an adjustable angle feature can provide a range of angles.

The electro-optic effect typically in lithium niobate waveguide devices can be used to facilitate TE–TM mode conversion. However, to allow the transformation of an arbitrary input polarization, not just TE or TM, it is necessary to control the relative phase between the TE and TM components. Such polarization transformers which operate as TE–TM mode converters can be employed as elements within intensity modulators (when combined with a polarizer), optical filters or polarization controllers. A basic example of the latter device is shown in Fig. 18. It comprises two phase modulators and a single TE–TM mode converter on X-cut lithium niobate.

The first phase modulator is required to adjust the phase difference between the incoming TE and TM modes to be $\pi/2$ so that the polarization controller can operate with all incoming polarization states. When this condition is satisfied the central phase-matched mode converter is operated as a linear polarization rotator. Although a linear output polarization of either TE or TM is sufficient in some applications, for full polarization control a second phase shifter is required to adjust the output phase to a desired value of elliptical output polarization.

A number of electro-optic waveguide devices can be used to provide wavelength conversion of an optical signal. A common technique is to employ a phase modulator in a serrodyne configuration to alter the optical frequency by a linearly increasing voltage applied to the device electrodes. In practice a continuously increasing ramp signal voltage cannot readily be produced and hence a sawtooth voltage waveform is used.

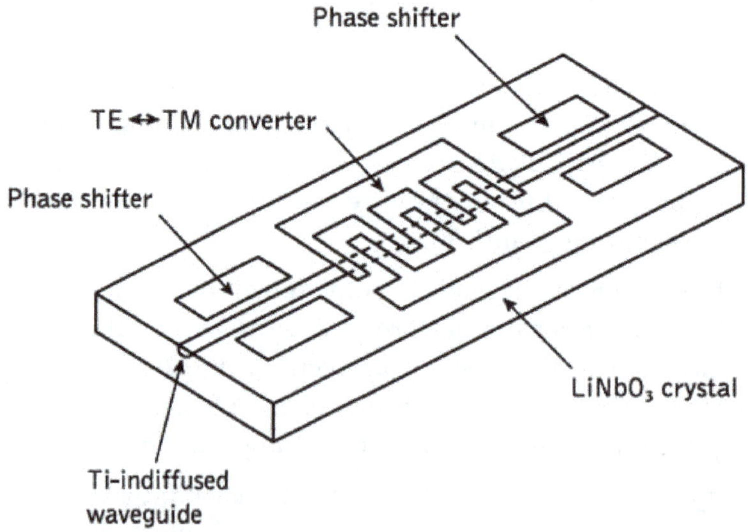

Fig. 18 An IO polarization controller

However, sawtooth waveforms with instantaneous fall times can be generated and hence additional frequency components tend to be produced. This factor, combined with the need to vary the rate of change of the applied voltage to alter the extent of the frequency shift, limits the use of this device to applications where a small constant frequency shift is required. In applications where large frequency translations are necessary and where the device is used as a control element in a feedback loop (e.g. coherent optical receivers), alternative frequency translators have been utilized.

Devices based on TE–TM mode conversion are also capable of generating frequencytranslated or wavelength-converted optical signals. When the region where the mode conversion takes place is made to move relative to the direction of the optical wave, the source of the converted signal appears to be moving to a stationary observer and the light is therefore Doppler shifted.

To generate the effect of a moving coupling grating in practice, a mode converter is divided into several sections and each is driven with a correctly phase-shifted sinusoidal signal which has a frequency equal to the desired up or down frequency translation. In principle, this technique should be highly efficient and generate no unwanted optical signals. However, significant unwanted sidebands have been observed with such devices which appear to arise from parasitic electrical fields.

Careful device design is therefore necessary to maintain these signals at an acceptable level.

Mach–Zehnder interferometric Y-junction modulators can also be used to generate double-sideband wavelength-converted optical signals when they are modulated with an intensity-modulated waveform. In this case the optical wavelength shift is proportional to the wavelength of the modulating signal. However, in simple device structures the charging of the electrode capacitance limits the maximum modulation frequency and thus the magnitude of the frequency translation that can be obtained. To overcome this problem Mach–Zehnder interferometers (MZIs) with traveling-wave electrode structures have been designed which provide multigigahertz bandwidths.

When a nonlinear medium (e.g. a semiconductor optical amplifier) is inserted into the arms of an MZI, however, the absorption characteristics of the nonlinear medium determine the refractive index variation with respect to change in the intensity of an input signal which produces phase variation in the output signal emerging from the output port of the MZI. This intensity– phase variation produces changes in the wavelength of an optical signal and therefore by controlling the optical signal power in each MZI arm, it is possible to produce any desired wavelength at the output of the wavelength converter. Integration of the aforementioned electro-optic devices into a single lithium niobate substrate, for use in coherent optical fiber communication systems, is an area of interest to reduce both losses between individual devices as well as system cost.

For example, the configuration of a potential coherent optical receiver device is illustrated in Fig. 19.

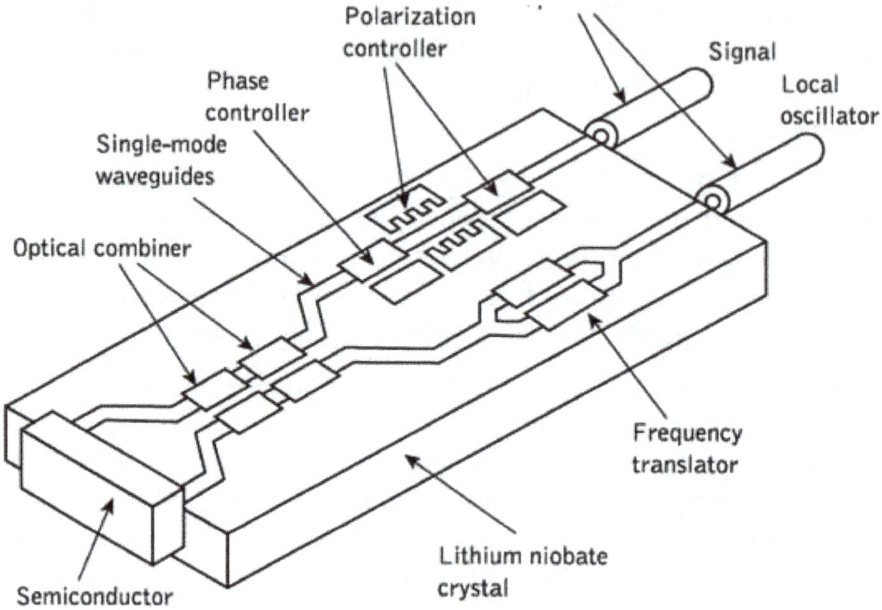

Fig. 19 Coherent optical receiver device

It was fabricated on Z-cut lithium niobate and comprises a polarization controller with output phase controller, and a frequency translator together with a directional coupler for mixing the two optical signals. Successful operation of this integrated device was demonstrated and a similar X-cut lithium niobate device requiring a lower operating voltage has been proposed. More recently, a coherent integrated optical receiver based on an optical phased locked loop implemented using InP technology has been demonstrated experimentally, which comprised a pair of balanced uni-traveling carrier photodetectors, a 2 × 2 multimode interference coupler and phase modulators located on a 1 × 2 mm substrate.

CHAPTER

11 OPTICAL FIBER SYSTEMS 1: INTENSITY MODULATION/DIRECT DETECTION

INTRODUCTION

Optical communication is an important part of modern communication techniques due to the excessive bandwidth of the light spectrum. Theoretically, optical communication has much higher system throughput than its radio frequency (RF) communication counterpart. Therefore, it finds many applications and facilitates our lives.

Some typical optical communication scenarios include optical fiber communication, free-space optical communication, and visible light communication. In those communication scenarios, intensity modulation and direct detection (IM/DD) is a cost-effective communication scheme compared to coherent ones. In IM/DD, the intensity, or power, of the light beam from a laser or a light-emitting diode (LED) is modulated by the information bits and no phase information is needed. Due to this nature, no local oscillator is required for IM/DD communication, which greatly eases the cost of the hardware.

In IM/DD channel, there are still some non-ideal factors that may deteriorate the quality of communication. One key factor is the multipath effect. This effect is caused by several mechanisms. First, in wireless communications, the light could be reflected at multiple locations and by many times by the surroundings before arriving at the receiver side. Second, the modulation bandwidth of LED is limited, typically below 100 MHz. When the bandwidth of signal exceeds the modulation bandwidth of LED, multipath effect occurs.

Third, in fiber communication, light components of different wavelength propagate through different paths, which also cause multipath effect. Therefore, effective means of mitigating the multipath effect are necessary in IM/DD optical communications.

In RF communication, orthogonal frequency division multiplexing (OFDM) is a powerful multi-carrier modulation scheme to combat the multipath effect. Compared to the single-carrier modulation schemes, OFDM avoids the usage of a complicated high-order time-domain equalizer. Instead, it employs frequency domain equalizer that only has a single tap. This greatly simplifies the equalization task and can perfectly resolves the multipath effects without any residual errors at high signal-to-noise ratio (SNR) region.

Thus, introducing OFDM to IM/DD optical communication is a natural choice. However, different from RF communication, IM/DD requires that the transmitted signal must be real and positive, which imposes strict constraint on the modulation scheme and the original OFDM transceiver must be modified carefully to satisfy the new scenario. In addition, different applications may have diverse emphasis such as spectral efficiency, power efficiency, detection capability, as well as computational complexity.

Within such perspectives, the purpose of this chapter was to analyze the potential forms of OFDM that are suitable for IM/DD transmission as well as various receiver designs in optical communication. We first study the concepts and basic modulation schemes of OFDM systems in IM/DD optical communication. They can be generally classified into three categories: direct-current-biased optical OFDM (DCO-OFDM), non-DC-biased optical OFDM, and hybrid optical OFDM. We will elaborate the system models and explain the validity of some fancy designs in those systems through analysis. Second, we investigate the preliminary receivers of those OFDM systems.

Besides, we will propose a new receiver that is capable of improving the detection performance based on the inherent signal structures of the specific transmitted signal. Third, the spectral efficiencies and computational complexities of different systems and receivers are analyzed and compared. Finally, the bit error rate (BER) performance of different systems is compared through computer simulations to give the reader a whole picture of different candidate OFDM systems in IM/DD optical communication.

INTENSITY MODULATORS

There are various kinds of optical modulators with which one can modulate the intensity – or more precisely the optical power – of light. In many case, the input light is delivered in the form of a free-space laser beam, in other cases, through a waveguide, e.g. an optical fiber. Most modulators are controlled with an input voltage varying in a certain range.

Intensity modulators are required for many purposes, e.g. in optical data transmission, in laser printers, in laser material processing or for the accurate stabilization of the optical power of a beam.

Intensity modulators can be realized based on a number of different physical principles. The most important types are described in the following sections. They differ substantially in various characteristics which can be important for applications:

- The input light is usually either a free-space beam or propagates in an optical fiber (fiber-coupled modulators).
- Some intensity modulators, e.g. electro-optic ones, can have a very high modulation bandwidth, i.e., they can be used for very fast modulations as required e.g. in optical communications.
- Some modulators can allow for transmission nearly in the full range of 0 to 100%, while others exhibit high transmission losses or do not allow for strong attenuation. Specifications may include an extinction ratio or a dynamic range in decibels. A high extinction ratio is important, for example, when producing pulses from a continuous-wave input.
- Some intensity modulators work only with polarized beams, while some others work in a polarization-independent way.
- Different types need very different amounts of drive voltage. Particularly high drive voltages are often required by electro-optic modulators, see the article on Pockels cell drivers.
- Some modulators simply absorb the non-transmitted part of the optical radiation (convert that energy to heat), while others release that power in the form of a second output beam.
- Depending on the type of the modulator, the permissible input optical power can be quite different. To high input powers may lead to damage of the modulator or only to temporarily degraded performance.
- Other aspects of practical importance may be the compactness, the preservation of beam quality, and the sensitivity to changes of optical wavelength and temperature.

Acousto-optic Modulators

Acousto-optic modulators (AOMs) naturally modulate the optical power both of the diffracted beam and the transmitted non-diffracted beam. With zero acoustical signal applied, the full incident power is transmitted in the non-diffracted beam, apart from some small amount of parasitic losses. If one applies some acoustical power, the power of the reflected beam is approximately proportional to the squared sign of a coefficient which itself is proportional to the square root of the acoustical power. Usually, the maximum diffraction efficiency stays well below 100% – often it is of the order of 50%.

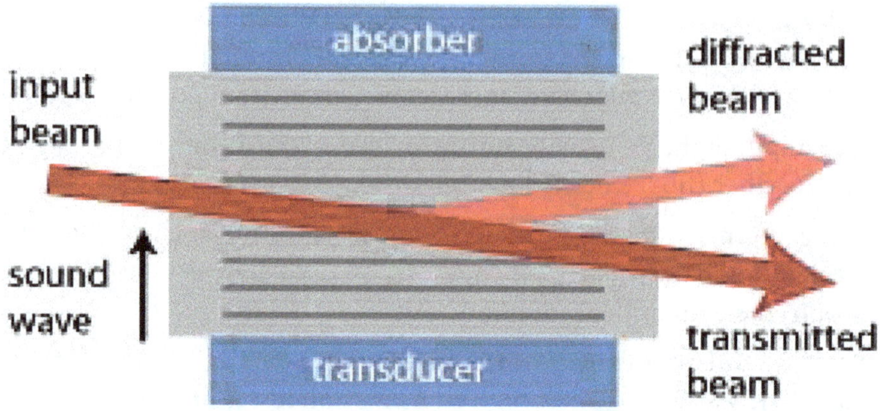

Fig. 1: Schematic setup of a non-resonant acousto-optic modulator. A transducer generates a sound wave, at which a light beam is partially diffracted.

If it is important to reach zero output power, one usually has to use the diffracted beam, which however limits the maximum transmission. Nearly complete transmission is possible for the non-diffracted beam, but in that case the transmission can be reduced only to e.g. 50%. In both cases, the possible transmission bandwidth is limited by the propagation time of the sound wave in the crystal. Therefore, small-aperture devices are tentatively suitable for higher modulation bandwidths.

Acousto-optic intensity modulators are used for various applications, e.g. for laser printers, for noise eaters, and for the active mode locking of lasers.

Typically, acousto-optical intensity modulators are designed for a certain polarization direction of the input beam, but there are also largely polarization-independent devices.

Electro-optic Intensity Modulators

And electro-optic modulator is based on a Pockels cell, which acts as a phase modulator. The phase modulation is then typically transformed into an intensity modulation by sending the light through a polarizer. Typically, the output power is proportional to $\cos^2 (C\ U)$ or $\sin^2 (C\ U)$ where C is a constant and U is the applied voltage. The constant C is related to the half-wave voltage $U_{À}$ of the Pockels cell: $C = À / (2\ U_{À})$. (See the article on Pockels cells for more details.) Unfortunately, the half-wave voltage is often quite high – hundreds of volts or even more, particularly for devices with large aperture. This increases the demands on the driver electronics if the full transmission range (0 – 100%) is required.

Unfortunately, the half-wave voltage is often quite high – hundreds of volts or even more, particularly for devices with large aperture. This increases the demands on the driver electronics if the full transmission range (0 – 100%) is required.

The modulation bandwidth can be very high, if an optimized electronic driver is used, which also needs to cope with the capacitance of the Pockels cell.

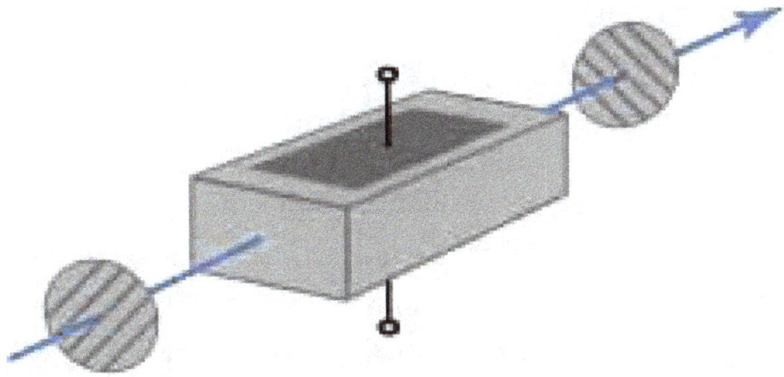

Fig. 2: Electro-optic intensity modulator, containing a Pockels cell between two polarizers.

Relatively small drive voltages are possible for fiber-coupled modulators, where the optical radiation propagates in a well-defined region of the crystal and thus allows the use of a Pockels cell with rather small aperture.

Particularly for waveguide devices, it is also possible to transform the phase modulation intern intensity modulation using an interference effect instead of a polarizer. For example, there are Mach–Zehnder modulators where a phase modulation is applied to one arm of a Mach–Zehnder interferometer.

Often, electro-optic intensity modulators are employed when the modulation needs to be rather fast. Most of them work only for a certain polarization direction.

Electroabsorption Modulators

Electroabsorption modulators are semiconductor devices working based on the Franz–Keldysh effect. They are usually operating on light in a waveguide and are coupled to optical fibers or placed on a chip together with other components, e.g. together with a laser diode to form a telecom transmitter. They are limited in terms of optical powers but can have a very high modulation bandwidth. Compared with electro-optic modulators, they require much smaller drive voltages.

MODULATORS BASED ON SEMICONDUCTOR OPTICAL AMPLIFIERS

It is also possible to use a semiconductor optical amplifier as an intensity modulator. Without drive current, it provides some degree of attenuation (negative gain), while some positive gain is achieved when the device is supplied with pump current. The modulation speed can be fairly high, with sub-nanosecond rise and fall times for the generation of rectangular pulses, for example. It can of course also be useful to have amplification at the same time.

In contrast to other optical modulators, such devices exhibit gain saturation and introduce amplifier noise, but this is not relevant for all applications. For example, a SOA-based modulator may be used to produce short seat pulses for fiber amplifier systems. The achievable extinction ratio can be more than 50 dB, i.e., far better than that of an acousto-optic or electro-optic modulator, for example.

Liquid Crystal Modulators

It is possible to modulate the polarization of light by applying a voltage to a liquid crystal material, and to obtain intensity modulation by adding a polarizer. Such liquid-crystal modulators can work with large apertures and low drive voltages, but they are seriously limited in terms of modulation bandwidth.

Devices can be made where different parts of the beam area are modulated independently (spatial light modulators).

OPTICAL OFDM SYSTEMS FOR IM/DD CHANNEL

Preliminaries

In IM/DD channel, there is a key difference with RF channel: the transmitted signal must be real and positive. This results from the fact that the intensity of light must be a real and positive quantity. Therefore, the structure for OFDM shown in Figure 1 cannot be directly used in IM/DD optical channel. Necessary changes must be made instead. A common approach is to generate a real time-domain signal first. This can be realized by imposing *Hermitian symmetry* on the frequency domain signal X, which is defined as follows:

$$X(N - k) = X^*(k),\ k = 1, \ldots, N - 1,\ X(0) = X(N/2) = 0. \quad (1)$$

It can be easily shown that the IFFT of X having property Eq. (1) is a pure *real*-valued signal x. Based on this real signal, one can further generate a positive signal to drive the optical source by various means. Those resultant new OFDM systems are typically referred to as *optical OFDM* systems.

DC-biased optical OFDM

The most straightforward approach to generate a positive signal from a real signal is to impose a proper DC bias. In optical communication, the DC bias is typically chosen such that the mean value of the positive signal just lies on the center point of the linear range of the optical source. This system is called direct current-biased optical OFDM, or DCO-OFDM, whose transmitter is shown in Fig. 3.

The clipping module shown in Fig. 3 is necessary. Since x(n)x(n) is Gaussian distributed, it is possible that x(n)x(n) plus a DC bias is still out of the linear range of the optical source. For example, if an LED accepts driving current within the range of [*a*, *b*], where 0d"a<b0d"a<b, then the clipping is needed to confine x(n)x(n) plus a DC bias into this range. Otherwise, the LED could not be illumined due to under-driving or even be damaged due to over-driving.

Non-DC-biased optical OFDM

Besides DCO-OFDM, there are many forms of optical OFDM systems that are not relying on DC bias. The most famous ones are introduced in this subsection.

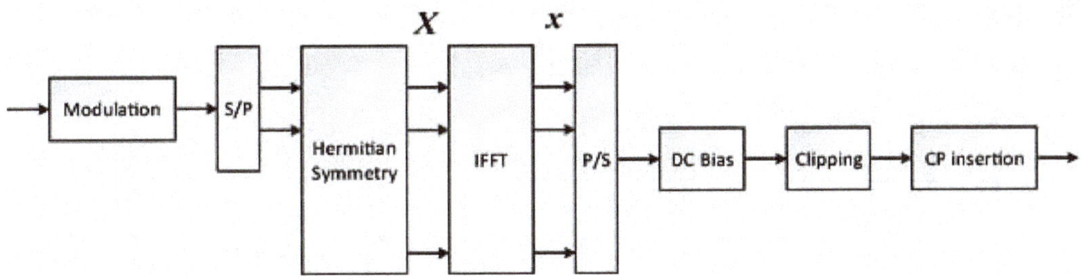

Fig. 3. Diagram of DCO-OFDM transmitter.

ACO-OFDM

Asymmetrically clipped optical OFDM (ACO-OFDM) is the most famous non-DC-biased optical OFDM system and has been extensively studied in literature. The basic idea of ACO-OFDM is to generate an asymmetrically structured time-domain signal such that direct clipping at zero (without adding DC bias) is allowed without any information loss. To do so, the frequency-domain input symbol X has a special structure besides satisfying Hermitian symmetry. Specifically, the *odd* components of X contain useful information U but the *even* components of X are set to zeros. This is shown in Fig. 4.

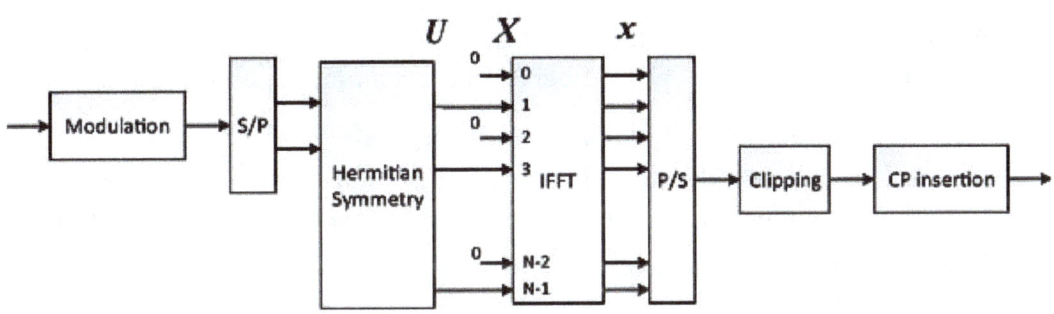

Fig. 4. Diagram of ACO-OFDM transmitter.

OPTICAL TRANSMITTER CIRCUIT

Introduction

Optical transmitter generates the optical signal. Transmitter imposes the information-bearing signal and transmits the modulated signal into the optical fiber.

Block diagram of the Transmitter

- The role of an optical transmitter is to convert the electrical signal into optical form and to launch the resulting optical signal into the optical fiber.
- Fig. 5 shows the block diagram of an optical transmitter. It consists of an optical source, a modulator, and a channel coupler

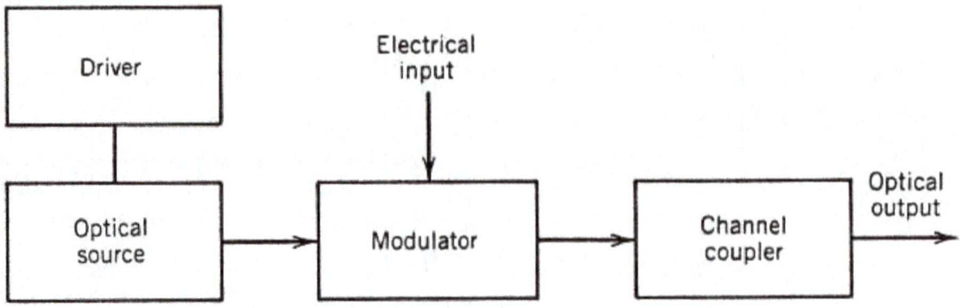

Fig. 5 Components of an optical transmitter

- Semiconductor lasers or light-emitting diodes are used as optical sources because of their compatibility with the optical-fiber communication channel
- The optical signal is generated by modulating the optical carrier wave.
- Although an external modulator is sometimes used, it can be dispensed with in some cases, since the output of a semiconductor optical source can be modulated directly by varying the injection current.
- Such a scheme simplifies the transmitter design and is generally cost-effective.
- The coupler is typically a microlens that focuses the optical signal onto the entrance plane of an optical fiber with the maximum possible efficiency.
- The launched power is an important design parameter.
- One can increase the amplifier spacing by increasing it, but the onset of various nonlinear effects limits how much the input power can be increased.
- The launched power is often expressed in "dBm" units with 1 mW as the reference level.
- The general definition is

$$\text{power (dBm)} = 10\log_{10}\left(\frac{\text{power}}{1\,\text{mW}}\right)$$

- The launched power is rather low (< "10 dBm) for light-emitting diodes but semiconductor lasers can launch powers <" 10 dBm.
- With proper design, optical transmitters can be made to operate at a bit rate of up to 40 Gb/s.

Source limitations

Power

- The electrical power required to operate both injection lasers and LEDs is generally similar, with typical current levels of between 20 and 300 mA and voltage drops across the terminals of 1.5 to 2.5 V.

- The optical output power against current characteristic for the two devices varies considerably, as indicated in Fig. 6.
- The injection laser is a threshold device which must be operated in the region of stimulated emission (i.e. above the threshold) where continuous optical output power levels are typically in the range 1 to 10 mW.
- Much of this light output may be coupled into an optical fiber because the isotropic distribution of the narrow-linewidth, coherent radiation is relatively directional.
- Injection lasers are capable of launching between 0.5 and several milliwatts of optical power into a fiber.
- LEDs are capable of similar optical output power levels to injection lasers depending on their structure and quantum efficiency, as indicated by the typical characteristic for a surface emitter shown in Fig. 6

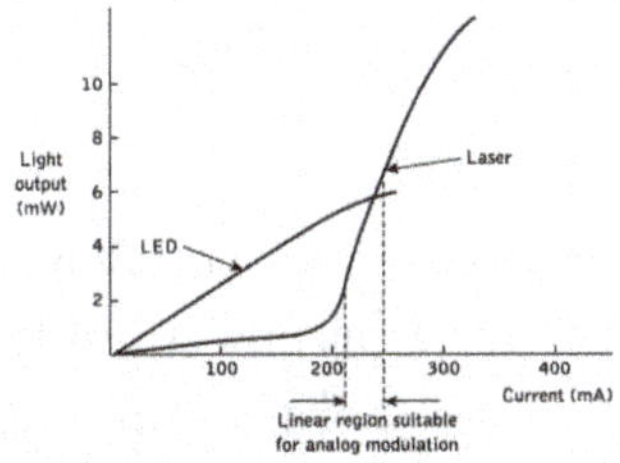

Fig. 6 the curves exhibit non-linearity at high currents due to junction heating

- The optical power coupled into a fiber from an LED can be 10 to 20 dB below that obtained with a typical injection laser.
- The power advantage gained with the injection laser is a major factor in the choice of source, especially when considering a long-haul optical fiber link.

Linearity

- Linearity of the optical output power against current characteristic is an important consideration with both the injection laser and LED.
- It is especially pertinent to the design of analog optical fiber communication systems where source nonlinearities may cause severe distortion of the transmitted signal.
- Firstly LED may appear to be ideally suited to analog transmission as its output is approximately proportional to the drive current.
- But most of LEDs display some degree of nonlinearity in their optical output against current characteristic because of junction heating effects which may either prohibit their use or necessitate the incorporation of a linearizing circuit within the optical transmitter.

- Certain LEDs do display good linearity, with distortion products between 35 and 45 dB below the signal level.
- An alternative approach to obtaining a linear source characteristic is to operate an injection laser in the light-generating region above its threshold
- This may prove more suitable for analog transmission than would the use of certain LEDs.
- However, gross nonlinearities due to mode instabilities may occur in this region.
- These are exhibited as kinks in the laser output characteristic.
- Therefore, many of the multimode injection lasers have a limited use for analog transmission.
- Digital transmission, especially that utilizing a binary (two-level) format, is far less sensitive to source nonlinearities and is therefore often preferred when using both injection lasers and LEDs

LED drive circuits

Circuit configurations that may be used to convert the information voltage signal at the transmitter into a modulation current suitable for an LED source. LED drive circuits are typically designed to control the current running through a string of LEDs independent of the voltage drop across the LED string.

Digital transmission:

- The operation of the LED for binary digital transmission requires the switching on and off of a current in the range of several tens to several hundreds of milli-amperes.
- This must be performed at high speed in response to logic voltage levels at the driving circuit input. A common method of achieving this current switching operation for an LED is shown in Fig. 7.
- The circuit illustrated uses a bipolar transistor switch operated in the common emitter mode.
- This single-stage circuit provides current gain as well as giving only a small voltage drop across the switch when the transmitter is in saturation.
- The maximum current flow through the LED is limited by the resistor R_2 while independent bias to the device may be provided by the incorporation of resistor R_3.
- The switching speed of the common emitter configuration is limited by space charge and diffusion capacitance; thus bandwidth is traded for current gain.

- This may be compensated by overdriving (pre-emphasizing) the base current during the switch-on period. In the circuit shown in Fig. 7 pre-emphasis is accomplished by use of the speed-up capacitor C.

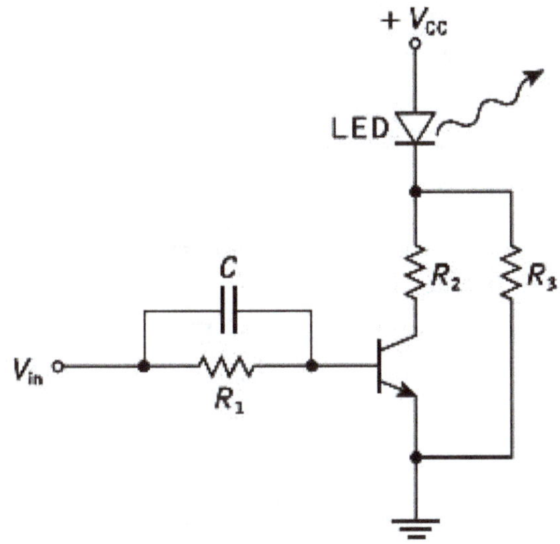

Fig. 7. A simple drive circuit for binary digital transmission consisting of a common emitter saturating switch

Low-impedance drive circuit:

- Increased switching speed may be obtained from an LED without a pulse shaping or speed-up element by use of a low-impedance driving circuit, whereby charging of the space charge and diffusion capacitance occurs as rapidly as possible.
- This may be achieved with the emitter follower drive circuit shown in Fig. 8.
- The use of this configuration with a compensating matching network (R_3C) provides fast direct modulation of LEDs with relatively low drive power.
- A circuit, with optimum values for the matching network, is capable of giving optical rise times of 2.5 ns for LEDs with capacitance of 180 pF, thus allowing 100 Mbit s^{-1} operations.

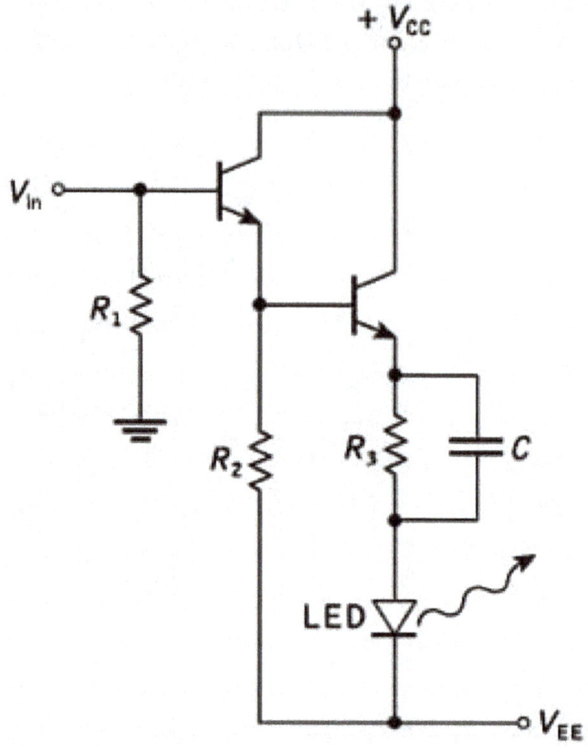

Fig. 8 Low-impedance drive circuit consisting of an emitter follower with compensating matching network

Laser drive circuits

Lasers are widely used as coherent light sources. Laser drive circuits are used to provide high speed operation. The mean optical power is maintained constant by varying the threshold current level by using the laser drive circuit.

Reasons for biasing:

- When digital transmission is considered, the laser is usually given a substantial applied bias, often referred to as prebias, in the off state.

Reasons for biasing the laser near but below threshold in the off state are as follows:

- It reduces the switch-on delay and minimizes any relaxation oscillations.
- It allows easy compensation for changes in ambient temperature and device aging.
- It reduces the junction heating caused by the digital drive current since the on and off currents are not widely different for most lasers.

Shunt drive circuit

- A simple laser drive circuit for digital transmission is shown in Fig. 9.

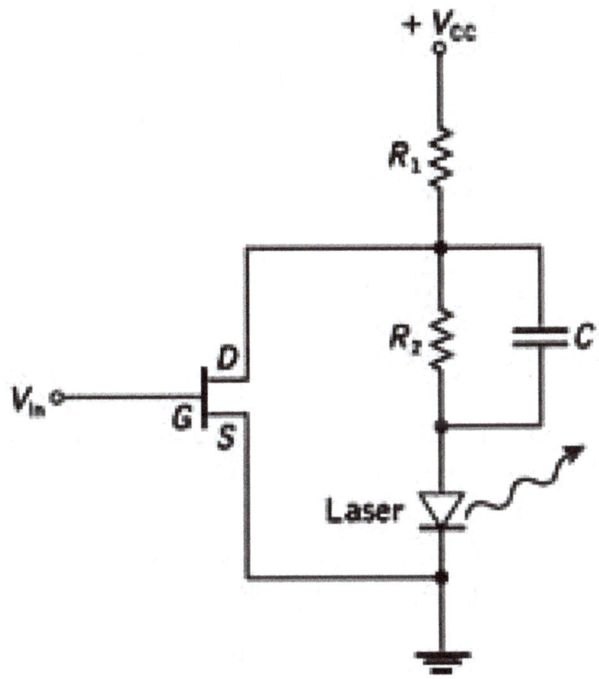

Fig. 9. A shunt drive circuit for use with an injection laser

- This circuit is a shunt driver utilizing a field effect transistor (FET) to provide high-speed laser operation.
- Sufficient voltage is maintained in series with the laser using the resistor R_2 and the compensating capacitor C such that the FET is biased into its active or pinch-off region.
- Hence for a particular input voltage V_{in} (i.e. VGS) a specific amount of the total current flowing through R_1 is diverted around the laser leaving the balance of the current to flow through R_2 and provide the off state for the device.
- Using stable gallium arsenide MESFETs the circuit shown in Fig. 9 has modulated lasers at rates in excess of 1 Gbit s^{-1}

ECL-compatible high-speed laser drive circuit

- An alternative high-speed laser drive circuit employing bipolar transistors is shown in Fig. 10.

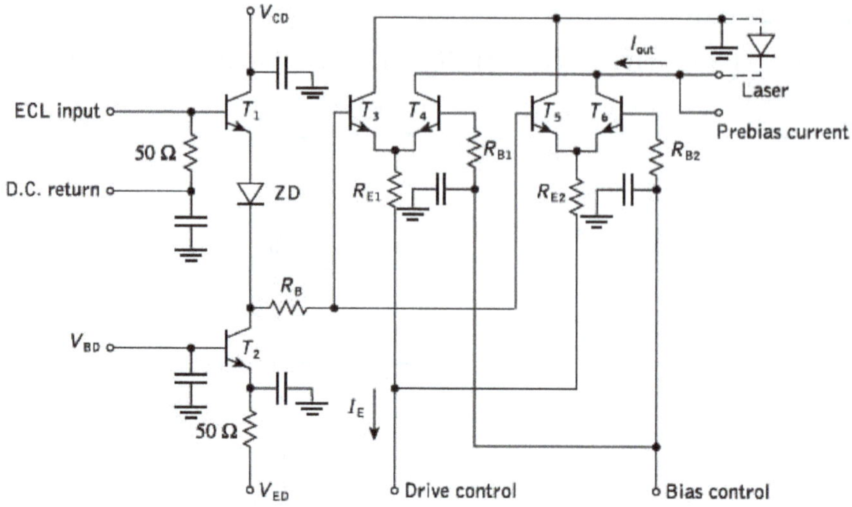

Fig. 10 An ECL-compatible high-speed laser drive circuit.

- This circuit configuration consists of two differential amplifiers connected in parallel.
- The input stage, which is ECL compatible, exhibits 50 Ω input impedance by use of an emitter follower T1 and a 50 Ω resistor in parallel with the input.
- The transistor T2 acts as a current source with the Zener diode ZD adjusting the signal level for ECL operation.
- The two differential amplifiers provide sufficient modulation current amplitude for the laser under the control of a d.c. control current IE through the two emitter resistors R_{E1} and R_{E2}.
- IE is provided by an optical feedback control circuit.

OPTICAL RECEIVER CIRCUIT

Receiver circuit of the optical communication system converts the optical signal into electrical signal to recover the transmitted data

Block diagram of the optical receiver:

- An optical receiver converts the optical signal received at the output end of the optical fiber back into the original electrical signal.
- Figure 11 shows the block diagram of an optical receiver.

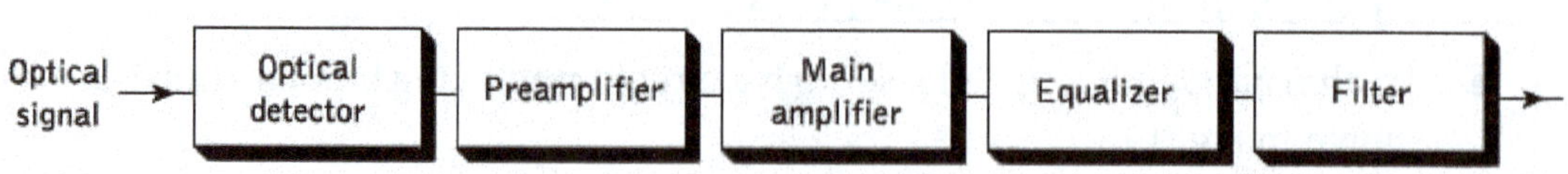

Fig. 11. elements of an optical fiber receiver

- Initial amplification is performed in the preamplifier circuit where it is essential that additional noise is kept to a minimum in order to avoid corruption of the received signal
- The main amplifier provides additional low-noise amplification of the signal to give an increased signal level for the following circuits.
- Although optical detectors are very linear devices and do not themselves introduce significant distortion onto the signal, other components within the optical fiber communication system may exhibit nonlinear behavior.
- The transfer function of the preamplifier–main amplifier combination may be such that the input signal becomes distorted.
- Hence, to compensate for this distortion and to provide a suitable signal shape for the filter, an equalizer is often included in the receiver.
- It may precede or follow the main amplifier, or may be incorporated in the functions of the amplifier and filter.
- The function of the last element in the receiver, the filter, is to maximize the received signal-to-noise ratio while preserving the essential features of the signal.
- In digital systems the function of the filter is primarily to reduce intersymbol interference, whereas in analog systems it is generally required to hold the amplitude and phase response of the received signal within certain limits.
- The filter is also designed to reduce the noise bandwidth as well as in-band noise levels.

Preamplifier

Preamplifier extracts the signal from the detector without significantly degrading the intrinsic signal-to-noise ratio

Low-input-impedance voltage preamplifier

- The simplest preamplifier structure is the low-input-impedance voltage amplifier.
- This design is usually implemented using a bipolar transistor configuration because of the high input impedance of FETs.
- The common emitter and the grounded emitter (without an emitter resistor) amplifier shown in Figure 12.
- They are favored connections, as they may be designed with reasonably low input impedance and therefore give operation over a moderate bandwidth without the need for equalization.

- This is achieved at the expense of increased thermal noise due to the low effective load resistance presented to the detector.

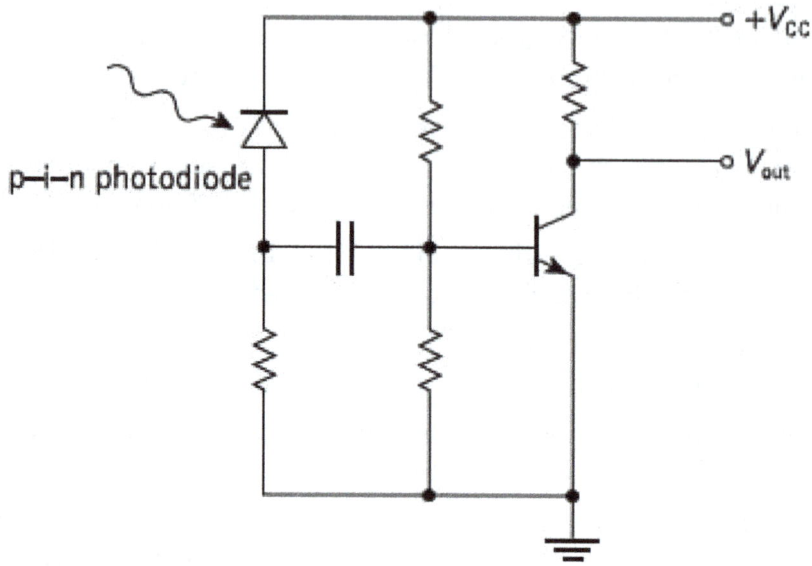

Fig. 12. A p–i–n photodiode with a grounded emitter, low-input-impedance voltage preamplifier

- It is possible to reduce the thermal noise contribution of this preamplifier by choosing a transistor with characteristics which give a high current gain at a low emitter current in order to maintain the bandwidth of the stage.
- An inductance may be inserted at the collector to provide partial equalization for any integration performed by the stage.
- The alternative connection giving very low input impedance is the common base circuit.
- This configuration has input impedance which gives insufficient power gain when connected to the high impedance of the optical detector.

FET common source preamplifier configuration:

- The preferred preamplifier configurations for low-noise operation use either high impedance integrating front-end or a transimpedance amplifier.
- Careful design employing these circuit structures can facilitate high gain coupled with low-noise performance and therefore enhanced receiver sensitivity.
- Although the bipolar transistor incorporated in the emitter follower circuit may be used to realize a high impedance front-end amplifier, the FET is generally employed for this purpose because of its low-noise operation.

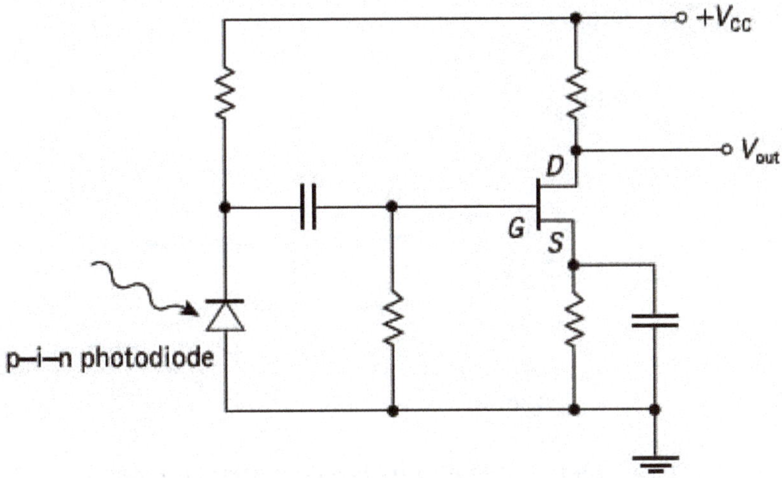

Fig. 13 An FET common source preamplifier configuration

- The grounded source FET connection is useful circuit to provide a high-impedance front-end amplifier.
- The same configuration with a source resistor (common source connection) shown in Figure 13 provides a similar high-input impedance and may also be used (often both configurations are referred to as the common source connection).
- When operating in this mode the FET power gain and output impedance are both high.
- This tends to minimize any noise contributions from the following stages.

Automatic gain control

This section explains the Automatic gain control in optical fiber which reduces the dynamic range of the signal applied to the preamplifier

Automatic gain control techniques:

- The multiplication factor can be held constant by some form of automatic gain control (AGC).

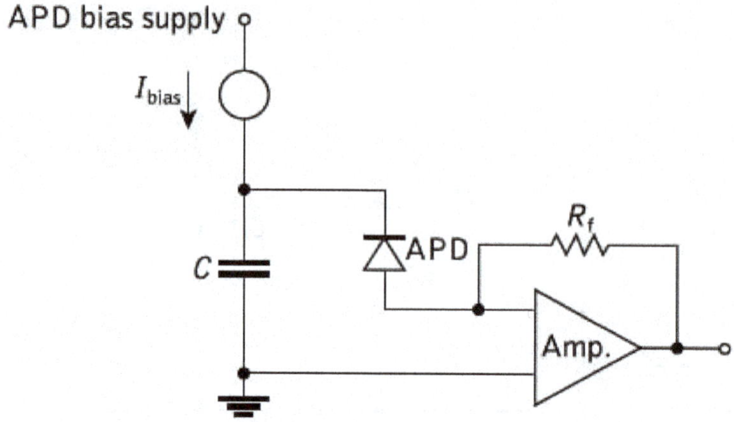

Fig. 14. Bias of an APD with a constant current source

- An additional advantage in the use of AGC is that it reduces the dynamic range of the signals applied to the preamplifier giving increased optical dynamic range at the receiver input.
- One method of providing AGC is simply to bias the APD with a constant d.c. current source I_{bias}, as illustrated in Figure 14.
- The constant current source is decoupled by a capacitor C at all signal frequencies to prevent gain modulation.
- When the mean optical input power is known, the mean current to the APD is defined by the bias which gives a constant multiplication factor (gain) at all temperatures.
- Any variation in the multiplication factor will produce a variation in the charge on C, thus adjusting the biasing of the APD back to the required multiplication factor.
- Therefore, the output current from the photo-detector is only defined by the input current from the constant current source, giving full AGC.
- However, this simple AGC technique is dependent on a constant, mean optical input power level, and takes no account of dark current generated within the detector.
- A more widely used method which allows for the effect of variations in the detector dark current while providing critical AGC is to peak-detect the a.c. coupled signal after suitable low-noise amplification, as shown in Figure 15.
- The signal from the final stage of the main amplifier is compared with a preset reference level and fed back to adjust the high-voltage bias supply in order to maintain a constant signal level.

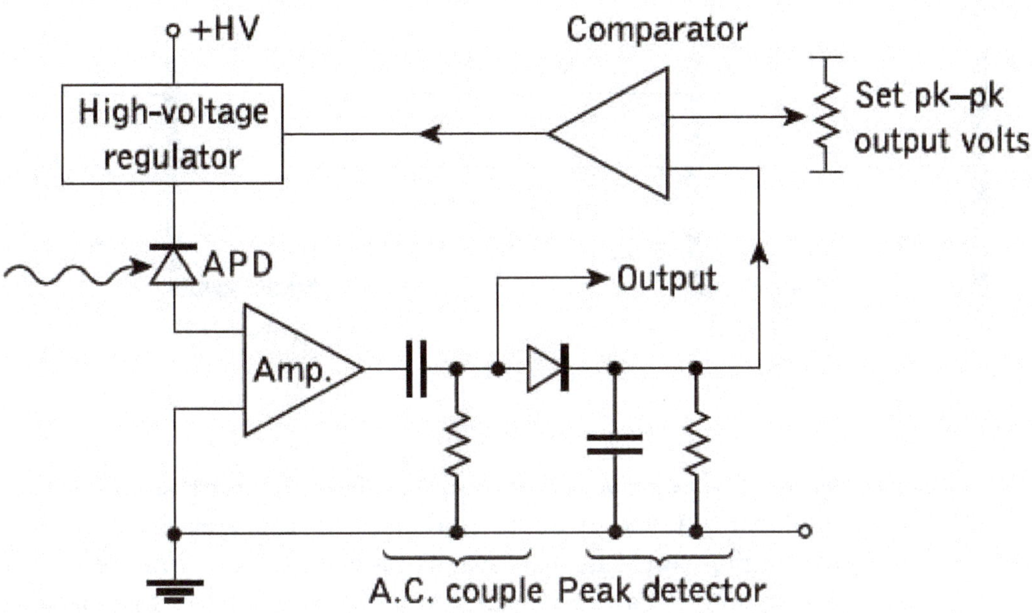

Figure 12.23 Bias of an APD by peak detection and feedback to provide AGC

- This effectively creates a constant current source with the dark current subtracted.
- A further advantage of this technique is that it may also be used to provide AGC for the main amplifier giving full control of the receiver gain.

SYSTEM DESIGN CONSIDERATIONS

For better communication it becomes necessary to consider the possible component selection in conjunction with the basic system requirements. For long-haul telecommunication systems maximum distance is covered without the use of much repeaters

Design considerations:

- Where the terminal equipment is more widely spaced than this maximum distance, as in long-haul telecommunication applications, it is necessary to insert repeaters at regular intervals, as shown in Figure 16.
- The repeater incorporates a line receiver in order to convert the optical signal back into the electrical regime where, in the case of analog transmission, it is amplified and equalized before it is retransmitted as an optical signal via a line transmitter.

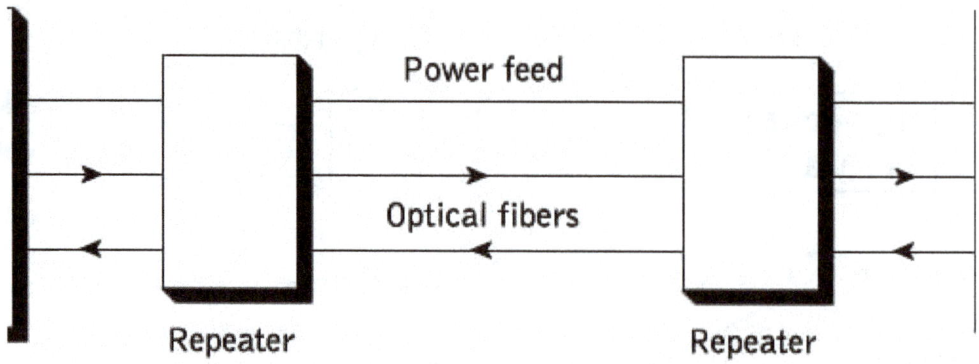

Fig. 16. The use of repeaters in a long-haul optical fiber communication system

- When digital transmission techniques are used the repeater also regenerates the original digital signal in the electrical regime (a regenerative repeater which is often simply called a regenerator) before it is retransmitted as a digital optical signal.
- In this case the repeater may additionally provide alarm, supervision and engineering order wire facilities.
- Optical or all-optical regeneration techniques now provide signal amplification with additional reshaping and retiming capabilities and they can be used to achieve longer transmission distances using optical amplifiers in comparison with optoelectronic repeaters
- The installation of repeaters substantially increases the cost and complexity of any line communication system.
- Hence a major design consideration for long-haul telecommunication systems is the maximum distance of un-repeatered transmission so that the number of intermediate amplification or regeneration stages may be reduced to a minimum.
- In this respect optical fiber systems display a marked improvement over alternative line transmission systems using metallic conductors.
- This major advantage of optical fiber communications is somewhat reduced when there is a requirement for electronic signal processing at the repeater.
- This necessitates the supply of electric power to the intermediate repeaters via metallic conductors

Before any system design procedures can be initiated it is essential that certain basic system requirements are specified. These specifications include

- Transmission type: digital or analog;
- Acceptable system fidelity, generally specified in terms of the received BER for digital systems or the received SNR and signal distortion for analog systems;
- Required transmission bandwidth;
- Acceptable spacing between the terminal equipment or intermediate repeaters;
- Cost;
- Reliability.

Component Choice

Component choice is optical fiber is very important for determining the system performance. For reliable and cost effective communication component choice becomes important.

Component Choice

The overall requirements of the system performance are ready availability of suitable reliable components, and cost.

Hence the major component choices are

- Optical fiber type and parameters:
- Multimode or single-mode; size, refractive index profile, attenuation, dispersion, mode coupling, strength, cabling, jointing, etc.
- Source type and characteristics:
- Laser or LED; optical power launched into the fiber, rise and fall time, stability, etc.
- Transmitter configuration:
- Design for digital or analog transmission; input impedance, supply voltage, dynamic range, optical feedback, etc.
- Detector type and characteristics:
- p–n, p–i–n, or avalanche photodiode; responsivity, response time, active diameter, bias voltage, dark current, etc.
- Receiver configuration:
- Preamplifier design (low impedance, high impedance or transimpedance front-end), BER or SNR, dynamic range, etc.
- Modulation and coding:

- Source intensity modulation; using pulse modulation techniques for either digital or analog so, encoding schemes for digital transmission such as biphase (Manchester) and delay modulation (Miller) codes.
- Analog transmission using direct intensity modulation or frequency modulation of the electrical subcarrier (subcarrier FM).
- In the latter technique the frequency of an electrical subcarrier is modulated rather than the frequency of the optical source, as would be the case with direct frequency modulation.
- The electrical subcarrier, in turn, intensity modulates the optical source.

Multiplexing

Multiplexing refer to technique in which multiple users share a medium with minimum or no interference

Requirement of Multiplexing:

- In order to maximize the information transfer over an optical fiber communication link it is usual to multiplex several signals onto a single fiber.
- It is possible to convey these multichannel signals by multiplexing in the electrical time or frequency domain, as with conventional electrical line or radio communication, prior to intensity modulation of the optical source.
- Hence, digital pulse modulation schemes may be extended to multichannel operation by time division multiplexing (TDM) narrow pulses from multiple modulators under the control of a common clock.
- Pulses from the individual channels are interleaved and transmitted sequentially, thus enhancing the bandwidth utilization of a single-fiber link.
- A number of baseband channels may be combined by frequency division multiplexing (FDM).
- In FDM the optical channel bandwidth is divided into a number of non-overlapping frequency bands and each signal is assigned one of these bands of frequencies.
- The individual signals can be extracted from the combined FDM signal by appropriate electrical filtering at the receive terminal.
- Hence, FDM in an IM/DD system is generally performed electrically at the transmit terminal prior to intensity modulation of a single optical source.
- It is possible to utilize a number of optical sources, each operating at a different wavelength on the single-fiber link.

- In this technique, often referred to as wavelength division multiplexing (WDM), the separation and extraction of the multiplexed signals (i.e. wavelength separation) is performed with optical filters (e.g. interference filters, diffraction grating filters, or prism filters).
- A multiplexing technique which does not involve the application of several message signals onto a single fiber is known as space division multiplexing (SDM).
- In SDM, each signal channel is carried on a separate fiber within a fiber bundle or multifiber cable form.
- The good optical isolation offered by fibers means that cross-coupling between channels can be made negligible.
- This technique necessitates an increase in the number of optical components required (e.g. fiber, connectors, sources, detectors) within a particular system and therefore has not been widely used to date.

CHAPTER

OPTICAL FIBER SYSTEMS 2: COHERENT AND PHASE MODULATED

12

INTRODUCTION

Optical communication refers to use optical fiber as a medium to transmit optical signal from the source to the destination. Since the 1990s the optical fiber has been widely installed for metropolitan and trans-ocean communications from 45 Mbit/s to now 100 Gbit/s. Depending on the detection methods, optical communication systems can be divided into two broad categories: direct detection and coherent detection. For direct detection system, the data is mapped into the intensity of the optical signals to simplify the system design and reduce the cost. Despite that the phase information of the optical signal can be modulated, the receiver complexity and cost will increase accordingly. To meet the bandwidth demand of the exponentially increasing internet traffic in the present and future, the capability of direct detection systems is extremely limited for upgrading the data rate from 40×10^9 Gbit/s up to 100×10^9 Gbit/s or even higher.

Recent advances in high-speed analog-to-digital converters (ADCs) have revived the coherent detection systems. When the received signals are sampled into digital waveforms using ADCs, its full information, such as phase, amplitude and polarization, can be preserved in the receiver such that higher-order modulation formats can be used to increase the transmission capacity. Besides, the bulky optical components are replaced by small and compact digital signal processing (DSP) processors to compensate for the fiber transmission impairments. This DSP functions make the digital coherent detection universal and independent of the modulation formats.

Polarization-multiplexed quadrature-amplitude modulation maximizes spectral efficiency and power efficiency, by utilizing all four available DOF, the two field quadratures in the two polarizations. Dual-polarization homodyne or heterodyne downconversion are linear processes that can fully recover the received signal field in these four DOF. When downconverted signals are sampled at the Nyquist rate, compensation of transmission impairments can be performed using digital signal processing (DSP).

Linear impairments, including chromatic dispersion and polarization-mode dispersion, can be compensated quasi-exactly using finite impulse response filters. Some nonlinear impairments, such as intra-channel four-wave mixing and nonlinear phase noise, can be compensated partially. Carrier phase recovery can be performed using feedforward methods, even when phase-locked loops may fail due to delay constraints. DSP-based compensation enables a receiver to adapt to time-varying impairments, and facilitates use of advanced forward-error-correction codes.

We discuss both single- and multi-carrier system implementations. For a given modulation format, using coherent detection, they offer fundamentally the same spectral efficiency and power efficiency, but may differ in practice, because of different impairments and implementation details. With anticipated advances in analog-to-digital converters and integrated circuit technology, DSP-based coherent receivers at bit rates up to 100 Gbit/s should become practical within the next few years.

BASIC COHERENT SYSTEM

Introduction

The most promising detection technique for achieving high spectral efficiency while maximizing power (or SNR) efficiency is coherent detection with polarization multiplexing. Coherent optical communication is one of the most promising ways to achieve highest receiver sensitivity, excellent spectral efficiency and longest transmission distance for the next generation of optical communication systems.

Transmitter of coherent system

- A block schematic of a generalized coherent optical fiber communication system is illustrated in Fig. 1.
- The dashed lines enclose the main elements which distinguish the coherent system from its IM/DD equivalent.
- At the transmitter a CW narrow-linewidth semiconductor laser is shown which acts as an optical frequency oscillator.
- An external optical modulator usually provides amplitude, frequency or phase shift keying of the optical carrier by the information signal.
- At present external modulators are generally waveguide devices fabricated from lithium niobate or the group III–V compound semiconductors.

- Internal modulation of the injection laser drive current may, however, also be utilized to produce either ASK or FSK.

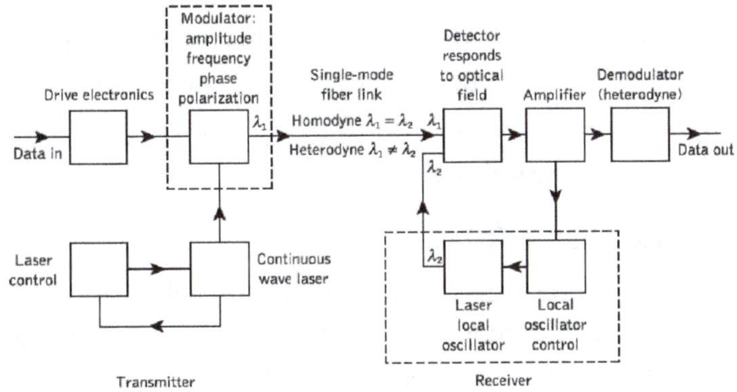

Fig. 1 The generalized coherent optical fiber system

Modulation techniques:

- Modulated carrier waveforms for the three standard modulation techniques with binary data are illustrated in Fig. 2.
- It may be observed from Fig. 2(a) why binary ASK is often referred to as on-off keying (OOK).
- Fig. 2(b) shows FSK in which the binary 1 is transmitted at a higher optical frequency than the binary 0 bit.

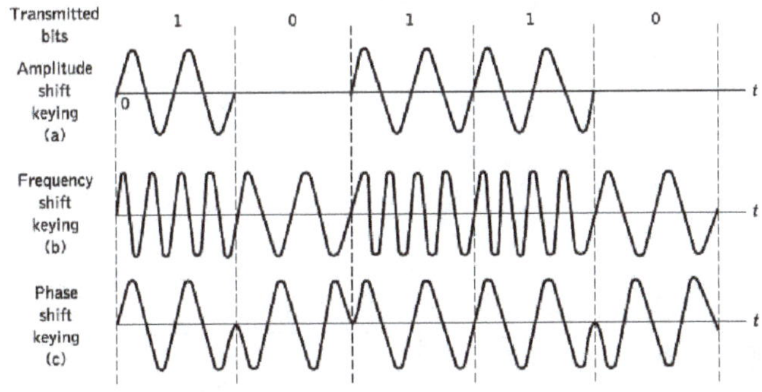

Fig. 2 Modulated carrier waveforms used for binary data transmission: (a) amplitude shift keying (ASK) or on-off keying (OOK); (b) frequency shift keying (FSK); (c) phase shift keying (PSK)

- The 180° phase shift between the binary 1 and 0 bits displayed in Fig. 2(c) depicts PSK.
- It should be noted that whereas with ASK the amplitude of the carrier waveform is effectively switched on and off, the amplitude of the optical carrier remains constant in the other two modulation schemes shown in Fig. 2.

- Variants on these standard modulation techniques exist, such as continuous phase FSK and differential PSK, have also been applied in experimental coherent optical fiber transmission systems.
- An alternative digital scheme based on the modulation of the polarization properties of the optical signal has more recently come under investigation.
- This strategy, known as polarization shift keying (PolSK), is discussed further along with the other modulation formats

COHERENT DETECTION PRINCIPLES

Coherent Detection is a receiver that measures the electric field strength of an incoming photon. Coherent detection potentially provides a substantial benefit for high-speed systems operating at longer wavelengths

Coherent Detection

- A simple coherent receiver model for ASK is displayed in Fig. 3.
- The low-level incoming signal field e_S is combined with a second much larger signal field e_L derived from the local oscillator laser.
- It is assumed that the electromagnetic fields obtained from the two lasers can be represented by cosine functions and that the angle $\phi = \phi_S - \phi_L$ represents the phase relationship between the incoming signal phase ϕ_S and the local oscillator signal phase ϕ_L defined at some arbitrary point in time.
- Hence the two fields may be written as

$e_S = E_S \cos(\omega_S t \mid \phi)$ (1)

and

$e_L = E_L \cos(\omega_L t)$ (2)

Where

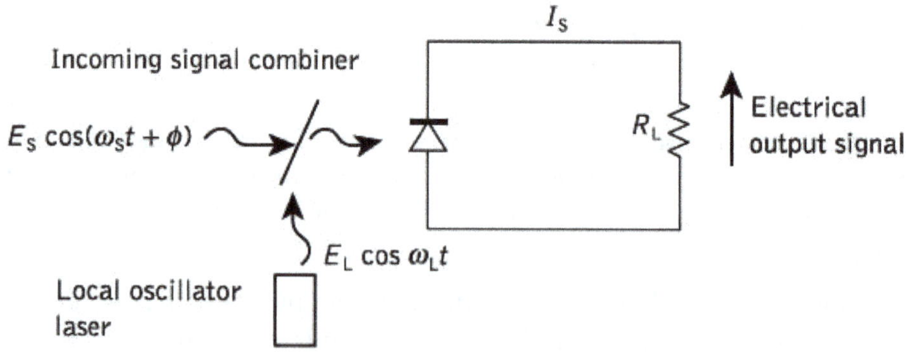

Fig. 3 Basic coherent receiver model

- E_S is the peak incoming signal field and

- ω_S is its angular frequency, and
- E_L is the peak local oscillator field and
- ω_L is its angular frequency.
- The angle $\phi(t)$ representing the phase relationship between the two fields contains the transmitted information in the case of FSK or PSK.
- With ASK $\omega(t)$ is constant and hence it is simply written as ω in Eq. (1), the information being contained in the variation of ES for ASK.
- For heterodyne detection, the local oscillator frequency ω_L is offset from the incoming signal frequency ω_S by an intermediate frequency such that

  $$\omega_S = \omega_L + \omega_{IF} \qquad (3)$$

Where

- ω_{IF} is the angular frequency of the IF
- The IF is usually in the radio-frequency region and may be a few tens or hundreds of megahertz.
- Within homodyne detection there is no offset between ω_S andω_L and hence $\omega_{IF} = 0$.
- In this case the combined signal is therefore recovered in the baseband.
- The two wave fronts from the incoming signal and the local oscillator laser must be perfectly matched at the surface of the photo detector for ideal coherent detection.
- This factor creates the normal requirement for polarization control of the incoming optical signal.

PRACTICAL CONSTRAINTS OF COHERENT TRANSMISSION

Noncoherent detection

In noncoherent detection, a receiver computes decision variables based on a *measurement of signal energy*. An example of noncoherent detection is direct detection of on-off-keying (OOK) using a simple photodiode (Fig. 4(a)). To encode more than one bit per symbol, multi-level amplitude-shift keying (ASK) – also known as pulse-amplitude modulation – can be used. Another example of noncoherent detection is frequency-shift keying (FSK) with wide frequency separation between the carriers. Fig. 4(b) shows a noncoherent receiver for binary FSK.

The limitations of noncoherent detection are: (a) detection based on energy measurement allows signals to encode only one degree of freedom (DOF) per polarization per carrier, reducing spectral efficiency and power efficiency, (b) the loss of phase information during detection is an irreversible transformation that prevents full equalization of linear channel impairments like CD and PMD by linear filters. Although maximum-likelihood sequence detection (MLSD) can be used to find the best estimate of the transmitted sequence given only a sequence of received intensities, the achievable performance is suboptimal compared with optical or electrical equalization making use of the full electric field.

Differentially coherent detection

In differentially coherent detection, a receiver computes decision variables based on *a measurement of differential phase between the symbol of interest and one or more reference symbol(s)*. In differential phase-shift keying (DPSK), the phase reference is provided by the previous symbol; in polarization-shift keying (PolSK), the phase reference is provided by the signal in the adjacent polarization. A binary DPSK receiver is shown in Fig. 5(a). Its output photocurrent is:

$$I_{DPSK}(t) = RR_e\{E_s(t)E * s(t - T_s)\} \quad (1)$$

where $E_s(t)$ is the received signal, R is the responsivity of each photodiode, and T_s is the symbol period. This receiver can also be used to detect continuous-phase frequency-shiftkeying (CPFSK), as the delay interferometer is equivalent to a delay-and-multiply demodulator. A receiver for M-ary DPSK, M>2, can similarly be constructed as shown in Fig. 5(b). Its output photocurrents are:

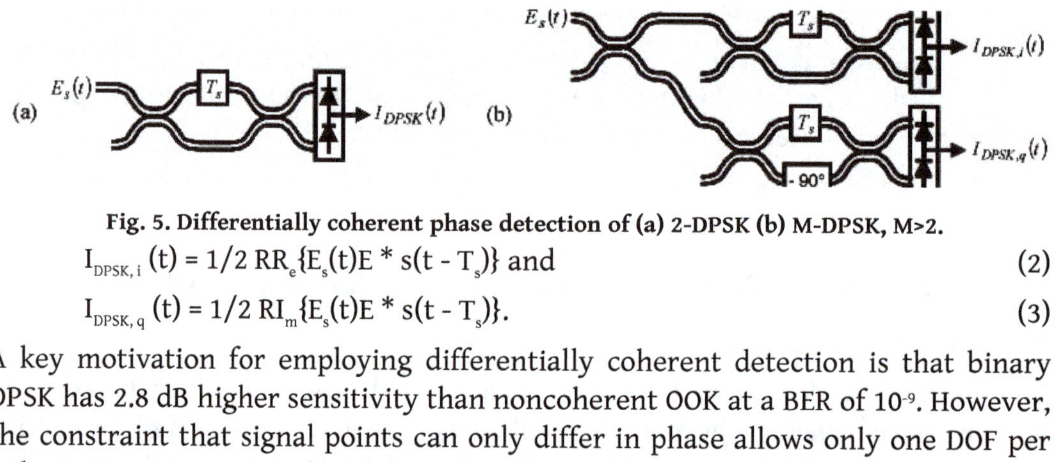

Fig. 5. Differentially coherent phase detection of (a) 2-DPSK (b) M-DPSK, M>2.

$$I_{DPSK,i}(t) = 1/2\ RR_e\{E_s(t)E * s(t - T_s)\} \text{ and} \quad (2)$$

$$I_{DPSK,q}(t) = 1/2\ RI_m\{E_s(t)E * s(t - T_s)\}. \quad (3)$$

A key motivation for employing differentially coherent detection is that binary DPSK has 2.8 dB higher sensitivity than noncoherent OOK at a BER of 10^{-9}. However, the constraint that signal points can only differ in phase allows only one DOF per polarization per carrier, the same as noncoherent detection. As the photocurrents in Eq. (1) to (3) are not linear functions of the E-field, linear impairments, such as CD and PMD, also cannot be compensated fully in the electrical domain after photodetection.

A more advanced detector for M-ary DPSK is the multichip DPSK receiver, which has multiple DPSK receivers arranged in parallel, each with a different interferometer delay that is an integer multiple of T_s. Since a multichip receiver compares the phase of the current symbol to a multiplicity of previous symbols, the extra information available to the detector enables higher sensitivity. In the limit that the number of parallel DPSK receivers is infinite, the performance of multi-chip DPSK approaches coherent PSK. In practice, the number of parallel DPSK receivers required for good performance needs to be equal to the impulse duration of the channel divided by T_s. Although multi-chip DPSK does not require a local oscillator (LO) laser, carrier synchronization and polarization control, the hardware complexity can be a significant disadvantage.

MODULATION FORMATS

The first step in the design of an optical communication systems is to decide how the electrical signal should be converted into an bit stream. There are two typical choices for the modulation format of the signal:

- return-to-zero (RZ)
- nonreturn-to-zero (NRZ)

Examples of the modulation formats are shown in Fig. 6.

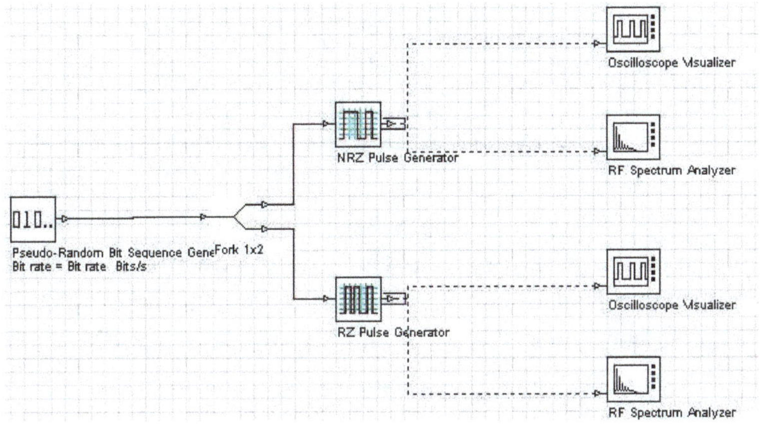

Fig. 6: Modulation Formats

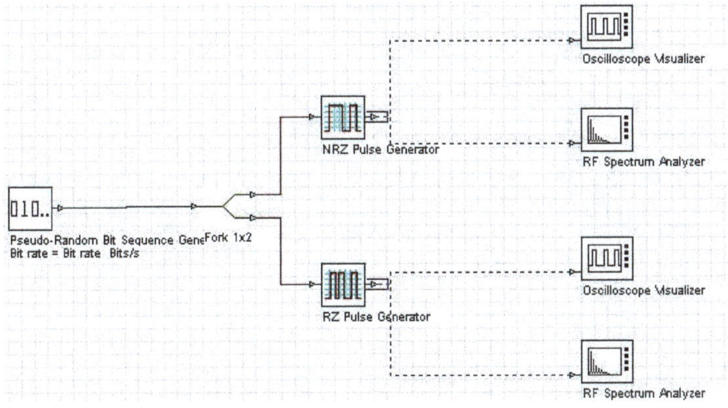

Fig. 7: NRZ – Time

In the RZ formats, each pulse representing bit 1 is shorter than the bit slot, and its amplitude returns to zero before the bit duration is over (see Fig 8). In the NRZ format, the pulse remains on throughout the bit slot and its amplitude does not drop to zero between two or more successive 1 bits (see Fig. 7). As a result, pulse width varies depending on the bit pattern, whereas it remains the same in the case of RZ format.

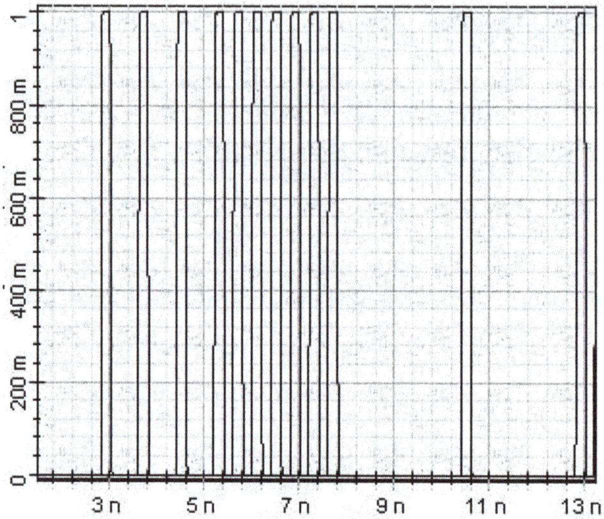

Fig. 8: RZ – Time

An advantage of the NRZ format is that the bandwidth associated with the bit stream is smaller than that of the RZ format by about a factor of 2, because on-off transitions occur fewer times (see Fig. 9 and Fig. 10).

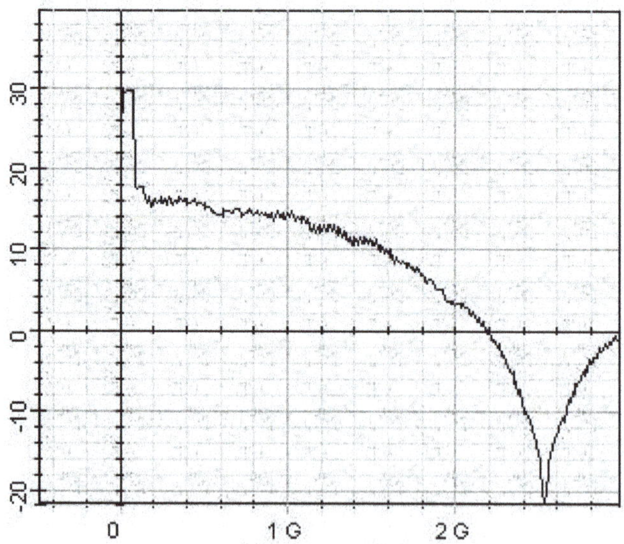

Figure 9: NRZ – Frequency

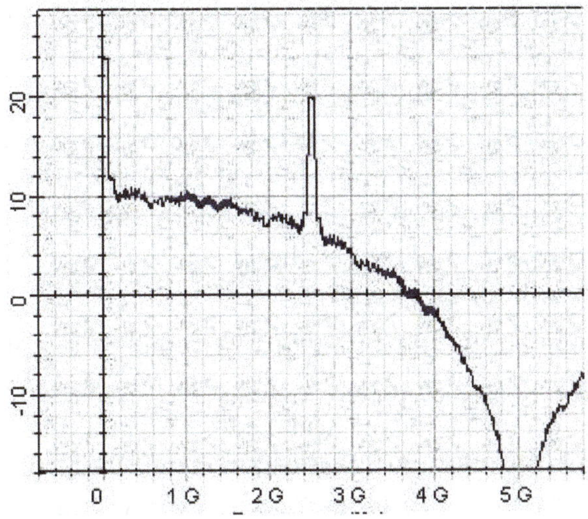

Figure 10: RZ – Frequency

DEMODULATION SCHEMES

Demodulation is defined as extracting the original information-carrying signal from a modulated carrier wave. A demodulator is an electronic circuit that is mainly used to recover the information content from the modulated carrier wave. There are different types of modulation, and so are demodulators. The output signal via a demodulator may describe the sound, images, or binary data.

Basic receiver configurations for optical heterodyne and homodyne detection are shown in Fig. 11. In both cases it has been assumed that some form of polarization control is required to match the incoming signal SOP to that of the local oscillator signal. This factor therefore implies the use of circularly symmetric standard single-mode fiber. For heterodyne detection (Fig. 11(a)), a beat-note signal between the incoming optical signal and the local oscillator signal produces the IF signal which is obtained using the square-law optical detector.

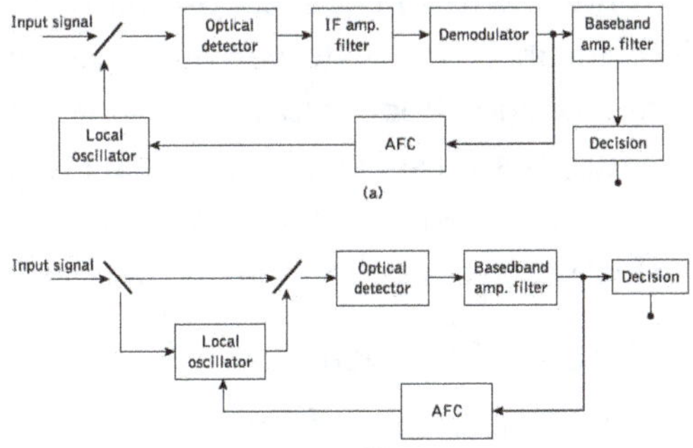

Fig. 11 Basic coherent receiver configurations: (a) optical heterodyne receiver; (b) optical homodyne receiver illustrating the phase locking between the local oscillator and incoming signals

The IF signal, which generally has a frequency of between three and four times the transmission rate, is then demodulated into the baseband using either a synchronous or asynchronous detection technique.* An optical receiver bandwidth several times greater than that of a direct detection receiver is therefore required for a specific transmission rate. Moreover, as IF frequency fluctuation degrades the heterodyne receiver performance, then frequency stabilization may be achieved by feeding back from the demodulator through an automatic frequency control (AFC) circuit to the local oscillator drive circuit.

In the case of homodyne detection in which the phase of the local oscillator signal is locked to the incoming signal, then, by definition, a synchronous detection scheme must be employed. Moreover, the result of the mixing process in the optical detector produces an information signal which is in the baseband and thus requires no further demodulation. An AFC loop is also shown within the homodyne receiver configuration of Fig. 11(b) to provide the necessary frequency stabilization between the two signals.

Hence any variant detection schemes based on homodyne detection, but in which the local oscillator laser is not phase locked to the incoming signal such as phase diversity or multiport detection, could be considered as a form of heterodyne rather than homodyne detection.

In both optical heterodyne and homodyne detection, where the incoming signal is demodulated using a local oscillator laser, FM noise in this device together with that resulting from the source laser causes SNR degradation in the receivers through FM to AM, or PM to AM, conversion which generally determines the lower limit of bit-error-rate performance. Frequency modulation noise which basically results from the spontaneous emission coupled to the lasing mode is, in the semiconductor laser, enhanced by AM noise caused by photon number fluctuation which is generated through the same mechanism.

Moreover, excess AM noise within the local oscillator laser due to its resonance characteristics also deteriorates the SNR performance and hence degrades the receiver sensitivity. To reduce the effect of local oscillator FM noise a semiconductor laser with a narrowed or suppressed spectral linewidth must be used. The excess AM noise in the semiconductor laser decreases with an increase in the bias level so that high-bias operation is effective in suppressing this mechanism.

OPTICAL HETERODYNE DETECTION

Heterodyne detection (also called *coherent detection*) is a detection method which was originally developed in the field of radio waves and microwaves. There, a weak input signal is mixed with some strong "local oscillator" wave in some nonlinear device such as a rectifier, and the resulting mixing product is then detected, often after filtering out the original signal and the local oscillator frequency. The frequency of the mixing product is the sum or the difference of the frequencies of the signal and the local oscillator.

Optical heterodyne detection involves optical signal and local oscillator waves, whereas the mixing product is an electrical signal. The mixing product is *not* obtained by mixing the signal and local oscillator wave in a nonlinear crystal, but rather simply by detecting the linearly superimposed waves with a square-law photodetector, typically a photodiode. For example, one uses a beam combiner (or beam splitter) as in Fig. 12 and aligns the two beams such that they are mode-matched. This means not only that their intensity profiles overlap, but also that their wavefronts have the same curvature on the detector, so that the interference conditions are uniform over the full detector area. Of course, this is possible only if the two beams are spatially coherent. In a fiber-optic setup, a fiber coupler would be used instead of the beam splitter, and all fibers would be single-mode fibers, possibly of polarization-maintaining type; the mode matching is then guaranteed without a special alignment.

The resulting photocurrent is propor-tional to the total optical intensity, thus to the square of the total electric field amplitude. If the signal and local oscillator powers and frequencies are constant, the photocurrent has two different frequency components:

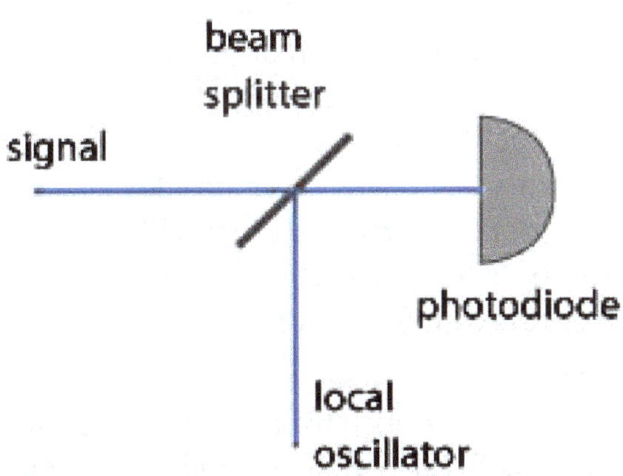

Fig. 12: Setup for optical heterodyne detection

- The constant (zero-frequency) part is proportional to the sum of local oscillator and signal power.
- The part oscillating with the difference frequency (the beat note) has an amplitude proportional to the product of the electric field amplitudes (*not* the optical powers) of signal and local oscillator.

The oscillating part can then be isolated and processed further with electronic means. Its electric power is proportional to the product of the optical powers of signal and local oscillator.

With a strong local oscillator, the heterodyne signal resulting from a weak input signal can be much more powerful than for direct detection. In that sense, heterodyne detection provides a signal gain, although there is no optical amplification involved.

Noise Limitations

The noise in the detected signal has different contributions:

- There is shot noise, the power spectral density (PSD) of which is proportional to the optical average power.
- The photodetector may add some noise, for example due to some dark current.
- Additional noise can be added by the electronics.
- It is also apparent that the beam combiner (see Figure 1) should have a high transmittance, so that most of the signal power is sent to the detector. The resulting high loss of local oscillator power at the detector can be compensated by increasing the local oscillator power incident on the beam combiner.
- If the beam combiner's transmission is high and the beam overlap is good ($\eta_m \approx 1$), the obtained signal-to-noise ratio is twice that for direct detection. In practice, heterodyne detection can offer a much greater advantage in terms of sensitivity, because direct detection reaches the shot noise limit only for rather high signal powers, whereas with heterodyne detection the standard quantum limit can be reached for much lower signal powers. Therefore, heterodyne detection is particularly suitable for very weak signals. Note, however, that there are cases where the standard quantum limit can not be reached because detector saturation limits the local oscillator power.
- An important detail is that the measured signal is not at zero frequency, but rather at some finite difference frequency, where the detector noise (e.g., $1/f$ noise) can be much weaker than around zero frequency.
- Using an amplitude-squeezed local oscillator beam, it is even possible to beat the standard quantum limit with heterodyne detection.

Spatial Single-mode Nature of Heterodyne Detection

- A strong heterodyne signal is obtained only if the beam overlap factor η_m is high. If the local oscillator beam is a perfect Gaussian beam, one may consider the total incident input signal light as a superposition of modes, using the mode system which is defined by the local oscillator beam. We see then that only the signal light in the lowest-order mode (the Gaussian) contributes to the heterodyne signal, because the overlap factors are zero for all higher-order modes. For any such higher-order mode, interference still occurs on the photodiode, but the local oscillations of intensity cancel out by integrating the power over the full detector area, provided that the detection efficiency is uniform over the full beam area.

- The single-mode nature of heterodyne detection is a crucial aspect. It can be very helpful, for example by suppressing influences of stray light in sensitive measurements. On the other hand, it would greatly reduce the detection efficiency, for example when light scattered on a large area needs to be detected.
- In principle, one could use a local oscillator beam consisting of different spatial modes, ideally each one being associated with a separate optical frequency. That would allow to simultaneously and independently measure signals in multiple modes. However, the power of each signal component in the heterodyne signal would then be proportional to the local oscillator power in the corresponding mode, whereas the noise level would be determined by the *total* local oscillator power. This shows again that heterodyne detection is essentially a single-mode detection technique.

Balanced Detection

A remaining problem of heterodyne detection in the simple form as shown in Fig. 12 is that excess noise of the local oscillator wave directly affects the signal. This is avoided with a *balanced* heterodyne setup as shown in Fig. 13. Here, the beam splitter must have a reflectivity of precisely 50%. With a simple electronic circuit, one can obtain the difference of the two photocurrents. That difference is to first order not influenced by noise of the local oscillator.

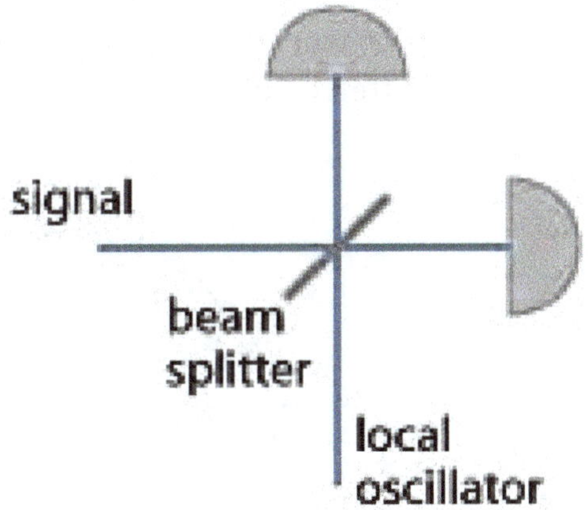

Fig. 13: Setup for balanced optical heterodyne detection.

TEMPORAL COHERENCE ISSUES

If the laser delivering the local oscillator beam is totally independent of the signal, it is hardly possible to measure the phase of the input signal, because the phase of the electronic signal is also dependent on the local oscillator phase, which is stable only over the typically short coherence time.

If phase measurements are required, the local oscillator and signal waves are usually generated from a single source, so that the *relative phase* of the two beams is fairly stable even if the source does not exhibit particularly low phase noise. For example, a signal wave may be generated from a local oscillator wave by frequency-shifting part of it with an acousto-optic modulator. A heterodyne measurement can then very sensitively reveal any phase changes to that signal.

By bandpass-filtering the electronic signal, it is also possible to sample a very narrow spectral portion of the signal. That detection bandwidth can easily be far below what is possible with optical bandpass filters. This can be useful for precision laser spectroscopy, for example.

HOMODYNE DETECTION

Homodyne detection is the standard method for measuring the quantum mechanical properties of light. Optical homodyne detection improves the receiver sensitivity.

Theory:

- The attraction of optical homodyne detection is not just the potential 3 dB improvement in receiver sensitivity but also that it can ease the receiver bandwidth requirement considerably.
- This factor is illustrated in Fig. 14 which compares the spectra at the output of the detector for PSK homodyne and PSK heterodyne detection.
- It may be observed that homodyne detection requires only the normal direct detection receiver bandwidth whereas heterodyne detection requires at least twice this bandwidth and often a factor of 3 or 4 times it.

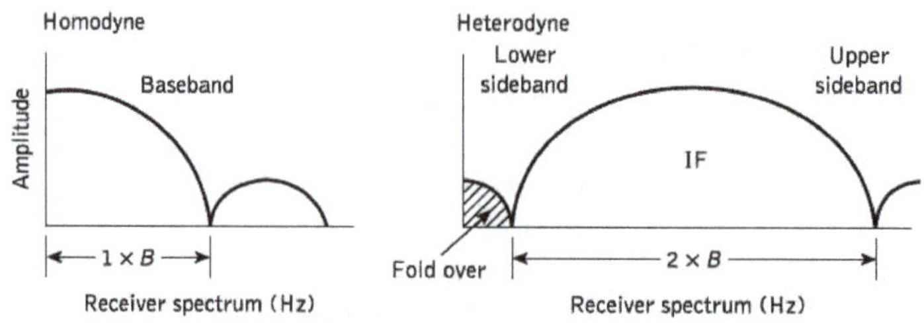

Fig. 14 Comparison of the electrical spectra at the optical detector output for homodyne and heterodyne detection of a PSK signal

- Optical homodyne detection using independent source and local oscillator lasers has proved quite difficult to achieve because of the problems associated with remotely optical locking a local oscillator laser to a low-level modulated signal.
- Such phase locking is essential because the phase difference ö must be held near zero for high-sensitivity reception.
- If ö drifts to $\pi/2$, then the output signal current IS will become zero and the detection process will cease.

Two homodyne demodulation strategies:

- Pilot carrier optical phase locked loop receiver
- Decision-driven optical phase locked loop receiver
- The optical phase difference between the incoming signal and the local oscillator signal, or the phase error signal, is detected by the photo detector prior to being fed back to correct the local oscillator frequency and phase.

Pilot carrier optical phase locked loop receiver:

- The optical PLL configuration shown in Fig. 15 employs a pilot carrier strategy for PSK homodyne detection.

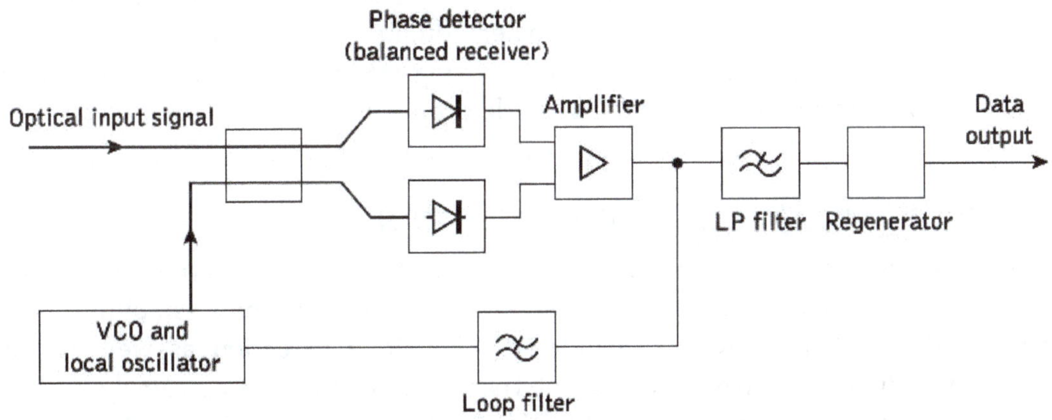

Fig. 15 Pilot carrier optical phase locked loop receiver

- In common with other pilot carrier techniques this carrier is generated by using incomplete (less than 180°) phase modulation.
- The pilot carrier signals, together with the incoming signal, are combined in a 3 dB fiber directional coupler and then detected using a balanced receiver.

HETERODYNE MEASUREMENTS WITH PULSED BEAMS

The heterodyne technique can also be applied to pulsed signals by using a pulsed local oscillator beam. It is even possible to use ultrashort pulses. As the heterodyne detector is essentially sensitive only while the local oscillator light is there (assuming a signal power well below the local oscillator power), it is possible to obtain a very high temporal resolution without requiring a very fast photodetector. This technique is called *heterodyne (or homodyne) linear optical sampling*.

Applications of Heterodyne Detection

Some examples for the application of optical heterodyne detection are:

- Heterodyne detection is used for *coherent Doppler LIDAR measurements* (and related laser radar techniques), where very weak light scattered in the atmosphere needs to be detected. It is possible, for example, to accurately measure wind speeds based on the Doppler shift. By deriving the signal and local oscillator beams from the same laser source, it is possible to resolve very small Doppler shifts.
- There are other applications in laser spectroscopy. Examples are solar radiometry and cavity ring-down spectroscopy.
- The linewidth of a laser, for example, can be measured with the self-heterodyne technique, where a frequency-shifted copy of the laser beam is superimposed with a temporally delayed copy on a photodetector.
- In optical fiber communications, the phase sensitivity of heterodyne and homodyne detection allows one to demodulate phase-encoded signals, based on techniques like frequency shift keying (FSK) (*coherent communications*). Also, it the basis for some forms of electronic dispersion compensation.
- Radio and microwave over fiber sometimes uses heterodyne detection.

HETERODYNE ASYNCHRONOUS DETECTION

Bboth ASK and FSK may be demodulated using asynchronous detection techniques, which puts the least demands on laser linewidth and phase stability. Such demodulation schemes, which include ASK envelope detection as well as FSK single- and dual-filter detection, do not therefore require the extremely narrow laser linewidths associated with synchronous binary PSK demodulation. Heterodyne envelope detection of an ASK signal may be achieved using an IF bandpass filter followed by a peak detector to recover the baseband signal as shown in Fig. 16(a). Such a scheme, however, incurs a receiver sensitivity penalty as a result of nonlinear filtering of the Gaussian-distributed noise in the peak detection process combined with the phase noise broadening of the signal spectrum such that a significant proportion of the signal energy can be translated outside the IF signal band. An optimum receiver bandwidth balances these factors and, for combined source and local oscillator linewidths of 10% of the transmitted data rate, the receiver penalty is around 3 dB.

By employing parallel filters with channels centered on the two transmitted frequencies it is possible to use envelope detection on each channel of a binary FSK signal. The configuration for this dual-filter demodulation technique is provided in Fig. 16(b). At the output it produces a differential ASK signal with a receiver sensitivity which is slightly better than asynchronous ASK demodulation in the presence of phase noise (i.e. a 2 dB penalty when source and local oscillator laser linewidths are 10% of the transmitted data rate. This improvement results from the complementary behavior of the dual-filter approach where, for large spacing between the two FSK signal channels, it is possible to have a significant spectral broadening of the signal but with insufficient energy in the complementary channel to register an error.

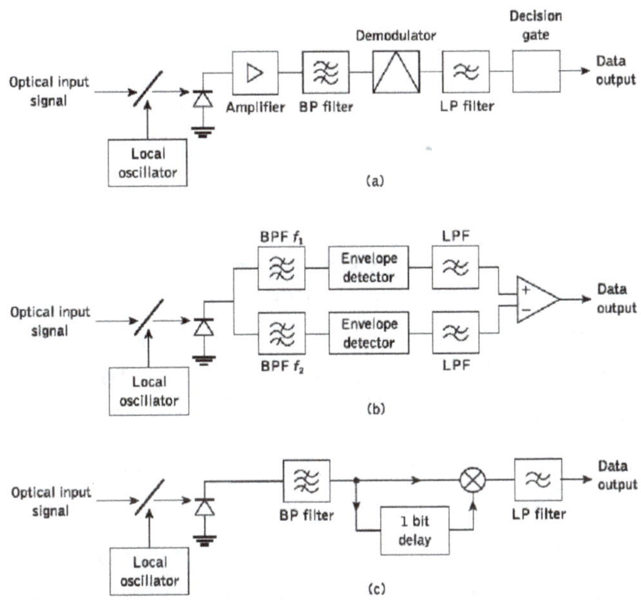

Fig. 16 Asynchronous heterodyne detection: (a) ASK single envelope detector receiver; (b) FSK dual-filter receiver; (c) asynchronous DPSK receiver

As mentioned earlier, asynchronous detection cannot be employed in the case of the PSK modulation format primarily because the phases of the received optical signal and the local oscillator laser are not locked and they can therefore drift over time.

The use of the DPSK format, however, does allow asynchronous demodulation as there is no similar requirement for phase locking. An asynchronous DPSK receiver is depicted in Fig. 16(c) which utilizes a 1 bit delay scheme. Hence the received bit stream is multiplied by its replica that has been delayed by 1 bit period duration.

This phase comparison demodulation reduces the synchronization problems associated with PSK as phase stability is only required over a few bit periods and it can therefore be implemented using distributed feedback lasers.

INTRADYNE DETECTION

This section explains the Intradyne detection. An alternative coherent optical detection concept to the common heterodyne or homodyne detection processes is that of intradyning.

Intradyne Detection method

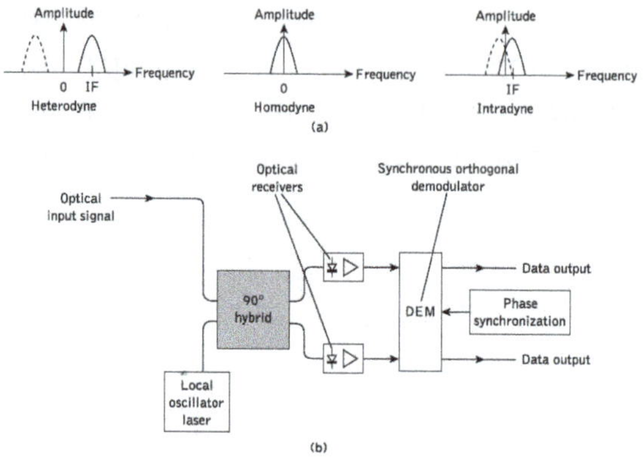

Fig. 17 Optical intradyne detection: (a) spectral comparison with heterodyne and homodyne detection; (b) two-phase QPSK intradyne receiver

- In the intradyne receiver the incoming signal is not precisely shifted to the baseband as in homodyne detection but shifted to a frequency much lower than the data transmission rate.
- Hence the electronic filtering can be undertaken using a baseband filter which is only slightly wider than that utilized for a homodyne receiver.
- The requirements on the automatic frequency control, however, for the local oscillator laser are similar to those of a heterodyne receiver.
- Use of a PLL can therefore be avoided with intradyne detection as there is a low IF generated and the IF is not zero as in homodyne receivers.

Optical intradyne detection

- Spectral comparison with heterodyne and homodyne detection;
- Two-phase QPSK intradyne receiver

Spectral comparison with heterodyne and homodyne detection

- A spectral diagram of the intradyne process for PSK comparing it with both heterodyne and homodyne PSK detection is provided in Fig. 17(a).
- It may be observed that the intradyne spectrum is down converted to produce an IF near the baseband as a consequence of the local oscillator laser only exhibiting approximately the same optical frequency as the incoming signal.

- The intradyne detection process was initially implemented in an experimental quadradature PSK receiver which employed purely electronic digital carrier recovery.

Two-phase QPSK intradyne receiver

- Fig. 17(b) depicts a two-phase intradyne receiver structure.
- There is no optical PLL shown as the receiver exhibits nonzero, albeit very low, IF.
- A minimum of two phases those being in-phase and quadrature (I and Q), is needed to avoid loss of information.
- This synchronous intradyne receiver incorporates a 90° hybrid and two balanced receivers using four identical photodiodes.
- The carrier recovery delivers the electronic carrier required to provide an input to the electronic PLL which facilitates the IF down conversion to convert the detected I and Q signals from the small IF signal into two data streams.
- This following synchronous demodulation is performed by an orthogonal demodulator.
- Although the intradyne approach removes the need for an optical PLL while exhibiting the nearly similar baseband, direct detection receiver bandwidth of homodyne detection, it can necessitate laser linewidth restrictions similar to those when an optical PLL is utilized.
- A feed forward carrier recovery QPSK/BPSK intradyne receiver strategy has, however, been explored which is predicted to be extremely tolerant to laser linewidths, thus enabling the use of distributed feedback lasers.

Phase Diversity Reception

This section explains the phase diversity reception in detail. The technique utilizes a fixed phase relationship between the ports of a multiport coupler to provide the direct demodulation to the baseband without the requirement for an optical PLL.

In-phase and quadrature phase diversity (two-phase) receiver:

- One variant of the optical phase diversity receiver known as the in-phase and quadrature (I and Q) receiver is shown in Fig. 18.
- In this two-phase scheme the incoming and local oscillator signals are combined in a 90° optical hybrid similar to the one.
- The 90° hybrid is connected to two detectors, the outputs of which are amplified and then passed through square-law devices prior to electrical recombination.

- The output signals from each receiver path prior to recombination can be written in terms of their voltages V1 and V2 as:

$$V_1^2 = k_1^2 m^2(t) \sin^2 \delta\varphi$$
$$V_2^1 = k_2^2 m^2(t) \cos^2 \delta\varphi$$

Where

- k_1 and k_2 are constants,

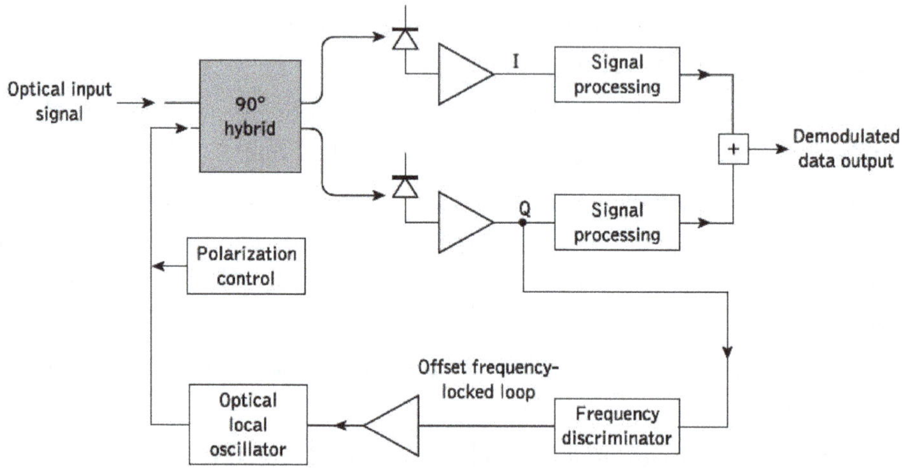

Fig. 18. In-phase and quadrature phase diversity (two-phase) receiver

- m(t) is the modulation
- $\delta\varphi$ is the phase error.
- Hence the output signal from the receiver is:

$V_1^2 = V_2^2 k^2[m(t)]^2$

Where we assume $k = k_1 = k_2$.

- It may be observed from Eq that for ASK the demodulated signal is constant irrespective of the relative phase between the incoming and local oscillator signals.
- With ASK modulation the laser linewidth requirements are comparable with heterodyne detection with asynchronous IF demodulation.
- For PSK modulation the signal may be differentially demodulated by the inclusion of a 1 bit delay in one of the inputs to the square-law mixer.
- Hence the change in phase during a single bit period only is of concern and any longer term phase drift is removed.

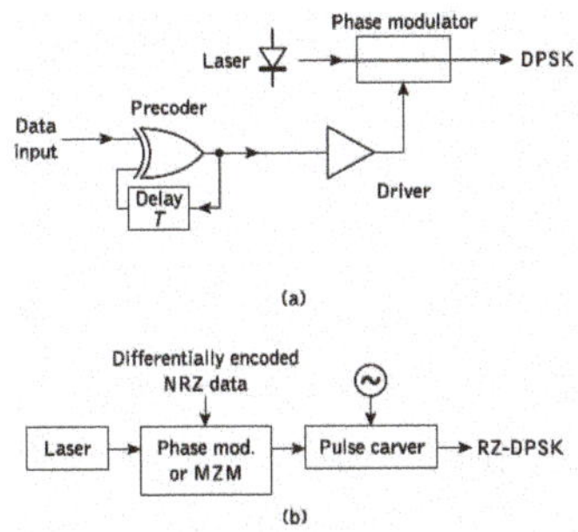

Figure 19 Differential phase shift keying transmitters: (a) DPSK transmitter; (b) RZ-DPSK transmitter

- This DPSK I and Q demodulation is expected to have similar laser linewidth requirements to DPSK heterodyne detection.
- The I and Q phase diversity receiver is more sensitive to fluctuations of the incoming SOP than a conventional optical heterodyne receiver.
- Polarization control should, however, be more straightforward in the I and Q case because it is possible to obtain electrical signals directly from the two receiver arms which provide exact information on the received polarization state.
- Problems do exist with two-port phase diversity reception, as in practice the electrical square-law demodulation is imperfect and produces additional terms which tend to appear in the baseband along with the demodulated signal.
- The two detected currents must be 90° out of phase, which may only be achieved at the expense of additional signal processing, and also the two arms of the receiver must be well matched.

DIFFERENTIAL PHASE SHIFT KEYING

Phase synchronization is eliminated using differential encoding. Optical DPSK signals carry the data in the phase difference between two consecutive symbols on the optical carrier as.

DPSK transmitter:

- A DPSK transmitter is illustrated in Fig. 19(a).
- The DPSK transmitter also requires a precoder to enable the generation of the DPSK signal prior to its application to the phase modulator.
- Although a straight line phase modulator is displayed in Fig. 19(a), a Mach-Zehnder modulator can be used in its place to produce the phase modulation.
- The RZ pulse format has proved of most interest for the provision of both spectral efficiency and improved tolerance to fiber nonlinearities, enabling

it to be utilized for long-haul transmission with optical amplifiers boosting the optical power levels

RZ-DPSK TRANSMITTER:

- An optical RZ-DPSK transmitter is shown in Fig. 19(b).
- In common with the DPSK transmitter in Fig. 19(a), the non return-to-zero (NRZ) data signal is differentially encoded using a precoder before it is applied to the phase modulator or the Mach–Zehnder modulator (MZM).
- The phase modulator only modulates the phase of the optical field resulting in a constant amplitude optical field, but it also introduces chirp across the bit transitions as a consequence of its non-instantaneous operation.
- An MZM always produces precise π phase transitions at the expense of residual optical intensity dips at the phase transition locations.
- Since exact phase modulation is more important for both DPSK and RZ-DPSK than constant optical intensity, such transmitters are more frequently implemented using an MZM to perform the phase modulation.
- A pulse carver, usually comprising an MZM controlled by a synchronized sinusoidal drive with the same amplitudes and of opposite phase, provides the RZ-DPSK output

Two types of DPSK receiver:

Differential phase shift keying receivers

- Heterodyne DPSK receiver;
- Direct detection DPSK receiver
- Two types of DPSK receiver are shown in Fig. 20.
- In both cases a balanced optical receiver is employed incorporating a fiber coupler and two matched photodetectors.

Heterodyne receiver:

- The heterodyne receiver depicted in Fig. 20(a) then utilizes an electrical delay of bit period T in the phase comparison system together with the un-delayed signal which then combines in a multiplier circuit to provide the differential phase.
- The output from the multiplexer is then integrated to eliminate residual noise, a positive voltage being obtained if the phases of the combined signals are the same.

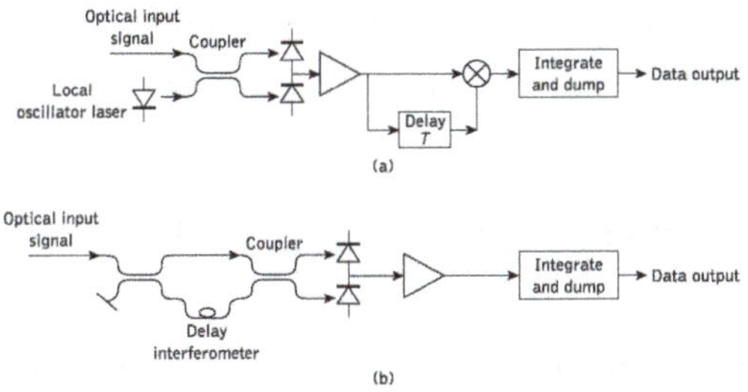

Fig. 20 Differential phase shift keying receivers: (a) heterodyne DPSK receiver; (b) direct detection DPSK receiver

- Alternatively, when the two phases differ by 180°, a negative voltage is obtained at the data output.
- It should be noted that while frequency locking may be required, phase locking is not necessary for optical heterodyne DPSK demodulation.

ASK HETERODYNE DETECTION

ASK heterodyne detection system in which the effect of laser phase noise is negligible. In a heterodyne receiver the analyses of signal and noise phenomena are more complicated

Heterodyne Detection method for ASK:

- The ASK or OOK modulation format has similarities to digital transmission in an IM/DD optical fiber system
- In a heterodyne receiver the analyses of signal and noise phenomena are more complicated than in the IM–DD case because the optical detector output appears as an IF signal and not as a baseband signal.
- Hence the IF output current from the photo detector IS(t) which corresponds to the input current to the preamplifier can be written as:

$$I_s(t) = \begin{cases} I_{SH}\cos(\omega_{IF}t + \phi) & \text{for a 1 bit} \\ 0 & \text{for a 0 bit} \end{cases} \tag{1}$$

Where:

$$I_{SH} = \frac{2\eta c}{hf}\sqrt{P_s P_I} \tag{2}$$

- Obtain the IF noise current, two assumptions can be made.
- Firstly, it is assumed that the local oscillator signal power is much larger than the incoming signal power so that the total noise current is approximately equal to and this applies for both the 1 and the 0 bit.

- Secondly, it is assumed that this IF noise current N(t) can be considered as narrowband noise which can be expressed as

  $N(t) = x(t) \cos \omega_{IF}t + y(t) \sin \omega_{IF}t$ (3)

- Where x(t) and y(t) are functions of time which vary at a much slower rate than the IF signal.
- It should be noted that the first and second terms in Eq. (3) represent the I and Q components respectively.
- Hence the mean square values of x(t) and y(t) may be written as:

  $$\overline{X}^2(t) = \overline{y}^2(t) = \overline{i}_{SL}^2 \quad (4)$$

- For heterodyne synchronous detection, the IF amplifier is followed by a demodulation circuit which has a phase synchronized reference signal proportional to $\cos\omega_{IF}t$.
- The probability density functions of $V_d(t)$ for the ASK signal $I_S(t)$ represented by Eq. (1) are illustrated in Fig. 21.

PSK HETERODYNE DETECTION

An optical phase-shift-keying (PSK) heterodyne receiver using synchronous detection by means of a decision directed phase locked loop (PLL) has been investigated taking the effect of shot noise, phase noise, adjacent channel interference and loop propagation delay. The evaluation of phase-error variance and error probability calculations are performed without affecting the loop stability. The plots of error probability with PNB are presented as a function of normalised optical domain channel spacing for several linewidth-to-bit-rate ratios. A relative improvement in error probability of 4 dB has been obtained with normalised control parameter of 0.4, $D = 10R_b$ & $\delta f = 0.001R_b$.

When a phase modulated optical signal at the wavelength of 1.55 μm is transmitted through a 1.3-μm zero dispersion fiber, phase modulation (PM) is converted to amplitude modulation (AM) by fiber chromatic dispersion.1 In optical heterodyne detection, the PM-AM conversion can be compensated by a delay equalizer.2 However, in a repeatered transmission system in which the fiber input power is large after in-line amplifiers. i.e., Er+3-doped optical fiber amplifiers (EDFAs) are used as in-line amplifiers, the refractive index of the fiber is modulated by the amplitude modulation resulting from the PM-AM conversion. This self-phase modulation (SPM) degrades the receiver sensitivity.

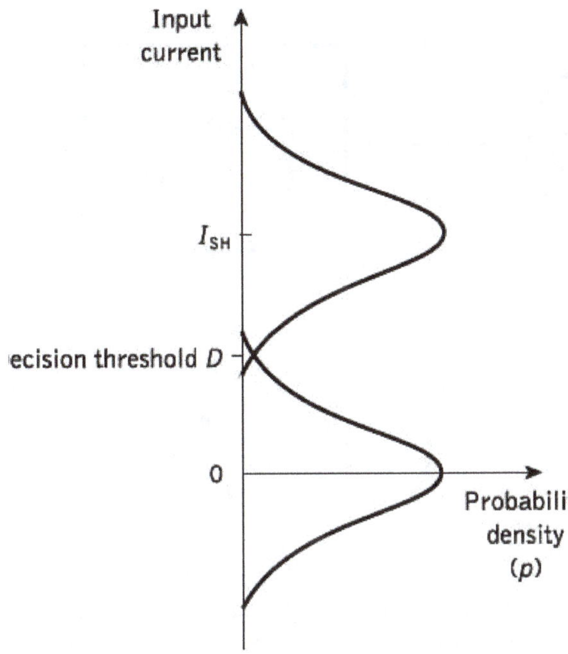

Fig. 21. Probability density functions for ASK heterodyne synchronous detection

Phase Shift Keying PSK is the digital modulation technique in which the phase of the carrier signal is changed by varying the sine and cosine inputs at a particular time. PSK technique is widely used for wireless LANs, bio-metric, contactless operations, along with RFID and Bluetooth communications.

PSK is of two types, depending upon the phases the signal gets shifted. They are -

Binary Phase Shift Keying BPSK

This is also called as 2-phase PSK or Phase Reversal Keying. In this technique, the sine wave carrier takes two phase reversals such as 0° and 180°.

BPSK is basically a Double Side Band Suppressed Carrier DSBSC modulation scheme, for message being the digital information.

Quadrature Phase Shift Keying QPSK

This is the phase shift keying technique, in which the sine wave carrier takes four phase reversals such as 0°, 90°, 180°, and 270°.

If this kind of techniques are further extended, PSK can be done by eight or sixteen values also, depending upon the requirement.

BPSK Modulator

The block diagram of Binary Phase Shift Keying consists of the balance modulator which has the carrier sine wave as one input and the binary sequence as the other input. Following is the diagrammatic representation (Fig. 22).

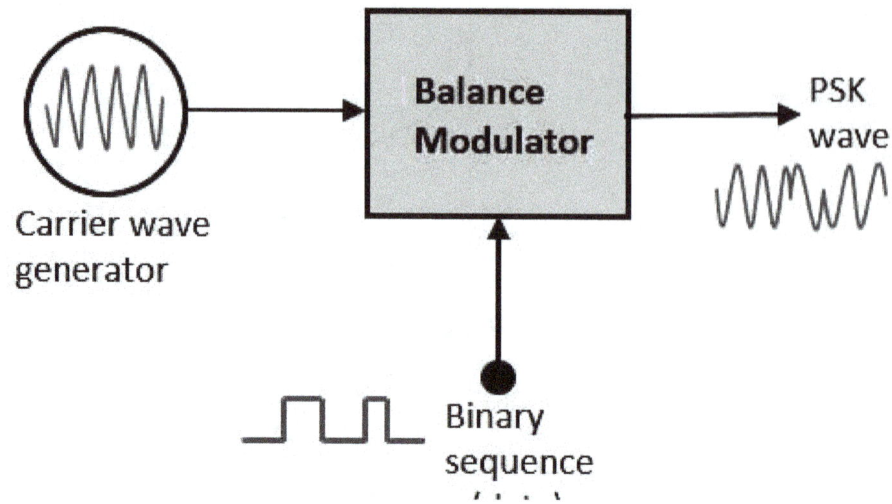

Fig. 22. BPSK Modulator

The modulation of BPSK is done using a balance modulator, which multiplies the two signals applied at the input. For a zero binary input, the phase will be 0° and for a high input, the phase reversal is of 180°.

Following (Fig. 23) is the diagrammatic representation of BPSK Modulated output wave along with its given input.

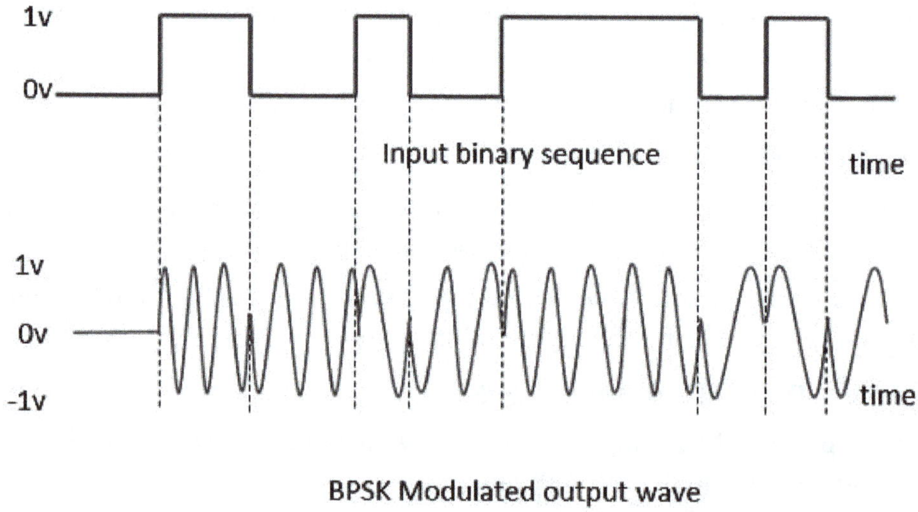

Fig. 23. Diagrammatic representation of BPSK Modulated output wave

The output sine wave of the modulator will be the direct input carrier or the inverted 180°phaseshifted input carrier, which is a function of the data signal.

BPSK Demodulator

The block diagram of BPSK demodulator consists of a mixer with local oscillator circuit, a bandpass filter, a two-input detector circuit. The diagram is as follows.

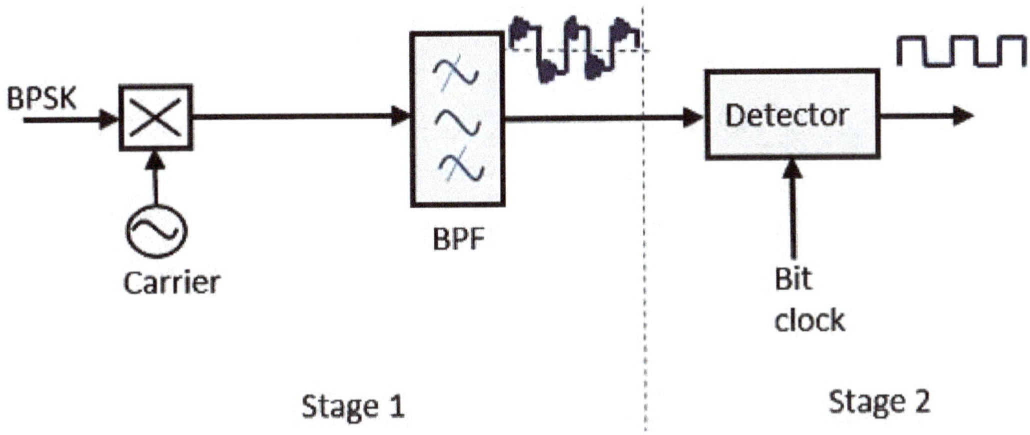

Fig. 24. BPSK Demodulator

By recovering the band-limited message signal, with the help of the mixer circuit and the band pass filter, the first stage of demodulation gets completed. The base band signal which is band limited is obtained and this signal is used to regenerate the binary message bit stream.

In the next stage of demodulation, the bit clock rate is needed at the detector circuit to produce the original binary message signal. If the bit rate is a sub-multiple of the carrier frequency, then the bit clock regeneration is simplified. To make the circuit easily understandable, a decision-making circuit may also be inserted at the 2nd stage of detection.

CHAPTER

13 OPTICAL NETWORKS

An Optical Network is basically a communication network used for the exchange of information through an optical fiber cable between one end to another. It is one of the quickest networks used for data communication.

Optical networks are telecommunications network of high capacity. They are based on optical technologies and components, and are used to route, groom, and restore wavelength levels and wavelength-based services. This tutorial is divided into distinct chapters, which explains the structural features of optical fibers and their connections in networks. The nature of optical networks along with the recent developments in the Optical and Networking systems using optical sources and devices is also dealt with.

The fiber cables in an optical network are made of a core surrounded by a glass cladding, which keeps light within the cable. Without optical networks, connections to distant data centers or other sources of data wouldn't be as fast as they are today. The internet's global reach relies on optical networking to ensure a user in the United States can access a server in Japan as fast as physics allows.

Optical networking is also used for connections in large-scale data centers. However, this article will focus more on the technology supporting long-haul connections.

Some basic vocabulary for fiber optic networking includes: cables, fiber, core, and cladding.

- **Cables**: bundles of optical fibers
- **Fiber**: what light travels through, similar to a wire but made of a core surrounded by cladding
- **Core**: the innermost layer of the fiber that contains the light, typically made of glass

- **Cladding**: material wrapped around the core, made of glass or plastic of a different density

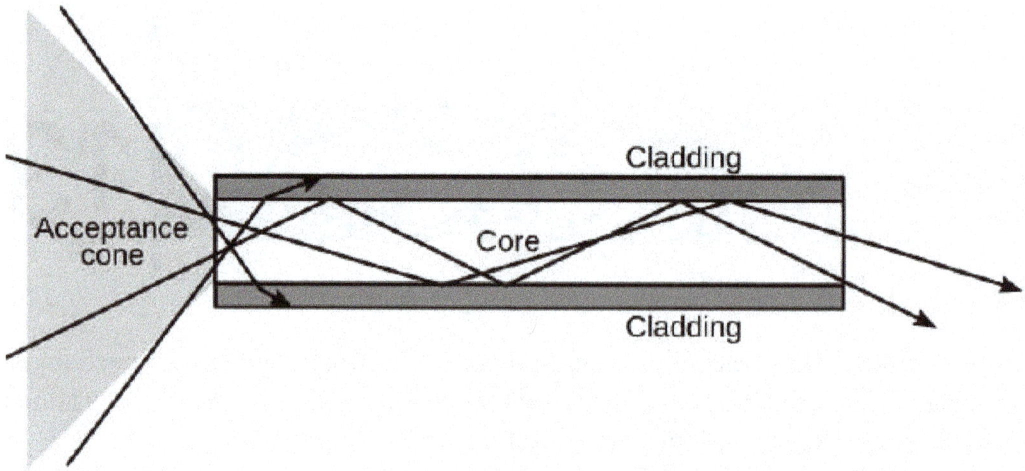

Fig. 1. Fiber optic networking includes: cables, fiber, core, and cladding

The cladding keeps the light from escaping the fiber. Light is sent into the fiber core at an angle of 42 degrees or less. When the light hits the cladding, it is reflected off of the cladding so it remains in the core. This is called total internal reflection.

The method for turning binary code from electric signals into light signals is fairly direct. The light source will send a pulse to indicate a one, while the lack of a light pulse indicates a zero.

The current thinking about IP over WDM by outlining a path to optical data networking, that includes multiple data networking protocol coupled with a protocol-neutral optical networking infrastructure is challenged. This tutorial discusses the diversity of data networking protocols and network architectures for optical data networking.

The bandwidth explosion ushered in by the popularity of the Internet has led to a paradigm shift in the telecommunication industry from voice-optimized circuit-switched services to data-optimized packet-switched services. The notation of supporting "data directly over optics" has been fueled by the promise that elimination of unnecessary network layers will lead to a vast reduction in the cost and complexity of the network.

In this view of reduced or collapsed network layers, existing TDM systems such as Synchronous Digital Hierarchy (SDH) plays a diminishing role, and optical transport networking emerges as the underlying transport infrastructure for the resultant "network of networks".

OPTICAL NETWORK CONCEPTS

Network Concepts

- A network utilizing optical fiber as a transmission medium provides a connection between many users to enable them to communicate with each other by transporting information from a source to a destination.
- It may also require an intermediate stage to process the data for control operation.
- Thus each stage of information transfer is required to follow the fundamentals of optical networking.
- These fundamentals essentially involve the methodology for the interconnection of various optical devices together with the operational procedures for successful transportation of optical signals between the source and the destination nodes.
- Figure 2 shows the structure of a simple optical network.

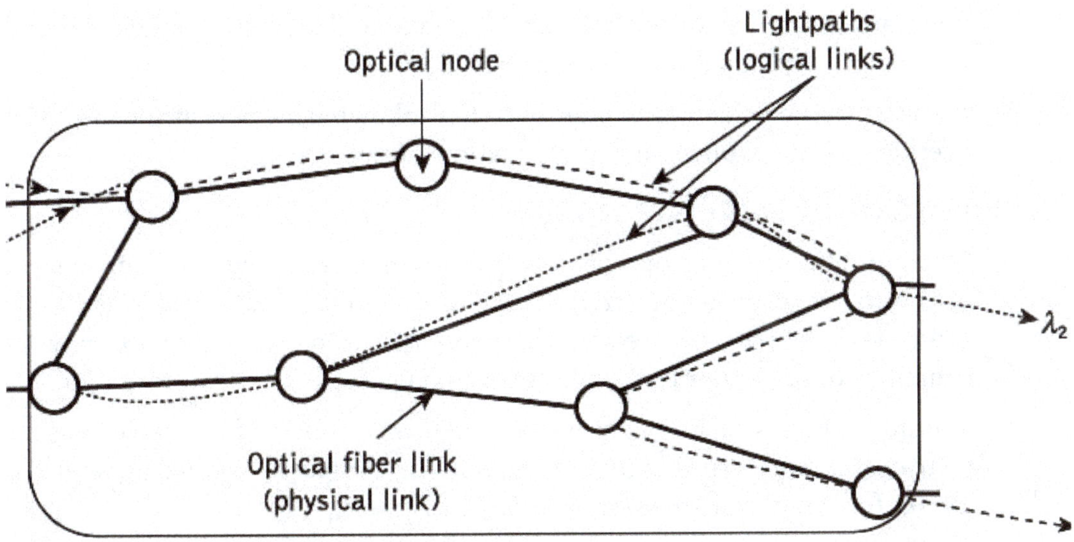

Fig. 2 Optical network structure

- It consists of optical nodes interconnected with optical fiber links.
- An optical node is a multifunctional element which basically acts as a transceiver unit capable of receiving, transmitting and processing (if required) the optical signal.
- Optical fibers provide point-to-point physical connections between network nodes.

- The point-to-point fiber links can be used to establish logical links where the destination node can be reached by traveling through one (or more intermediate nodes) in a single or multiple hops.
- Ideally, for an uninterrupted optical signal to reach a destination node using multi hops each channel is assigned a specific signal wavelength from source to destination.
- A signal carried on a dedicated wavelength from source to a destination node is known as a light path (i.e. ë1 and ë2 are two light paths in this particular case).
- The data can be sent over these light paths once the connections are set up.
- In addition, a controlling mechanism is also required to provide for data flow during its transmission in order to authenticate the entire data transportation between each transmitting and receiving node.
- Based upon this simple optical network scenario, the fundamentals can be divided into three broad areas: namely, optical networking terminology, the functions and types of optical network node and switching elements, and wavelength division multiplexed optical networks.
- These aspects are further explained in the following sections prior to a brief overview of the public telecommunications networks

ELEMENTS OF OPTICAL NETWORK

- **Stations:** Stations in an optical network serves as the source and destination of the information being transmitted and received. Stations are basically those devices that are used by the users of the network. For example, a computer or any other telecommunication device.
- **Trunk:** A trunk is basically a transmission line i.e., optical fiber cable in order to transmit the optical signal. A network is composed of one or multiple trunks for signal transmission over large distance.
- **Node:** Node is nothing but acts as a hub for multiple transmission lines inside the network. In case of a single transmission line, an optical network does not require nodes, as in this case stations at both the ends can be directly connected to the fiber cables.
- **Topology:** When multiple fiber cables are employed in an optical network, then these are connected through nodes. But the way in which the multiple nodes are connected together denotes the topology of the network.
- **Router:** A router is basically placed inside an optical network that provides a suitable path for signal transmission.

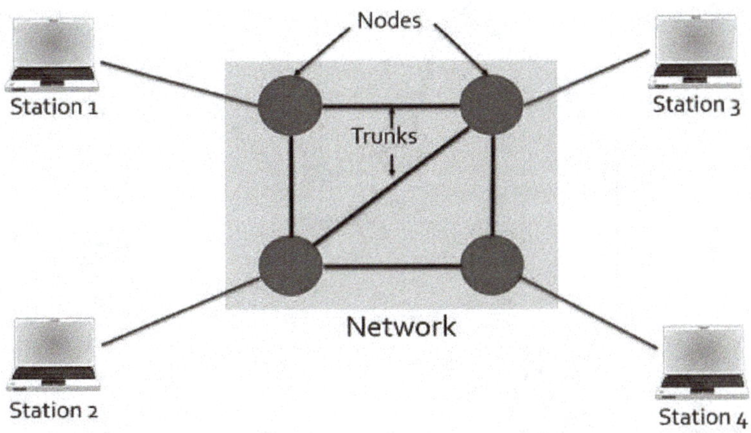

Fig. 3. Elements of optical network

Optical Network Topologies

As we have already discussed in the previous section that topology is the arrangement of multiple optical fiber transmission lines in an optical network. So, let us now move forward to understand the various topology configurations:

Bus Topology: In a bus topology (Fig. 5), the various nodes are connected through a single trunk line with the help of optical couplers. This allows a convenient as well as a cost-effective method to transmit the signal. However, in a bus topology, it is difficult to determine the faulted node as well as it also takes time to restore the transmitted signal from that particular node.

Ring Topology: In a ring topology (Fig. 6), one single node is joined to its neighbouring node thereby forming a closed path. So, the transmitted information in the form of light is sent from one node to another. Also, optical couplers are installed within the network in order to the couple the transmitted optical signal from one node to another.

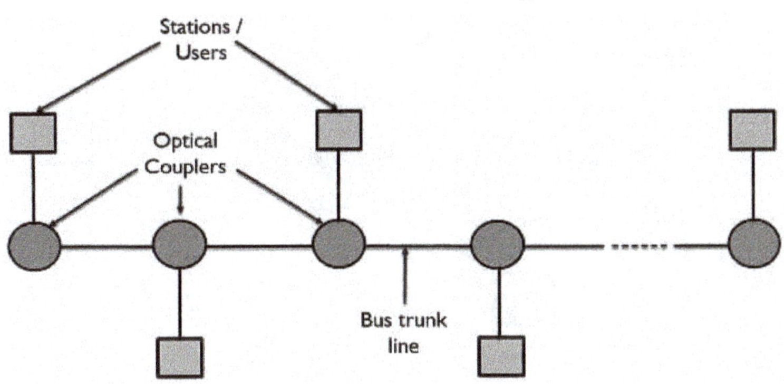

Fig. 5. Bus Topology

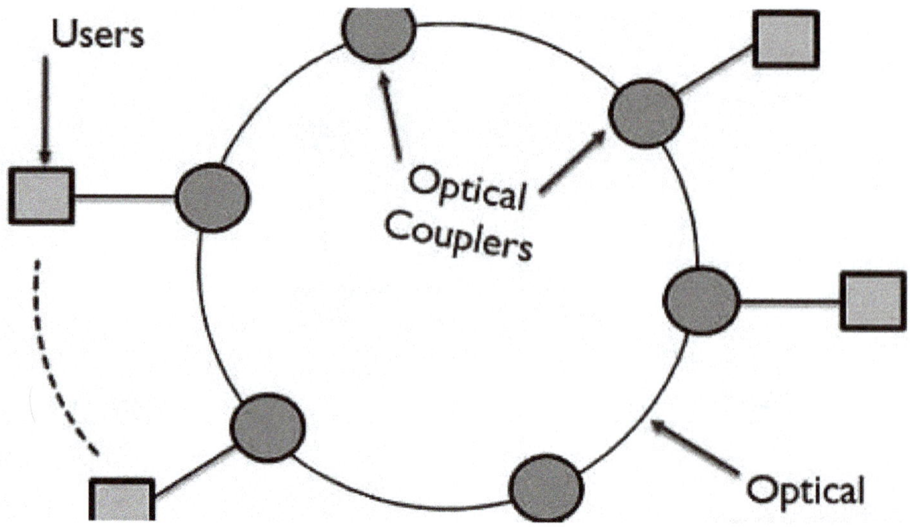

Fig. 6. Ring Topology

Star Topology: In star connection (Fig. 7), the various nodes of the network are connected together with a single central hub. This central hub can be active or passive network. This central hub then controls and directs the transmitted optical signal inside the optical network.

Mesh Topology: In a mesh topology (Fig. 8), an arbitrary connection is formed between the nodes in the network. This point to point connection can be changed according to the application. This shows the flexible nature of star topology as in case of failure of one node, others can be used for signal transmission.

Basically, in mesh connection, failure of any link or node is generated then firstly that particular failure is detected and then the signal traffic is diverted from failed node to another link inside the connection.

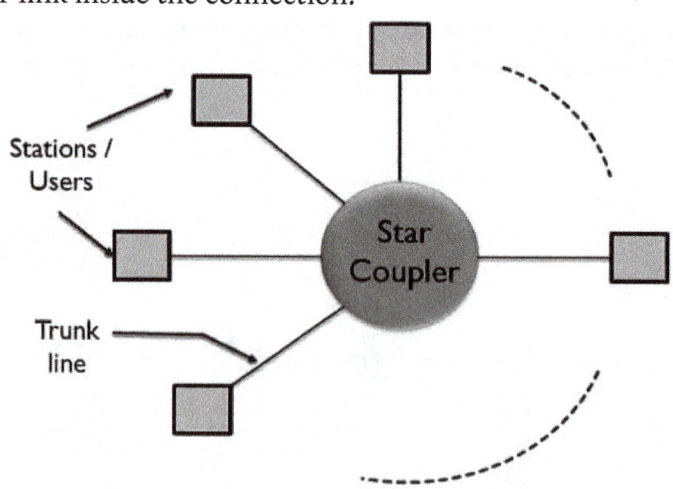

Fig. 7. Star Topology

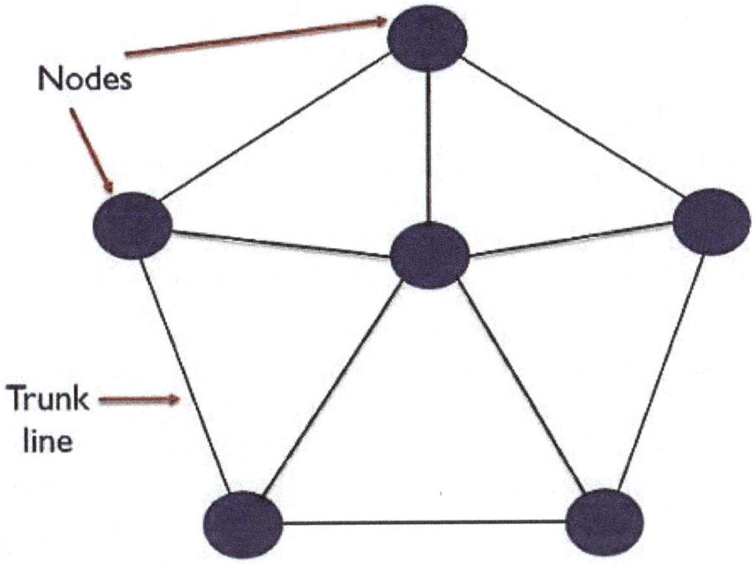

Fig. 8. Mesh Topology

CATEGORIES OF OPTICAL NETWORK

The categories of optical network are based on the area that connects the user of the network. These are classified as:

- **Local Area Network (LAN)**: Basically a LAN connection provides the interconnection of users that are present in localized areas like a building, a department or an office etc.
- The example of networking topology of LAN is Ethernet. As in LAN, users are permitted to share the resources together like servers etc. These are personally owned by an organization. It is quite inexpensive.
- **Campus network**: This network category is formed by the interconnection of multiple LAN's. This is basically extended to a large level but is still confined within a localized area. It is also governed by a single organization.
- The examples of campus network are university campus, a government organization, or a medical centre etc.
- **Metropolitan Area Network (MAN)**: It is also known as a metro network and covers a greater area than a campus network. It permits the interconnection of several buildings that are present in different cities.
- Due to its large operating area, MAN is controlled by several communication organizations.
- **Wide Area Network (WAN)**: Unlike MAN, a WAN provides inter-connection of users from neighbouring cities as well as cross-country regions. It is employed to establish communication over a large geographical distance

and is controlled and maintained by some private organizations or telecommunication service providers.

OPTICAL INTERNET

Optical internet working, for example, as defined by the Optical Interworking Forum (OIF), is a data-optimized network infrastructure in which switches and routers have integrated optical interfaces and are directly connected by fiber or optical network elements, such as Dense Wavelength-Division Multiplexers (DWDMs).

At present, however, the notion of IP directly over WDM is little more than cleverly disguised marketing. Almost invariably, IP over WDM is IP packets mapped into SDH, coupled with SDH based point-to-point DWDM systems. SDH standalone elements, often referred to as Time-Division Multiplexer (TDMs), are not required, but SDH remains an integral element of the data networking equipment interface.

Ever-increasing reliance on the presence of SDH in DWDM systems limits technological innovation. For example, it may inhibit packet over fiber applications such as Asynchronous Transfer Mode (ATM), Gigabit Ethernet (GbE) and 10 GbE over DWDM. Nor does it bring us any closer to realizing the ultimate vision of optical transport networking.

As compared to the present view of IP over WDM, there is a more balanced view of data/transport network evolution. This balanced view is based on two fundamental principles "

- Every data network is unique, in a marketplace governed by differentiation.
- The Optical Transport Network (OTN), as the underlying infrastructure "network of networks" should be capable of transporting a wide variety of client signals, independent of their format.

Together, these fundamental principles form the basis for the notion of optical data networking.

CONVERGENCE NETWORKS

Today's TDM-based transport networks have been designed to provide an assured level of performance and reliability for the predominant voice and based-line services. Proven technologies, such as SDH, have been widely deployed, providing high-capacity transport, scalable to gigabit per second rates, for voice and leased-line applications.

SDH self- healing rings enable service-level recovery within tens of milliseconds following network failures. All of these features are supported by well-established global standards enabling a high degree of multivendor interoperability.

TODAY'S NETWORK

In contrast to today's TDM-based transport networks (and, to some extent, with ATM networks), "best-effort" IP networks generally lack the means to guarantee high reliability and predictable performance. The best-effort service provided by most legacy IP networks, with unpredictable delay, jitter, and packet loss, is the price paid to achieve maximum link utilization through statistical multiplexing. Link utilization (e.g. the number of users per unit of bandwidth) has been an important figure of merit for data networks, since the links are usually carried on leased circuits through the TDM transport network.

Given the inherently bursty nature of data traffic, the fixed-bandwidth pipes of TDM transport may not be an ideally efficient solution. However, this inefficiency has traditionally been considered of less importance important than the network reliability and congestion isolation features of a TDM-based transport network provider.

The surging demand for high bandwidth and differentiated data services is now challenging this dual architecture model of TDM-based transport and best effort packet networks. It is not cost- effective to extend the usefulness of best-effort networking by over provisioning network bandwidth and keeping the network lightly loaded.

Furthermore, this approach cannot always be achieved or guaranteed due to spotty demand growth, and is a particular issue for the network access domain, which is most sensitive to the economic constraints of underutilized facilities. As a result, in general, data service providers today do not have the network infrastructure support to provide customer- specific differentiated service guarantees and corresponding service-level agreements.

NEXT GENERATION NETWORK

Next generation network architectures for cost-effective, reliable, and scalable evolution will employ both transport networking and enhanced service layers, working together in a complementary and interoperable fashion. These next-generation networks will dramatically increase, and maximally share, backbone network infrastructure capacity, and provide sophisticated service differentiation for emerging data applications.

Transport networking enables the service layers to operate more effectively, freeing them from constraints of physical topology to focus on the sufficiently large challenge of meeting service requirements.

Hence, complementing the many service-layer enhancements, optical transport networking will provide a unified, optimized layer of high-capacity, high reliability bandwidth management, and create so-called optical data networking solutions for higher capacity data services with guaranteed quality.

OPTICAL TRANSPORT NETWORKING: A PRACTICAL VIEW

Visions of optical networking have captured the imagination of researchers and network planners alike, since the rapid and successful commercialization of WDM. In the original vision of optical transport networking, a flexible, scalable, and robust transport network emerges, catering to an expanding variety of client signals with equally varied service requirements (flexibility, scalability, and survivability coupled with bit rate and protocol independence).

The promise of a transport infrastructure capable of meeting the burgeoning bandwidth demands well into this new century, wherein wavelengths replace timeslots as the medium for providing reliable transfer of high-bandwidth services across the network, is indeed tantalizing. But what is optical networking ? The answer varies widely, and in fact has evolved over recent years. Early attempts at optical networking focussed on an optical transparency and the design of optically transparent networks on a global scale.

Practical Solution

In the absence of viable "all-optical" solutions more practical solutions for optical networking accommodate the need for opto-electronics to support optical signal regeneration, and optical signal performance monitoring. In what is termed all-optical networking, signals traverse the network entirely in the optical domain, with no form of opto-electronic processing. This implies that the all signal processing-including - signal regeneration, routing, and wavelength interchange - takes place entirely in the optical domain.

Due to limitations of analog engineering (e.g. limiting factor in a properly designed digital system is an one accuracy of the conversion of the original analog message waveform into digital form) and considering the current state- of- the- art in all-optical processing technology, the notion of global or even national all optical networks is not practically attainable.

In particular, opto-electronic conversion may be required in opto network elements to prevent the accumulation of transmission impairments - impairments that result from such factors areas fiber fibre chromatic dispersion and nonlinearities, cascading of non-ideal flat-gain amplifiers, optical signal crosstalk, and transmission spectrum narrowing from cascaded non-flat filters. Opto-electronic conversion can also support wavelength interchange, which is currently a challenging feature to realize in to the all optical domain.

In short, in the absence of commercially available devices that perform signal regeneration to mitigate impairment accumulation and support wavelength conversion in the all-optical domain, some measure of opto-electronic conversion should be expected in near-term practical optical networking architectures. The resulting optical network architectures can be characterized by optically transparent (or all-optical) subnetworks, bounded by feature-enhanced opto-electronics, as shown in Fig. 9.

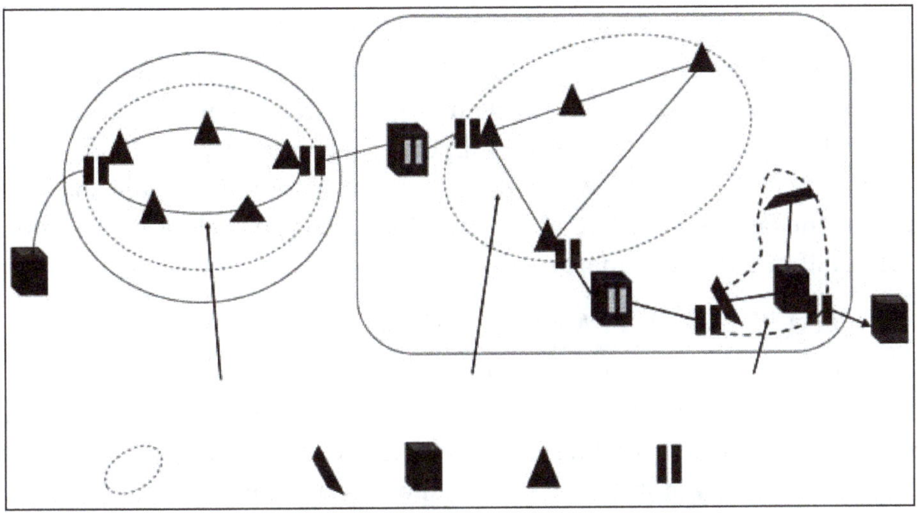

Fig. 9. Optically transparent subnetworks, bounded by feature-enhanced optoelctronics

CLIENT SIGNAL TRANSPARENCY

Beyond analog network engineering, practical considerations will continue to govern the ultimate realization of the OTN. Paramount among these considerations is the network operator's desire for a high degree of client signal transparency within the future transport infrastructure.

What is meant by "Client signal transparency"? Specifically, for the desired set of client signals targeted for transport on the OTN, individual mappings are defined for carrying these signals as payloads of optical channel (OCh) server signals. Signals expected in the OTN include legacy SDH and PDH signals, and packet-based traffic such as Internet Protocol (IP), ATM, GbE and Ssimple Ddata Llink (SDL). Once a client signal has been mapped into its OCh server signal at the ingress of the OTN, an operator deploying such a network need not have detailed knowledge of (or access to) the client signal, until it is demapped at the network egress.

The optical network ingress and egress points should delimit the domain of OTN client signal transparency. Hence, the most important factor in realizing client signal transparency is to eliminate all client-specific equipment and processing between OTN ingress and egress points. Fortunately, it is easier to accept client-dependent equipment at the ingress/egress, since it is generally dedicated on a per-service basis.

OPTICAL TRANSPORT NETWORKING VIA DIGITAL WRAPPERS

The widespread use of DWDM technology has presented service providers with a new challenge : how to cost- effectively manage the increasing number of wavelengths to provide fast, reliable services to their end customers. To effectively manage wavelength or OChs, requires that optical networks support per wavelength or OCh-level operations, administration and maintenance (OAM) functions.

ITU(T) Rec. G872 defines some functionality for OCh-level OAM implemented in the form of overheads without specifying how this overhead is to be carried. Until now, the only feasible way to support signal regeneration and to monitor, analyze, and manage OChs (wavelengths) was to rely on SDH signals and equipment throughout the network. This requires that the signals on each of the wavelengths in the WDM system be SDH formatted.

AN OPTICAL CHANNEL (WAVELENGTH)

Taking advantage of the existing opto-electronic regeneration points in DWDM systems, the notion of using digital wrapper technology will provide functionality and reliability similar to SDH, but for any client signal, bringing us one step closer to realizing the original vision of optical transport networking.

Digital wrapper technology provides the network management functions outlined in ITU(T) Rec. G.872 to enable OTNs. These include optical-layer performance monitoring, Fforward Eerror Ccorrection (FEC), and ring protection and network restoration on a per-wavelength basis, all independent of the input signal format as is shown in Fig. 10.

The notion of using a digital (or TDM) wrapper per "around" the OCh client to support channel-associated OCh overhead has recently been proposed, and has in fact, been adopted as the basis for definition of the OCh. This scheme will take advantage of the need for OCh regeneration to add additional capacity to the OCh client. Of course, once we have a means of adding overhead to the OCh client signal digitally, it makes sense to use this to support all of the OCh-level OAM requirements.

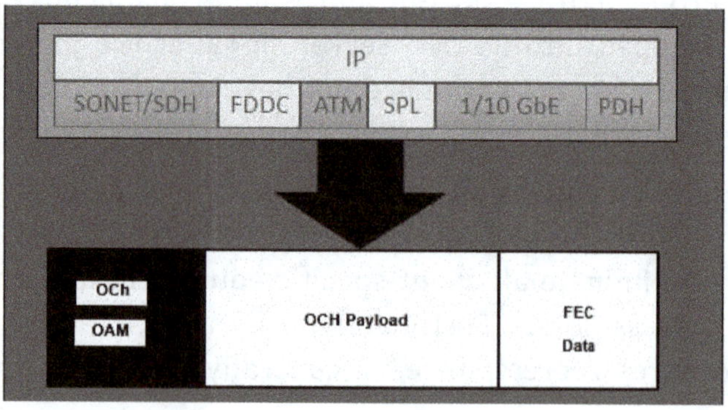

Fig. 10. Ring protection and network restoration on a per-wavelength basis, all independent of the input signal format

In particular, digitally added overhead makes it almost trivial to solve the major performance monitoring problem of the OTN, namely providing access to Bbit Eerror Rrate (BER) in a client-independent manner. BAnd by optionally using FEC, the digital wrapper method can significantly enhance the BER performance of the client signal, further minimizing the requirement for opto-electronic conversion.

One method of enhancing the performance of the transport network is through the use of FEC, which is currently provided in some equipment. Hence, an added benefit of the digital wrapper technique is the ability to optionally support FEC for system margin enhancement.

OCH FRAME STRUCTURE

In functional terms, the OCh payload and OAM should be separable from the FEC mechanism. This allows for carrying the payload and OAM end to end across the network, while using different FEC schemes on different links. An obvious example of where this might occur is between submarine and terrestrial links. In the former, new FEC codes are under investigation for the next generation of systems.

Following figure Below Fig. illustrates the proposed basic frame structure of the OCh, and the types of functions that may be carried in the OCh frame structure. While it might be argued that this proposal is inconsistent with long-term goals of the all optical networking, we should not expect the need for regeneration to disappear.

The distance between regeneration points will continue to increase; however, the need for regeneration at signal handoff points will remain. Coupled with the use of the Ooptical Ssupervisory Cchannel (OSC) to manage OChs within optically transparent subnetworks, digital wrappers will support end- to- end management of OChs (wavelength) across national or global OTNs.

3R-regeneration (Reshaping, Retiming, and Regeneration) is provided by means of optical-to-electrical conversion and vice versa, and the digital wrapper proposal takes advantage of this. Would the picture change should all-optical 3R-regeneration becomes available? If all-optical regeneration is capable of adding overhead, the argument is unchanged; only the regenerator implementation would change.

If Should optical regenerators are be unable to add overhead, the need for OChs overhead will not disappear. ; Optical regenerators would then simply increase the potential distance between the opto-electronic regeneration points, and the digital wrapper would pass transparently through them. The implications of the use of digital wrappers on the evolution of optical transport networking may be profound, especially when taken in the context of data networking trends.

PROTOCOL STACK CHOICES

The IP protocol is clearly the convergence layer in today's data communication networks, and it is foreseeable that it will expand this role to multi- service networks in the years to come. IP can be transported over a broad variety of data link layer protocols and underlying networking infrastructures. Fig. 11 shows some of the possible protocol stacks, or mappings, of IP into a WDM network infrastructure.

What is IP over WDM ?

The protocol stacks labelled a, b and d in Fig. 11 are the most commonly deployed today. They use the classical IP over ATM over SDH mapping as shown in Fig 11(a); packet over SDH (POS) as shown in Fig. 11(b); or the classical and well-extended IP over Ethernet as shown in Fig. 11(d).

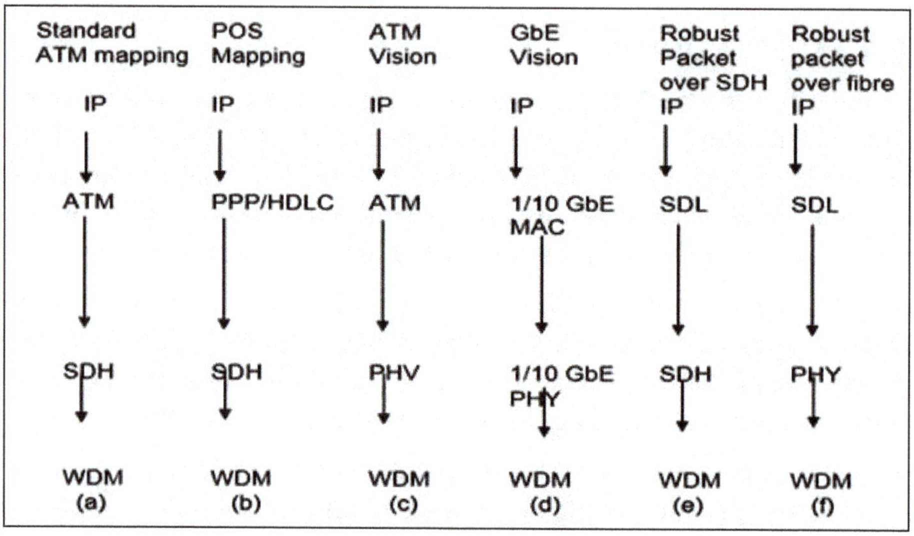

Fig. 11. Possible protocol stacks, or mappings, of IP into a WDM network infrastructure.

Cases (e) and (f) use Simple Data Link (SDL), a new data link layer recently proposed as an alternative to POS. The protocol stack labeled (c) is an alternative to case (a), where the intermediate SDH layer is eliminated and a direct mapping of ATM cells into WDM is performed.

These different protocol stacks provide different functionality, in terms of bandwidth overhead, rate scalability, traffic management, and QOS. To state that any one particular mapping represents IP over WDM is extremely disingenuous.

This diversity of data link layer protocols and mappings of IP into different underlying network infrastructures is one of the major strengths of IP, and it is a characteristic that will not disappear.

On the contrary, it is quite very possible that new, innovative, and more efficient protocol mapping will be proposed for the transport of IP packets. This is already the case for low-bandwidth and low-reliability networks, and will also be so for high-bandwidth and highly-reliable optical networks. This view also fits within the vision of "everything on IP and IP on everything".

OPTICAL NETWORK NODE AND SWITCHING ELEMENTS

The following section explains the network node and switching element in detail. An optical node is considered as a multifunctional element that performs several tasks depending upon its type and the network requirements

Optical Network Node

- Essentially it sends, receives and resends or redirects optical signals to its neighboring connected nodes.
- The resending or redirecting of an optical signal to the desired networking nodes requires the node to perform either a routing or switching function.
- It should be noted that when several optical signals travel in a multiplexed form, the networking node becomes transmission system dependent since different optical multiplexing techniques can be used such as time division or wavelength division multiplexing.
- An optical node can also function as a router directing an input signal wavelength to a specified output port.
- A router essentially comprises optical couplers where the desired signal from an incoming multiplexed signal can be isolated (i.e. de-multiplexed) and then directed or routed to the output port.
- For this reason it is commonly referred to as a wavelength router.
- It is also possible to change the wavelength of the signal before sending it to the specified output port.
- In this case the router will be called a wavelength converting router and since the device is used to switch the wavelength by changing it to the compliant signal wavelength it is also known as a wavelength routing switch or simply an optical switch
- It should be noted that unlike semiconductor switches an optical switch performs wavelength operational functions (i.e. multiplexing/de-multiplexing and switching) and the physical medium remains fixed with only the information contained in the messages being changed to send (or route/reroute) it to the corresponding destination node through the assigned route. Such wavelength switches can be constructed using integrated optical or optoelectronic devices

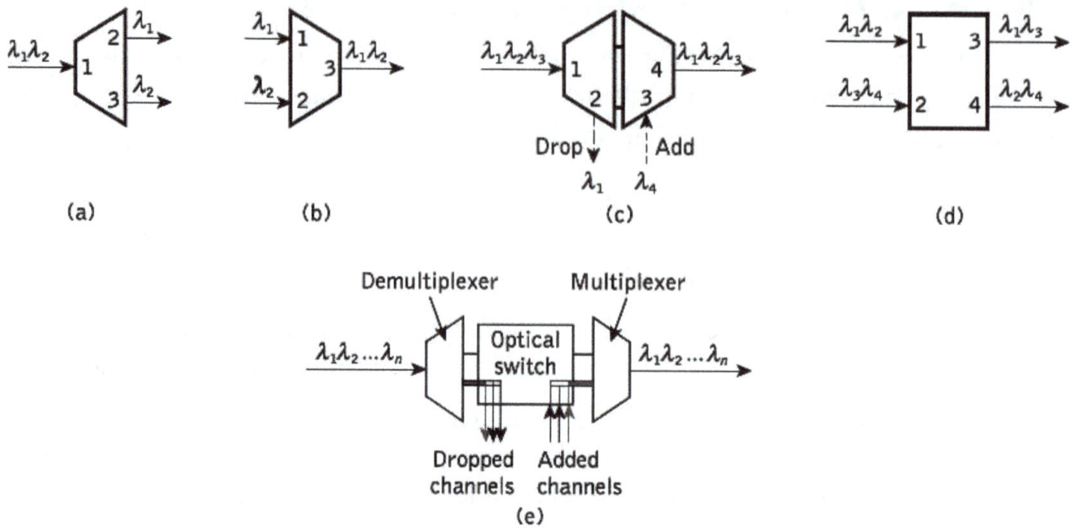

Fig. 12. Optical networking node elements: (a) wavelength demultiplexer; (b) wavelength multiplexer; (c) optical add/drop multiplexer; (d) 2 × 2 optical switch; (e) reconfigurable optical add/drop multiplexer

Functions of an optical router:

- The four different functions of an optical router are depicted in Fig. 12.
- A 1 × 2 wavelength de-multiplexer is shown in Fig. 12(a) which illustrates the splitting of an optical signal present at input port 1 containing two signal wavelengths (i.e. λ_1 and λ_2) and routing them to ports 2 and 3, respectively.
- A three-port wavelength multiplexer combining two wavelength signals is indicated in Fig. 12(b).
- An optical add/drop multiplexer (OADM) which also comprises a wavelength add/drop device (WADD) is shown in Fig. 12(c) where an incoming multiplexed signal comprising three wavelengths (i.e. λ_1, λ_2 and λ_3) is partially de-multiplexed by dropping the λ_2 signal at an intermediate port 2.
- The OADM further adds another wavelength signal λ_4 at intermediate port 3 which is them multiplexed with the transmitted signal wavelengths so that the combined signal leaving port 4 contains wavelengths λ_1, λ_3 and λ_4. Devices which are configured to drop aparticular wavelength (in this case λ_2) and add another specific wavelength (in this case λ_4) are also known as fixed OADM
- Fig. 12(d) shows a simple 2 × 2 optical switch with two input and two output ports.
- The two optically multiplexed signals comprising wavelengths $\lambda_1\lambda_2$ and $\lambda_3\lambda_4$ are present at input ports 1 and at 2, respectively.

- At the output ports the wavelengths are required to be switched and multiplexed as $\lambda_1\lambda_3$ and $\lambda_2\lambda_4$ emerging at the output ports 3 and 4, respectively.
- Using this approach it is possible to produce an optical switch with a greater number of ports (i.e. an N × N optical switch, where N is a large number).

WAVELENGTH DIVISION MULTIPLEXED NETWORKS

Optical wavelength division multiplexing (WDM) technology, partitions the available bandwidth into a number of manageable smaller bandwidth channels. WDM has undoubtedly become the solution adopted in order to increase the capacity of long-haul wide area networks.

Types of WDMA

Optical fiber networks using wavelength division multiplexing (WDM) techniques can be classified as

1. Broadcast-and-select networks
   - i. Single-hop
   - ii. Multi-hop
2. Wavelength routing networks
   - i. Single-hop
   - ii. Multi-hop

Broadcast-and-select network

- A broadcast-and-select network strategy based on a star coupler is shown in Fig. 13.
- The optical transmission is broadcast to all other nodes using fixed transmitters and a tunable receiver at the destination node extracts the desired signal from the entire group of wavelength multiplexed transmitted signals (i.e. $\lambda_1, \lambda_2, \lambda_3 \ldots \lambda_n$).
- All transmissions are broadcast to all network nodes and hence most of the transmitted power is depleted on the receivers which do not use it.
- Consequently, as the number of nodes increases, each station receives a small fraction of the overall transmitted power.

Wavelength routing network

- A wavelength routing network can be used to avoid this wastage of transmitted power where each node within the network is provided with restricted connection(s) to the receiver(s).

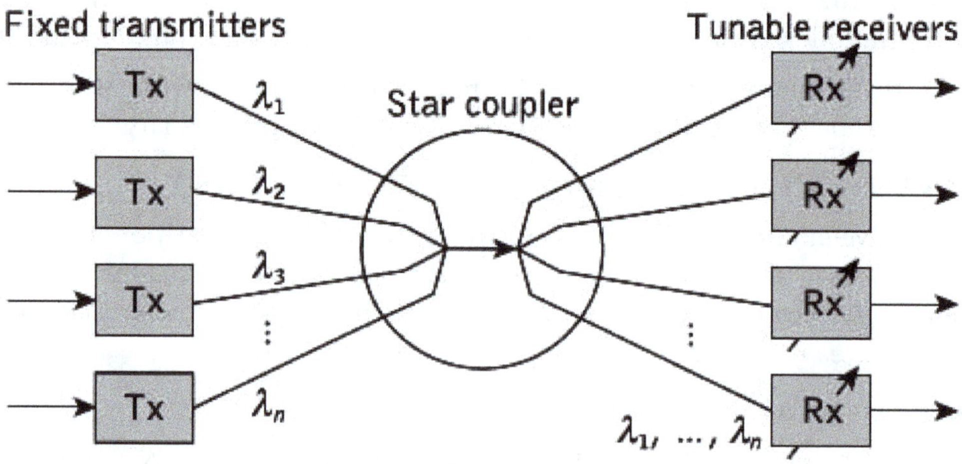

Fig. 13. Broadcast-and-select network

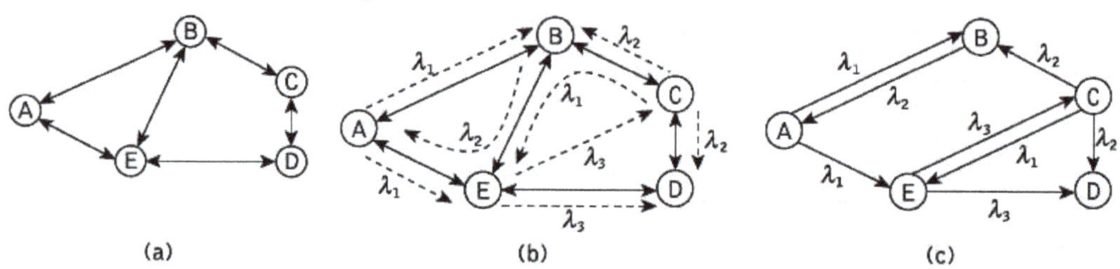

Fig. 14. Wavelength routing: (a) physical connection; (b) wavelength assignments; (c) virtual topology

- In wavelength routing, instead of distributing the message over the entire network, the signal is routed to the specific destination through either a single node or using multiple nodes
- The concept of wavelength routing is illustrated in Fig. 14 where the physical bidirectional interconnections between five nodes (i.e. A, B, C, D and E) are shown in Fig. 14(a).
- Using three wavelengths (i.e. λ_1, λ_2 and λ_3) any network node can transmit or receive a signal from another node within the network.
- This strategy of wavelength implementation or path selection shown in Fig. 14(b) is known as routing and wavelength assignment (RWA).
- Node A can transmit to node B using wavelength λ_1 and it can simultaneously receive from node B using wavelength λ_2 only when the signal is routed through node E.
- To simplify such routing and wavelength assignments a virtual topology is generally used to describe only the enabled wavelength paths.

- The virtual topology is indicated Fig. 14(c) where only wavelength signals identify the possible interconnections between nodes as given by the RWA.
- It can be observed by comparing Fig. 14(a) and (c) that there exists no physical connection between nodes C and E but both nodes can communicate via the virtual connection set up using wavelengths λ_1 and $\lambda_{3.}$

Both broadcast-and-select and wavelength routing networks can be further classified into single-hop or multi-hop.

Single-hop network

- Single-hop networks allow direct communication between any two nodes and the data remains entirely in the optical domain (i.e. without opto-electrical conversion) until it reaches its destination.

Multi-hop network

- In a multi-hop network a transmission may take place through intermediate nodes before reaching its destination.
- At each intermediate node the data can be switched electronically to the next possible node and it is then retransmitted as an optical signal.
- Although this conversion process is inefficient, it is necessary if there is no common wavelength path between two nodes in order to establish a direct connection

PUBLIC TELECOMMUNICATIONS NETWORK

The telecommunications network providing services in the public domain is known as the public telecommunications network where the service providers (or carriers) offer a variety of services for the provision of voice, data and video transmission.

Optical public telecommunications network hierarchy

A simple block hierarchy for the optical public telecommunications network is illustrated in Fig. 15, which is divided into three tiers: long-haul, metropolitan and access networks.

- The long-haul network, also known as the core or sometimes the backbone network, provides national or global coverage with a reach of thousands of kilometers.

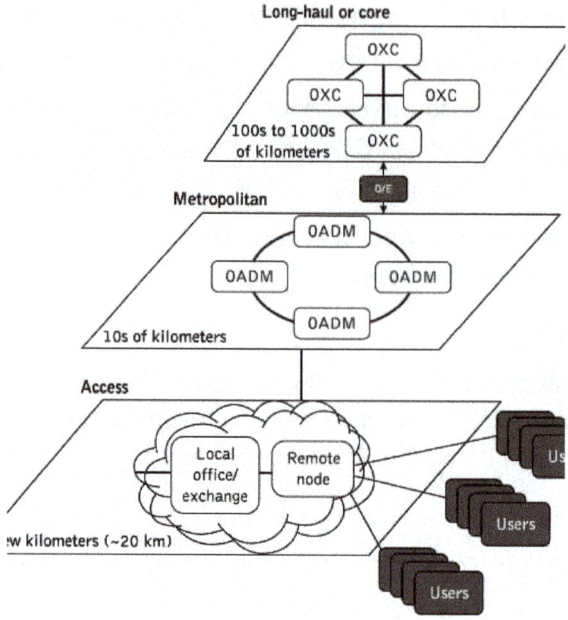

Fig. 15 Optical public telecommunications network hierarchy

- These networks connect many metropolitan networks within a country or interconnect together several country-wide long-haul networks.
- Interconnection between optical nodes is generally accomplished by means of optoelectrical conversion and/or optical switches employing OXCs.
- At the next lower hierarchal level resides the metropolitan area network (MAN), often called the metro, or sometimes the back-haul network.
- These networks offer a multiservice platform and may be confined to a region spreading to tens of kilometers.
- At present they are largely implemented using the ring topology.
- The interconnections between optical nodes in metro networks are also achieved using OADMs while the larger metro networks can also incorporate OXCs.
- The lowest tier in the hierarchy is the local access network which may be extended from a few hundreds of meters to 20 kilometers or so.
- Hence the access network provides the initial interface to the telecommunications network for residential and business customers.
- Although these networks can be configured based on the bus, star or ring topologies, the fiber access network configuration illustrated in Fig. 15 is currently gaining favor.

- The users are connected to a branching node known as a remote node (RN) which interconnects the users with the local office/ telephone exchange.
- In order to provide larger interconnectivity, several local offices/ exchanges are connected using a metropolitan area network

SYNCHRONOUS NETWORKS

SONET is ideal particularly for network providers, as it delivers an efficient, economical network management system that can be easily adapted to accommodate the demand for "bandwidth-hungry" applications and services. The SONET standard as ultimately developed by ANSI defines a digital hierarchy with a base rate of 51.840 Mbit s^{-1}

Pleisochronous structure

- A schematic of the way in which the European hierarchy is multiplexed up to the 140 Mbit s^{-1} rate from the constituent 2 Mbit s^{-1} (30-channel) signals is shown in Fig. 16(a).
- Difficulties arise with this multiplexing strategy in that each 2 Mbit s^{-1} transmission circuit has its own independent clock to provide for timing and synchronization.

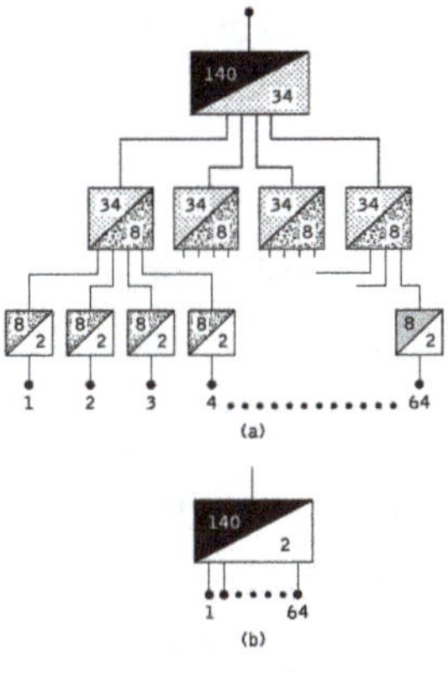

Fig. 16. European multiplexing hierarchy: (a) existing pleisochronous structure; (b) synchronous multiplexing

- This results in slightly different frequencies occurring throughout a network and is referred to as pleisochronous transmission.
- Although this strategy is well suited to the transport of bits it suffers a major drawback in that in order to multiplex the different levels extra bits need to be inserted (bit stuffing) at each intermediate level so as to maintain pleisochronous operation
- The presence of bit stuffing in the existing pleisochronous digital hierarchy

(PDH) makes it virtually impossible to identify and extract an individual channel from within a high-bit-rate transmission link.

- Thus to obtain an individual channel the whole de-multiplexing procedure through the various levels (Fig. 16(a)) must be carried out.

Synchronous multiplexing

- Pleisochronous structure is complex and uneconomic, particularly when considering future telecommunication networking requirements such as drop and insert where individual channels are extracted or inserted at particular stages.
- A substantial saving in electronic hardware together with increased reliability could be achieved by having a straight and multiplexing/de-multiplexing capability as illustrated in Fig. 16(b)
- The synchronous optical network recommendations tend to be referred to as SONET in North America and SDH in Europe
- The SONET standard as ultimately developed by ANSI defines a digital hierarchy with a base rate of 51.840 Mbit s^{-1}.
- SONET defines a hierarchy of electrical signaling levels called synchronous transport signals (STSs). SDH specifies a similar system called a synchronous transport module (STM).

STS-1 frame structure

- The OC notation refers to the optical carrier level signal.
- Hence the base rate signal is OC-1.
- The STS level in brackets refers to a corresponding synchronous transport signal from which the optical carrier signal is obtained after scrambling and electrical to optical conversion.
- Thus STS-1 is the basic building block of the SONET signal hierarchy.
- Higher level signals in the hierarchy are obtained by byte interleaving (where a byte is 8 bits) an appropriate number of STS-1 signals
- The STS-1 frame structure shown in Fig. 17 is precisely 125 μs and hence there are 8000 frames per second.
- This structure enables digital voice signal transport at 64 kbit s^{-1} (1 byte per 125 μs) and the North American DS1-24 channel (1.544 Mbit s^{-1}), as well as the European 30-channel (2.048 Mbit s^{-1}) signals to be accommodated.
- The basic STS-1 frame structure illustrated in Fig. 17 comprises nine rows, each of 90 bytes, which therefore provide a total of 810 bytes or 6480 bits per 125 μs frame.

- This result in the 51.840 Mbit s^{-1} base rate.

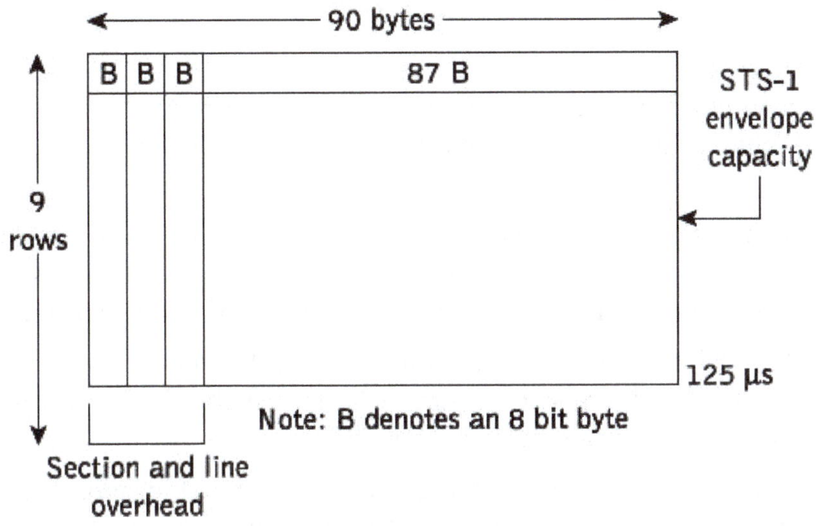

Fig. 17. STS-1 frame structure

- The first 3 bytes in each row of the STS-1 frame contain transport overhead bytes, leaving the remaining 783 bytes to be designated as the synchronous payload envelope (SPE).
- Apart from the first column (9 bytes) which is used for the path overhead, the remaining 774 bytes in the SPE constitute the SONET data payload
- A 'payload pointer' within the transport overhead is employed to designate the beginning of the SPE within that frame.
- This provides the flexibility required in order to accommodate different bit rates and a variety of services.
- To accommodate sub-STS-1 signal rates a virtual tributary (VT) structure is defined comprising four rates:
    - 1.728 Mbit s^{-1} (VT 1.5);
    - 2.304 Mbit s^{-1} (VT 2);
    - 3.456 Mbit s^{-1} (VT 3);
    - 6.912 Mbit s^{-1} (VT 6)..
- The higher order multiplexing of a number of STS-1 signals is obviously important in order to achieve the higher bit rates required for wideband services.

ASYNCHRONOUS TRANSFER MODE

ATM (asynchronous transfer mode) is a dedicated-connection switching technology that organizes digital data into 53-byte cell units and transmits them over a physical medium using digital signal technology. ATM system uses high-speed transmission (155 Mbps) and is a connection-based system.

ATM cell

- Asynchronous transfer mode (ATM) is a packetized multiplexing and switching technique which seeks to combine the benefits of packet switching and circuit switching.
- Asynchronous transfer mode transfers information in fixed size units called cells where each cell contains the information identifying the source of the transmission but which generally contain less data than packets.
- Unlike the fixed time division multiplexed technique where each user waits to send in the allocated time slot, ATM is asynchronous and therefore the time slots are made available on demand.
- To enable correct segmentation and assembly of different cells at the destination, each cell contains significant information in addition to data.
- An ATM cell comprises a header and payload data as shown in Fig. 18.

	Layers	Function	Data unit	
7	Application	Network process to application		
6	Presentation	Data representation and encryption	Data	Host layers
5	Session	Interhost communication		
4	Transport	End-to-end connections and reliability	Segments	
3	Network	Path determination and logical addressing	Packets	
2	Data link	Physical addressing	Frames	Media layers
1	Physical	Media, signal and binary transmission	Bits	

Fig. 18 Format of an ATM cell

- It contains 48 bytes of data with 5 bytes of header information.
- Each single byte in the header field includes different information to identify destination, path, channel and the error control bits.
- Before sending ATM cells carrying user data, a virtual connection between source and destination has to be established.
- All connections follow the same path within the network.
- During the connection setup each control bit (1 or 0) generates an entry in the virtual path identifier (or virtual channel identifier) translation table to inform the destination to receive the incoming packet.

OPEN SYSTEMS INTERCONNECTION REFERENCE MODEL

The OSI model is a layered framework for the design of network systems that allows communication between all types of computer systems. An understanding of the fundamentals of the OSI model provides a solid basis for exploring data communications

OSI Model structure

- Open Systems Interconnection (OSI) describes a standard architecture to be used for designing networks.
- The OSI model was specified jointly by the International Organization for Standardization (ISO)

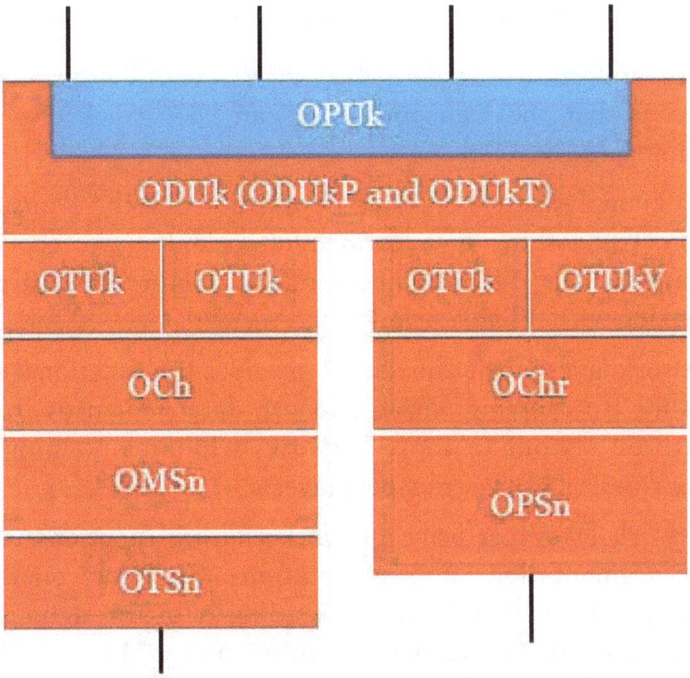

Fig. 19. Description of the layers in the OSI reference model

- Fig. 19 shows the structure of the OSI network reference model and identifies the functions of each of the seven levels or layers.
- The application layer which sits at the top of the hierarchy as the seventh layer provides a means for a user to access information on or utilize the network by receiving a service (e.g. database management, network management).
- This layer is the main interface for the users to interact through applications with the network.
- The sixth level is the presentation layer which transforms data to provide a standard interface for the application layer.

- The session or fifth layer controls the dialogs between intelligent devices.
- It establishes, manages and terminates the connections between the local and remote application.
- It also provides for either duplex or half-duplex operation and establishes check pointing, adjournment, termination and restart procedures.
- The top three layers (i.e. 5, 6 and 7) are therefore data-driven levels.
- The fourth level is the transport layer which provides transparent transfer of data between end users, thus relieving the upper layers from any transport concern while providing reliable data transfer.
- The transport layer controls the reliability of a specific link through flow control, segmentation/de-segmentation and error control.
- Some protocols are both state and connection oriented.
- This means that the transport layer can keep track of the packets and retransmit those that fail.
- The best known example of a layer four protocol is the Transmission Control Protocol (TCP) which is a connection oriented protocol that provides basic data transfer between nodes offering interconnection, flow control, multiplexing with reliable priority and security
- The network layer at the third level provides the functional and procedural method for transferring variable length data sequences from a source to a destination via one or more networks while maintaining the quality of service requested by the transport layer
- The network layer performs network routing functions, and might also perform segmentation/ de-segmentation and report on delivery errors. Routers operate at this layer for sending data throughout the extended network, thus making Internet transmission possible.
- The remaining two layers are the data link and physical media.
- The data link layer provides the functional and procedural mechanisms to enable transfer of data between network entities and to detect, and potentially correct, errors that may occur in the physical layer.
- It is the layer at which network bridges and switches operate and that connectivity is provided among locally attached network nodes forming layer 2 domains for unicast or broadcast forwarding.
- Other protocols can be implemented on the data frames where each frame is logically addressed.

- The best known example of such a protocol is carrier sense multiple access with collision detection (CSMA/CD) or the Ethernet protocol.
- Other examples of data link layer protocols are High-level Data Link Control (HDLC) and the Advanced Data Communications Control Procedure (ADCCP) for point-to-point or packet-switched networks
- The physical layer (PHY) is located at the bottom of the OSI reference model hierarchy and it defines all the electrical, optical and physical media specifications for devices (e.g. the fiber type, cable specifications, hubs, repeaters, network adaptors). It both establishes and terminates a connection to the communications medium.
- Furthermore, it also ensures that communication resources are effectively shared among multiple users.
- For example, it implements contention resolution and flow control while also carrying out data conversion between the equipment and the signal requirements of the communications channel.

OPTICAL TRANSPORT NETWORK

The transmission of information over optical media in a systematic manner is an optical transport network. The optical transport network consists of the networking capabilities and the technology required to support them.

It is regarded as the set of facilities using optical fiber interconnections to carry data between network elements that switch or route the data from different customers into the network

OTN Hierarchy Overview

It is specified in ITU-T Recommendation G.873.1 and it provides for the transport, multiplexing, routing, supervision and survivability for client signals.

The OTN has been proposed as the successor to SDH/ SONET which is the current standard for voice and data traffic in the public switched telecommunications network.

OTN hierarchy

ITU-T Recommendation G.709 provides information about the hierarchical structure and interface functionalities of the OTN by describing the four layers

These are:

- Optical channel (OCh);
- Optical channel transport unit (OTU);

- Optical channel data unit (ODU);
- Optical channel payload unit (OPU).

FULL OTN HIERARCHY

The OTN hierarchy to support these four layers may also include three sections as indicated in Fig. 19 given below, which are:

These sections are further divided into a number of subsections n.

The layers and sections in an OTN produce a confluence of:

- inter- domain
- intra-domain

These interfaces enabling optical signal transmission to reach from one end to the other using electronic, optoelectronic or all-optical interfaces.

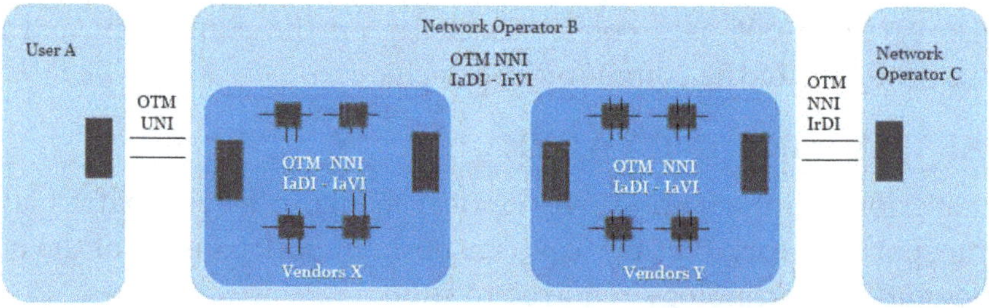

Fig. 20. How OTN maps the client payload

OTN ports

The OTN standards define types of interfaces (ports) where users attach to the network and interfaces between elements in the network, and where different domains interconnect. Figure 21 shows the architecture of the interfaces.

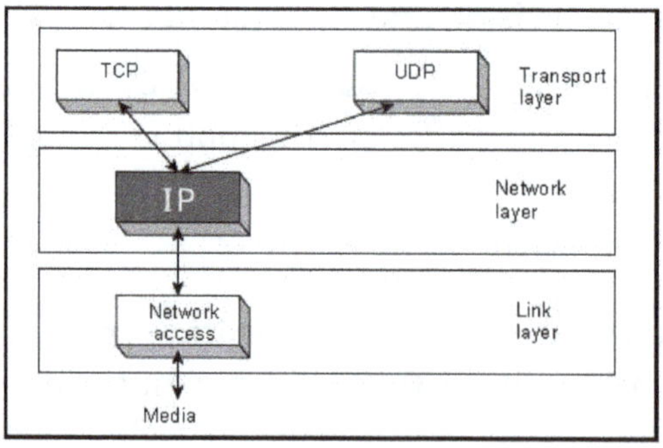

Fig. 21. The interconnections between users and the OTN network

Equipment functions

The OTN standards define functions for different network elements. This helps standardize the design of vendor equipment and ensures interworking and interoperability, which makes it easier for operators to make informed, optimal choices. The functions include:

01. Supervision

- Alarm reporting control
- Defects
- Consequent actions
- Defect correlations
- Performance filters

02. Information flow across reference points

03. Generic processes

- Scrambling processes
- Alignment processes
- Signal quality supervision
- BIP correction
- OTUk forward error correction processing
- Trail trace identifier (TTI) processing
- Payload structure indication (PSI) acceptance processes
- Status information (STAT) acceptance process
- Generic AIS generation and detection
- Generic layer fault processing
- Optical signal processing

04. Optical transmission

- Connection functions
- Termination functions
- Adaptation functions

05. Optical multiplex section (OMS) layer functions

06. Optical physical section (OPS) layer functions

- Connection functions
- Termination functions
- Adaptation functions
- Sub-layer functions

07. OCh (layer) functions

- Connection functions
- Termination functions
- Adaptation functions

08. Layer functions

- OTU (layer) functions
- Connection functions
- Termination functions
- Adaptation functions
- Sub-layer functions

09. ODU (layer) functions

- Connection functions
- Termination functions
- Adaptation functions
- COMMS functions
- Sub-layer functions
- Virtual concatenation functions

INTERNET PROTOCOL

The Internet Protocol (IP) is a network-layer (Layer 3) protocol that contains addressing information and some control information that enables packets to be routed.

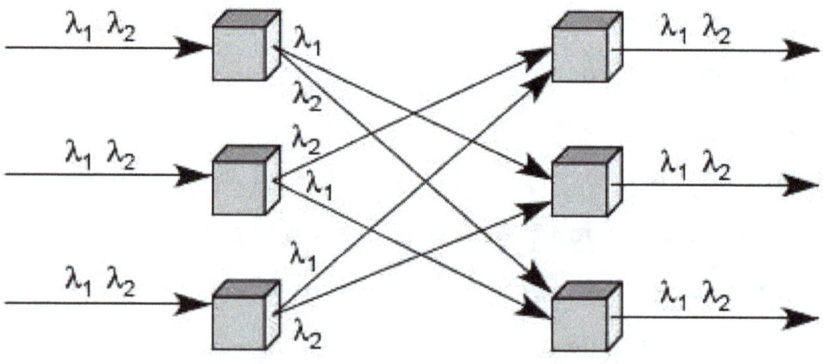

Fig. 22. Internet Protocol for optical networks

Internet Protocol for optical networks

- The IP provides the means of communication between the link and transport layers.
- A Virtual connection is established between nodes requiring communication when IP is combined with a specific higher level protocol such as the Transmission Control Protocol (TCP) or the User Datagram Protocol (UDP)
- The IP provides protocols for both the functions of signaling and routing required to carry the entire signal operation necessary to transmit and receive from optical nodes.

The signaling protocols include

- Multiprotocol Label Switching (MPLS);
- Generalized Multiprotocol Label Switching (GMPLS);

Routing protocols include the

- The Open Shortest-Path First (OSPF);
- The Intermediate-System-to-Intermediate System (IS–IS);
- The Border Gateway Protocol (BGP)

WAVELENGTH ROUTING NETWORKS

The idea at the basis of the wavelength routing network consists of maximum reuse the same wavelengths. Figure 23 illustrates a node of a routing network wherein wavelengths of the same wavelength are used repeatedly.

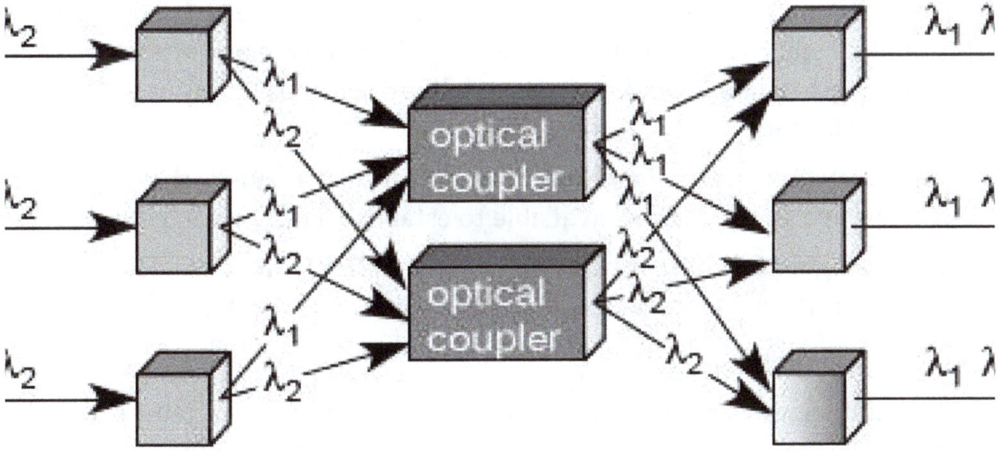

Fig. 23. Node of a routing network

This architecture is a fixed routing on wavelengths. We can also develop in wavelength routed networks with dynamic routing in time. For this purpose, it is necessary to insert optical or optoelectronic switches, according to the technology used, between transmit and receive ports. An example of this technique is illustrated in Fig. 24.

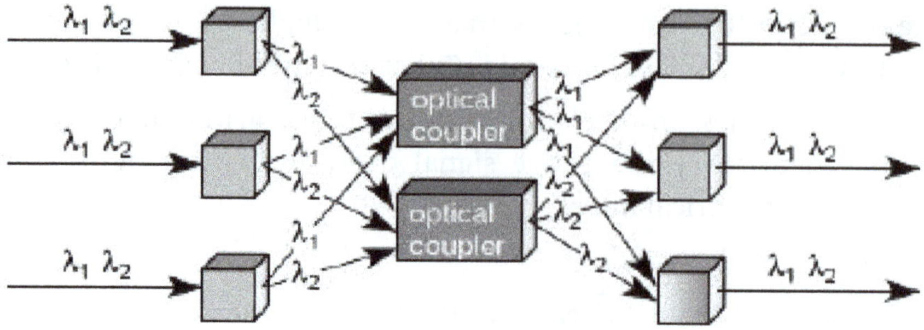

Fig. 24. Dynamic Routing[1]

Much research still takes place in the field of optics to optimize the use of wavelengths. This technique achieves particularly high flow rates, which are measured in terabits per second. The difficulties are still high cost of the wavelength-division multiplexing and especially optical switches. When we want to minimize the cost and increase the range, use optoelectronic switches. Certain fragility is visible at each passage of a light environment to an electrical environment. Considerable progress still needed to reamplify optically signals and adjust almost instantly emission or reception couplers on the right wavelength.

Fixed wavelength nodes

- Three network nodes are interconnected using two wavelength channels (i.e. λ_1 and λ_2) where the solid line connecting the nodes represents the available wavelength channel and the dashed line identifies that the wavelength channel is in use.
- If the network node 1 is required to connect with node 3 then there is no single wavelength channel available to establish a light path between them.
- When a light path cannot be established on a link using a single wavelength channel it is referred to as a wavelength continuity constraint.

Wavelength convertible node

- A methodology to reduce this wavelength continuity constraint is to switch the wavelength channel at node 2 by converting the incoming wavelength λ_2 to λ_1 (which is available between nodes 2 and 3) to enable a link between node 2 and 3 to be established.
- Wavelength conversion is required to convert from λ_2 to a compliant

1

wavelength (i.e. λ_1) at the output port of network node 2 (which functions as an intermediate node) in order to provide a path.

- Hence the newly set up path uses two wavelength stages (i.e. two hops) to interconnect nodes 1 and 3.
- Such networks which employ wavelength conversion devices (or switches) are known as wavelength convertible networks.

OPTICAL NETWORK DEPLOYMENT

Connecting a building to a fiber or coax network can be extremely complex. When planning the deployment, you need to take into account the environmental and topographical conditions, select the best installation methodology, and choose the right equipment for the job.

Then you have to implement the plan, essentially carrying out a civil engineering project to ensure the cable successfully reaches its destination. This can involve re-using existing ducts or creating completely new paths into, and then around, buildings.

However, this is not the end of the job, and perhaps the most vital part is yet to come – testing. This not only enables you to check that the connection works correctly, but, most importantly, that it is reliable, meets relevant industry standards, and is acceptable to the network owner. This should be required on all installs, even if the network owner hasn't mandated it.

TESTING IN FIBER NETWORK DEPLOYMENTS CAN BE SPLIT INTO THREE PHASES

Visual Inspection

To start the process, walk around and inspect the installation, checking the cable route (where visible), its start and termination point, pathways and all equipment rooms. Look for any issues with the install itself, check that the specified materials and equipment have been used (vital for meeting industry standards), and ensure that the design, specifications and relevant safety codes have been met. Any problems should be corrected, or if changes have had to be made, noted on the plan. Look at the placement of the cable, how and where it has been terminated, whether it has been bonded or grounded correctly, and check that the overall infrastructure is satisfactory.

For fiber optic cables, pay particular attention to the connector end faces as microscopic amounts of dust or dirt can dramatically impact performance. Clean every connector as a matter of routine, using either cleaning tools or high-grade, completely lint-free wipes (aim for clean room quality) and pure Isopropyl Alcohol (IPA).

Finally, make sure that every component has been correctly labeled, and logged on a permanent record – and that labels are positioned where they are secure, yet visible, in case of further inspections.

Technical Testing

Once the visual verification is completed, it is time to test that the network actually performs correctly. Obviously, the equipment used for this depends on the type of network itself. For a fiber network, the key parameter to measure is optical loss (often called dB loss or attenuation), as a high optical return loss (ORL) can cause the laser within the system to stop transmitting correctly. This is particularly crucial within systems that use wavelength-division multiplexing.

It is also good practice to test the loss readings (and of course record and file the results) for individual components, such as connectors, cables and splices, using an Optical Time Domain Reflectometer (OTDR). This not only helps uncover potential faults that may be missed by an Optical Loss Test Set (OLTS) that measures end-to-end readings, but, importantly, pinpoints issues with specific components, making it easier to identify where failures occur. It also means that you have an audit trail if the system starts to degrade over time. The TIA/TSB 140 standard requires that you test each fiber link with an OLTS kit. Use an optical power meter and a light source (laser for single mode, LED for multimode) to measure loss and then check to see if it is within acceptable parameters. If you have excessive loss readings look at the connectors first, and then check the cable route. Cables that have been pulled too tightly around a corner and don't meet minimum bend radius specifications can also lead to high loss readings.

Document the Job

The final part of testing is to document everything. After all, if there is no record of tests being completed, there's no way you can prove they were carried out.

Before beginning testing, agree on the format for the documentation with the customer, or adopt its corporate standard, so that it meets everyone's requirements. Ensure that documents are clear and show what has been tested, how, when, where, and by whom. Be particularly careful to include adherence to national and international standards (such as in the case of fire regulations), as this can be used for insurance purposes.

The increasing pace of FTTH rollouts does mean the pressure is on to complete deployments as quickly as possible. However, this shouldn't come at the expense of a proper testing regime that certifies the network, provides a record of compliance, and avoids potential future issues.

OPTICAL ETHERNET

Fiber optic Ethernet is a data service that provides connectivity between two or more locations using fiber optic cables comprised of thin glass or plastic strands. Fiber optic Ethernet transmits data via pulses of light, which can travel much faster than data sent as electrical signals over the copper wiring of traditional Ethernet cable.

- Fiber optic Ethernet transmits data signals over fiber cables rather than the copper cables of traditional Ethernet wiring.
- The benefits of fiber optic Ethernet include greater bandwidth, better signal quality over longer distances, increased reliability and improved security.

Fast Ethernet-Physical Layer

The physical layer in Fast Ethernet is more complicated than the one in Standard Ethernet. We briefly discuss some features of this layer.

Topology

- Fast Ethernet is designed to connect two or more stations together.
- If there are only two stations, they can be connected point-to-point.
- Three or more stations need to be connected in a star topology with a hub or a switch at the center.

Implementation

- Fast Ethernet implementation at the physical layer can be categorized as either two-wire or four-wire.
- The two-wire implementation can be either category 5 UTP (lOOBase-TX) or fiber-optic cable (lOOBase-FX).
- The four-wire implementation is designed only for category 3 UTP (l00Base-T4).

Gigabit Ethernet is more complicated than that in Standard or Fast Ethernet. The following section will explain some of the features of Physical layer of GIGABIT Ethernet.

Topology

- Gigabit Ethernet is designed to connect two or more stations.
- If there are only two stations, they can be connected **point-to-point.**
- Three or more stations need to be connected in a star topology with a hub or a switch at the center.
- Another possible configuration is to connect several star topologies or let a star topology be part of another.

An Ethernet service is offered to the end-customer (the enterprise or residential customer), runs end-to-end (customer premise-to-customer premise), and is one in which the traffic flow into/out of the system at the customer consists of Ethernet frames. An Ethernet service is thus the Ethernet connectivity between customer equipment. A carrier-grade Ethernet service is one that is scalable (to many MAC addresses and end points), offers QoS (traffic management), reliability (protection), and manageability (OAM and monitoring), and can span long distances (of MAN/WAN scope; typically tens to thousands of kilometers).

Ethernet transport refers to the ability to switch/route Ethernet frames (belonging to an Ethernet service) between network nodes, by setting up/using connection-oriented, traffic engineered paths in the network with deterministic performance (QoS, delay, jitter, loss, reliability). In other words, Ethernet transport refers to the setting up of the "pipe" through which the Ethernet frames travel, and to determining its routing within the cloud.

Ethernet transport makes it possible to realize connection-oriented Ethernet (COE). COE, in essence, refers to the collection of control-plane protocols and data-plane settings that create a connection- oriented capability for transferring the frames of an Ethernet service. We mention that Ethernet transport could be provided either by enhancing Ethernet technology (e.g. as is done in Provider Backbone Bridging with Traffic Engineering, PBB-TE, in the IEEE 802.1Qay standard) or by a different technology (e.g. using MPLS-TP technology being developed jointly by the IETF & the ITU-T). Both of these forms of transport involve switching/routing data frames and are, therefore, referred to as Layer 2 (or L2) transport.

It is also possible to embed Ethernet frames in a different transport networking layer, such as the one provided by the ITU-T's G.709 OTN (Optical Transport Network) standard. This form of transport involves switching/routing traffic at the optical channel data unit (ODU) level and is, therefore, referred to as Layer 1 (or L1) transport.

Ethernet PHY refers to the framing and timing of the actual bits of the Ethernet frame, and their transmission over a physical medium — copper wire, coaxial cable, or optical fiber — to connect switches at the physical layer. Some common Ethernet PHYs are the 1 GE (IEEE 802.3z), 10 GE (IEEE 802.1ae), and 100 GE (IEEE 802.3ba) Ethernet PHYs. Note that Ethernet frames can also be embedded in other PHY framing standards, such as those in the ITU-T's G.709 OTN (Optical Transport Network) standard.

Optical Ethernet Network

With this background, we may now define an Optical Ethernet Network as a network spanning a MAN/WAN that offers a carrier-grade Ethernet service, running over a connection-oriented Ethernet (COE) transport infrastructure over an optical PHY. The optical PHY could be provided either by the OTN's optical channel (OCh), or by an Ethernet PHY running over optics, and may be multiplexed onto a given fiber using CWDM/DWDM technology.

A key characteristic of optical Ethernet is that its scope is beyond the enterprise LAN, and spans a metropolitan- area or wide-area network.

"Carrier Ethernet" Vs Optical Ethernet

The term "Carrier Ethernet" was formalized by the work of the MEF (Metro Ethernet Forum) in the time frame, which defines Carrier Ethernet as "a ubiquitous carrier-grade Ethernet service that has the following five attributes: standardized services, scalability, reliability/ protection, hard QoS, and service management." The technical work of the MEF (as described in its specifications) together with the technical work of associated standards bodies (ITU-T, IEEE, IETF) enable the functionality and attributes of Carrier Ethernet.

The services defined by the MEF are in terms of an Ethernet Virtual Connection (EVC), which is defined as an association of two or more User Network Interfaces (UNIs) at the edge of a metro Ethernet network (MEN) cloud (i.e. subscriber sites), where the exchange of Ethernet service frames is limited to the UNI's in the EVC. The MEF defines three standardized services: E-Line (a point-topoint EVC), E-LAN (a multipoint-to-multipoint EVC), and E-Tree (a point-to-multipoint "rooted" EVC, where the root(s) can communicate with any of the leaves, but the leaves must communicate with each other only via the root).

Scalability refers to a service that scales to millions of UNIs (end-points) and MAC addresses, spanning access, local, national, and global networks, with the ability to support a wide bandwidth granularity and versatile QoS options. Reliability refers to the ability to detect and recover from errors/faults without impacting customers, typically with rapid recovery times, as low as 50ms. Hard QoS implies providing end-to-end performance based on rates, frame loss, delay, and delay variation, and the ability to deliver SLAs that guarantee performance that matches the requirements of voice, video, and data traffic over heterogeneous converged networks. Service management implies having carrier-class OAM, and standards-based, vendor-independent implementations to monitor, diagnose, and manage networks offering Carrier Ethernet service.

PACKET-OPTICAL TRANSPORT SYSTEMS

Packet-optical transport systems (P-OTS or P-OTP) are a new class of networking platforms that combine the functions and features of SONET/SDH/OTN ADMs or cross-connects, Ethernet switching and aggregation systems, and WDM/ROADM transport systems into a single network element, thus providing "data-aware optical networking."

A P-OTS network element typically will have ITU-T G.709 OTN support, a COE component, and support for WDM. These elements also offer transport of a wide range of client signals — Ethernet (dominant), legacy SONET/ SDH, SAN traffic, IP/ ATM, video traffic. Optical Ethernet Network with the service, transport and PHY components in operation. (WDM), sub-wavelength (or ODU) level, TDM level (SONET/ SDH), and packet level (Ethernet, MPLS). A P-OTS network element enables a carrier, especially in the MAN/WAN, to quickly and cost-effectively change connectivity and bandwidth in the network, without knowing about the actual services.

Key architectural features of P-OTS elements are:

- Universal switching architecture/fabric for switching traffic at different layers (OTN, TDM and packet)
- Ability to switch, groom, and manage traffic in its native format (i.e. SONET/ SDH traffic as TDM traffic, and IP or Ethernet traffic as packet traffic), thus, allowing for the percentage of each traffic type to vary dynamically (all Ethernet to all SONET/SDH and anything in between, for instance)
- Software-selectable ports that can switch between switching SONET/ SDH to switching Ethernet, depending on the traffic

Even as this definition is gaining industry consensus, according to research firm Heavy Reading, there are three architectures that are currently deemed to fall under the packet optical transport umbrella.

Thus, P-OTS platforms provide the transport and PHY components of optical Ethernet networks.

OPTICAL ETHERNET APPLICATIONS

So which applications/services are optical Ethernet being used for (or envisaged for) today?

As expected, it is the business or residential services with triple-play applications (voice, video, and data to the desktop), mobile backhaul applications (where the Ethernet PHY is used between the base-station and the first switching node, and regular optical Ethernet networks are used in the backhaul and backbone networks), and utility infrastructure networks (where oil, gas, water, and electric utilities are transforming their aged communication systems into "data-aware" systems that allow for automation of functions such as billing, monitoring, meter reading).

Applications such as software-as-a-service, VoIP, VoD, and hosted unified communications are driving demand, as are ICT trends such as virtualization, data center outsourcing, data replication, disaster recovery, remote backup, and IT outsourcing.

As optical Ethernet evolves over the next few years, there will be further reduction in the layers leading to a fused Ethernet-WDM packet transport layer with circuit-like capabilities, and to packet- optical systems optimized for it. This allows the providers to handle increasing volume of data traffic, while reducing the number of network elements by using Ethernet as the common packet technology in access, aggregation, and core networks.

DEFINITIONS AND TERMINOLOGY USED IN OPTICAL NETWORKS

Acceptance Angle: The half-angle of the cone (a) within which incident light is totally internally reflected by the fiber core. It is equal to sin-1(NA)

Active device: A device that requires a source of energy for its operation and has an output that is a function of present and past input signals. Examples include controlled power supplies, transistors, LEDs, amplifiers, and transmitters.

American National Standards Institute (ANSI): The American National Standards Institute (ANSI) is the official American standards body through which standards are published, and various other standards-setting committees are accredited.

Amplifier: A device, inserted within a transmission path that boosts the strength of an electronic or optical signal. Amplifiers may be placed just after the transmitter (power booster), at a distance between the transmitter and the receiver (in-line amplifier), or just before the receiver (preamplifier).

Analog: A continuously variable signal. Opposite of digital.

Angled physical contact (APC): A polishing technique for fibers/ferrules that minimizes reflective light. A style of fiber-optic connector with a 5 -15 degree able on the connector tip for the minimum possible back reflection. Usually used in AM or DWDM single-mode transmission systems using laser light sources. Typically 65-70 dB.

Application-specific optical fiber (ASOF): Fibers built for specific applications such as erbium fibers used in EDFA optical amplifiers. Other types include high N.A. fibers used for manufacturing filters and gratings, etc.

Arrayed waveguide grating (AWG): A device that allows multiple wavelengths to be combined and separated in a DWDM system. An array of planar waveguides diffracts light at angles that depend on the wavelength. The central element is an array of narrow, curved waveguides that run parallel to each other between a pair of mixing regions. The waveguides differ in length by an increment that is much larger than the wavelength. The input signal enters the first mixing region and is coupled into the waveguides to pass into the second mixing region.

Attenuation: The loss of optical power, whether caused intrinsically (absorption, scattering and microbends), or by extrinsic components such as connectors, splices, splitters and other optical components. External stresses such as microbends and macrobends result in fiber attenuation. The term can also be used synonymously with attenuation coefficient, which is expressed in dB/km.

Avalanche photodiode (APD): A photodiode designed to take advantage of avalanche multiplication of photocurrent. It converts one photon to multiple electrons.

Central Office Home-run : Type of Centralized FTTH architecture where splitters are located inside the central office & no splitting beyond.

Centralized Topology FTTH : The type of FTTH architecture where fiber optic cable from OLT is split only once before reaching customers.We mainly know the two types of Centralized FTTH topology namely: Central Office Homerun and Local Convergence Point.

CER: Communication Equipment Room : Buildings drop point to house indoor FDT cabinets, FAT and ONT's.

Cladding: The layer which surrounds the core of a fiber. The cladding is slightly less transparent than the core, and reflects light back in to the core.

Connector: The structure at the end of a fiber optic cable that allows connection to a fiber port on a patch panel or electronic device.

Core: The transparent center of a fiber optic cable, where light is internally transmitted.

dB: Abbreviation for decibel, the unit of measurement for optical power in a fiber optic cabling system.

Distributed / Cascaded : The type of FTTH architecture where the fiber is spliced from the OLT twice (one in / out of the exchange or one out of the exchange) before the customers arrive at the location. The division can be located in the exchange, FDT, manhole, handhole or FAT.

Distribution : In conventional method of FTTH deployment this is equivalent to the secondary cables in copper network. These are fiber cables that connect the FDT or cross connect cabinet to the final distribution point (a FAT or a splice in a handhole before the drop). In short, we can say that this is the cable between the feeder and the drops.

Drops : The optical fiber cable from the last distribution point up to customer's location.

Ethernet: A data communications protocol originally developed for premises and local access networks (IEEE 802.3) operating at speeds from 10 Mbps to 10 Gbps. It was originally developed for peer-to-peer communications using shared media over relatively short distances. Ethernet features variable length packets that allow data to be sent with less overhead.

FAT: Fiber Access Terminal : A distribution point in the distribution end that distribute the all drop fibers to each customer. It can also divide the house to distribute drops in homes and offices.

FDF: Fiber Distribution Frame : Exchange termination equipment where fibers from OLT are split and jointed to Feeder cable when there is mass FTTH deployment.

FDT: Fiber Distribution Terminal : A hub where all the fiber from the telecom provider's central office is split and distributed to the end users.

Feeder : Equivalent to primary cables in copper networks. Fiber cables that connect the exchange to a split point in the FDT or manhole serve as a cross connect to the distribution network.

Fiber (or, optical fiber): The glass construction of a fiber optic cable, consisting of the core, cladding, and buffer.

FTB: Fiber Termination Box : Customer Lead-In Unit, installed at the customer outer wall for termination of incoming & indoor fibers.

FTTH - Fiber to the Home : Access technology architecture that brings fiber optic cable from the exchange direct to customer's location at high bandwidth delivering triple play (voice, video and data) from a single fiber.

GBIC: Gigabit Interface Converter (pronounced "JEE-bic"). A transceiver device that plugs into a switch or other network device in order to provide a high-speed fiber optic interface.

GPON - Gigabit Passive Optical Network : A flexible optical fiber access network cable of supporting bandwidth requirements of subscriber and covers system with nominal line rates of 1.2 Gbps and 2.4 Gbps in downstream direction; and 155 Mbps up to 2.4 Gbps in the upstream direction.

Laser-Optimized Multi-Mode Fiber (LOMMF): Multi-mode optical fiber that is designed for use with VCSEL lasers. Normally has an aqua-colored jacket; also known as OM3.

LCP: Local Convergence Point : Type of Centralized FTTH architecture where splitters are concentrated in an FDT or a manhole outside the exchange.

Multimode: A fiber cable that transmits more than one mode, or paths, of light to travel down the core. May come in two standard core sizes — 62.5/125 (OM1) or 50/125 (OM2).

MDF: Main Distribution Frame : Location where copper OSP cables are terminated inside the exchange make access to Network switch.

ODF: Optical Distribution Frame : Location where fibers from OLT ports are terminated then interfaced to Feeder cable of FTTH.

OLT: Optical Line terminal : Located inside the exchange and serve as the termination point of OSP Feeder fiber cables. An FTTH access platform designed for the GPON standard provides RF video and TDM voice services with ease, while IP provides a smooth migration path to triple play services.

ONT: Optical Network Terminal : A media converter typically installed outside / indoors or designed to deliver multi-line POTS, data and video.

ONU: Optical Network Unit : A unit located outside customer's area to serve more customers, similar to RAN.

PON - Passive Optical Network : An architecture using optical transmission system designed to carry signals via fiber optic cable from the central office directly to end users such as businesses and multi dwelling units. Meaning of Passive is that there is no active or powered element or component in between the telecom exchange and the customer's location.

Sp: Passive Optical Splitter : Split the signal bidirectionally from a single fiber to multi fiber without any power application.

ABBREVIATION USED IN OPTICAL NETWORK

AON	:	Active Optical Network
APON	:	ATM Passive Optical Network
BPON	:	Broadband Passive Optical Network
CER	:	Communication Equipment Room
E2E	:	End To End
EBU	:	Enterprise Business Unit
ePON	:	Ethernet Passive Optical Network
FAT	:	Fiber Access terminal
FDH	:	Fiber distribution hub
FDT	:	Fiber distribution Terminal
FTTB	:	Fiber to the building
FTTH	:	Fiber To The Home
Gbps	:	Gigabits per second
GPON	:	Gigabit Passive Optical Network
HBU	:	Home Business Unit
I&R	:	Department of Installation & Repair
LCP	:	Local Convergence Point
Mbps	:	Megabits per second
MDU	:	Multi Dwelling Unit
NAP	:	Network Access Point
Ne	:	Network Engineer Tool System
ODF	:	Optical Distribution Frame

ODN	:	Optical Distribution Network
OLT	:	Optical Line Terminal OLT
ONT	:	Optical Network Terminal
ONU	:	Optical Network Unit
P2MP	:	Point To Multi Point
PON	:	Passive Optical Network
POS (Sp)	:	Passive Optical Splitter
PTP (P2P)	:	Point To Point
RF	:	Radio Frequency
SFU	:	Single Family Unit
SO/HO	:	Small Office/Home Office
TDM	:	Time Division Multiplexing in downstream
TDMA	:	Time Division Multiple Access in upstream
WDM	:	Wavelength division multiplexer

BIBLIOGRAPHY

Ackerman, D. A., J. E. Johnson, L. J. P. Ketelsen, L. E. Eng, P. A. Kiely and T. G. B. Mason, 'Telecommunication lasers', in I. P. Kaminow and T. Li (Eds), Optical Fiber Telecommunications IVA, pp. 587–665, Academic Press, 2002.

Beese, N. C. 'Light sources for optical communication', Infrared Phys., 1, pp. 5–16, 1961.

Belahlou, A. et al., 'Fiber design considerations for 40 Gb/s systems', J. Lightwave Technol., 20(12), pp. 2290–2305, 2002.

Bell, A. G. 'Selenium and the photophone', The Electrician, pp. 214, 215, 220, 221, 1880.

Bhattacharya, P., Semiconductor Optoelectronic Devices, Englewood Cliffs, NJ: Prentice-Hall, 1997.

Born, M. and E. Wolf, Principles of Optics (7th edn), Cambridge University Press, 1999.

Bube, R. H., Photoconductivity of Solids. New York: Wiley, 1960.

Chuang, S. L., Physics of Photonic Devices, New York: Wiley, 2009.

Davis, C. C., Lasers and Electro-Optics: Fundamentals and Engineering, Cambridge: Cambridge University Press, 2014.

DeSalvo, R. et al., 'Advanced components and sub-system solutions for 40 Gb/s transmission', J. Lightwave Technol., 20(12), pp. 2154–2181, 2002.

Donati, S., Photodetectors: Devices, Circuits, and Applications. Upper Saddle River, NJ: Prentice-Hall, 2000.

Gagliardi, R. M. and S. Karp, Optical Communications, Wiley, 1976.

Gambling, W. A. 'The rise and rise of optical fibers', IEEE J. Sel. Top. Quantum Elecbron, 6(6), pp. 1084–1093, 2000.

Garmire, E. 'Sources, modulators, and detectors for fiber-optic communication systems', in M. Bass and Eric W. Van Stryland (Eds), Fiber Optics Handbook, pp. 4.1–4.80, McGraw-Hill, 2002.

Goodwin, A. R., J. F. Peters, M. Pion and W. O. Bourne, 'GaAs lasers with consistently low degradation rates at room temperature', Appl. Phys. Lett., 30(2), pp. 110–113, 1977.

Hartman, R. L., J. C. Dyment, C. J. Hwang and H. Kuhn, 'Continuous operation of GaAs–GaxAl1–xAs, double heterostructure lasers with 330 °C half lives exceeding 1000 h', Appl. Phys. Lett., 23(4), pp. 181–183, 1973.

Haus, H. A., Waves and Fields in Optoelectronics. Englewood Cliffs, NJ: Prentice-Hall, 1984.

Hecht, E. and A. Zajac, Optics (4th edn), Addison-Wesley, 2003.

Huxford, W. S. and J. R. Platt, 'Survey of near infra-red communication systems', J. Opt. Soc. Am., 38, pp. 253–268, 1948.

Iizuka, K., Elements of Photonics for Fiber and Integrated Optics, Vol. II. New York: Wiley, 2002.

Kao, K. C. and G. A. Hockham, 'Dielectric fiber surface waveguides for optical frequencies', Proc. IEE, 113(7), pp. 1151–1158, 1966.

Kasap, S. O., Optoelectronics and Photonics: Principles and Practices, Upper Saddle River, NJ: Prentice-Hall, 2012.

Kraemer, A. R. 'Free-space optical communications', Signal, pp. 26–32, 1977.

Liu, J.M., Photonic Devices. Cambridge: Cambridge University Press, 2005.

Maiman, T. H. 'Stimulated optical radiation in ruby', Nature, 187, pp. 493–494, 1960.

Miller, S. E. 'Overview and summary of progress', in S. E. Miller and I. P. Kaminow (Eds), Optical Fiber Telecommunications II, pp. 1–27, Academic Press, 1988.

Nalwa, H. S., ed., Photodetectors and Fiber Optics. San Diego, CA: Academic Press, 2001.

Nayayama, K., M. Kakui, M. Matsui, T. Saitoh and Y. Chigusa, 'Ultra-low-loss (0.1484 dB/km) pure silica core fibre and extension of transmission distance', Electron. Lett., 38(20), pp. 1168–1169, 2002.

Oh, K., S. Choi, Y. Jung and J. W. Lee, 'Novel hollow optical fibers and their applications in photonic devices for optical communications', J. Lightwave Technol., 23(2), pp. 524–532, 2005.

Pedrotti, F. L., L. S. Pedrotti and L. M. Pedrotti, Introduction to Optics (3rd edn), Prentice Hall, 2006.

Rosencher, E. and Vinter, B., Optoelectronics. Cambridge: Cambridge University Press, 2002.

Saleh, B. E. A. and Teich, M. C., Fundamentals of Photonics. New York: Wiley, 1991.

Smith, F. Graham, T. A. King and D. Wilkins, Optics and Photonics: An Introduction (2nd edn), Wiley, 2007.

Smith, W. J. Modern Optical Engineering (3rd edn), McGraw-Hill, 2000.

Werts, A. 'Propagation de la lumière cohérente dans les fibres optiques', L'Onde Electrique, 46, pp. 967–980, 1966.

Yariv, A. and Yeh, P., Photonics: Optical Electronics in Modern Communications. Oxford: Oxford University Press, 2007.

INDEX

www.ingramcontent.com/pod-product-compliance
Lightning Source LLC
La Vergne TN
LVHW062218200926
848484LV00006B/1428